IMMUNOBIOLOGY OF PROTEINS AND PEPTIDES VI

Human Immunodeficiency Virus, Antibody Immunoconjugates, Bacterial Vaccines, and Immunomodulators

ADVANCES IN EXPERIMENTAL MEDICINE AND BIOLOGY

Recent Volumes in this Series

Volume 300
MECHANISMS AND SPECIFICITY OF HIV ENTRY INTO HOST CELLS
Edited by Nejat Düzgüneş

Volume 301
MECHANISMS OF ANESTHETIC ACTION IN SKELETAL, CARDIAC, AND SMOOTH MUSCLE
Edited by Thomas J. J. Blanck and David M. Wheeler

Volume 302
WATER RELATIONSHIPS IN FOODS: Advances in the 1980s and Trends for the 1990s
Edited by Harry Levine and Louise Slade

Volume 303
IMMUNOBIOLOGY OF PROTEINS AND PEPTIDES VI: Human Immunodeficiency Virus, Antibody Immunoconjugates, Bacterial Vaccines, and Immunomodulators
Edited by M. Zouhair Atassi

Volume 304
REGULATION OF SMOOTH MUSCLE CONTRACTION
Edited by Robert S. Moreland

Volume 305
CHEMOTACTIC CYTOKINES: Biology of the Inflammatory Peptide Supergene Family
Edited by J. Westwick, I. J. D. Lindley, and S. L. Kunkel

Volume 306
STRUCTURE AND FUNCTION OF THE ASPARTIC PROTEINASES: Genetics, Structures, and Mechanisms
Edited by Ben M. Dunn

Volume 307
RED BLOOD CELL AGING
Edited by Mauro Magnani and Antonio De Flora

IMMUNOBIOLOGY OF PROTEINS AND PEPTIDES VI

Human Immunodeficiency Virus, Antibody Immunoconjugates, Bacterial Vaccines, and Immunomodulators

Edited by

M. Zouhair Atassi

Baylor College of Medicine
Houston, Texas

PLENUM PRESS • NEW YORK AND LONDON

Library of Congress Cataloging-in-Publication Data

International Symposium on the Immunbiology of Proteins and Peptides
(6th : 1990 : Scottsdale, Ariz.)
Immunobiology of proteins and peptides VI : human immunodeficiency virus, antibody immunoconjugates, bacterial vaccines, and immunomodulators / edited by M. Zouhair Atassi.
p. cm. -- (Advances in experimental medicine and biology ; v. 303)
"Proceedings of the Sixth International Symposium on the Immunobiology of Proteins and Peptides, held October 26-30, 1990, in Scottsdale, Arizona"--T.p. verso.
Includes bibliographical references and index.
ISBN 0-306-44038-5
1. AIDS (Disease)--Immunological aspects--Congresses. 2. HIV (Viruses)--Congresses. 3. HIV antibodies--Congresses. 4. Antibody -drug conjugates--Congresse. 5. Bacterial vaccines--Congresses. 6. Radioimmunotherapy--Congresses. I. Atassi, M. Z. II. Title. III. Series.
[DNLM: 1. Adjuvants, Immunologic--congresses. 2. Antibodies, Monoclonal--Immunology--congresses. 3. Bacterial Vaccines--immunology--congresses. 4. HIV--immunology--congresses. W1 AD559 v. 303 / QW 166 1613i 1990]
QR201.A37I57 1990
616.97'92079--dc20
DNLM
for Library of Congress 91-31837
CIP

Proceedings of the Sixth International Symposium on the
Immunobiology of Proteins and Peptides,
held October 26–30, 1990,
in Scottsdale, Arizona

ISBN 0-306-44038-5

Printed in the United States of America

SCIENTIFIC COUNCIL OF THE SYMPOSIUM

M.Z. Atassi, *President*
Howard L. Bachrach
Eli Benjamini
Alec Sehon
Garvin Bixler, *Meeting Secretary*
Nickolas Calvanico, *General Secretary*

SESSION ORGANIZERS

Steven Gillis
Thomas Matthews
Peter Paradiso
Ellen S. Vitetta

MAJOR SPONSORS OF THE SYMPOSIUM

Praxis Biologics
United States Army Medical Research and Development Command

The following organizations also contributed to the Symposium:

Connaught Laboratories
Bristol-Myers Squibb
Sandoz Pharmaceuticals
Fisher Scientific

PREFACE

The articles in this volume represent papers delivered by invited speakers at the 6th International Symposium on the Immunobiology of Proteins and Peptides. In addition, a few of the abstracts submitted by participants were scheduled for minisymposia and some of the authors, whose presentations were judged by the Scientific Council to be of high quality, were invited to submit papers for publication in this volume.

This symposium was established in 1976 for the purpose of bringing together, once every two or three years, active investigators in the forefront of contemporary immunology, to present their findings and discuss their significance in the light of current concepts and to identify important new directions of investigation. The founding of the symposium was stimulated by the achievement of major breakthroughs in the understanding of the immune recognition of proteins and peptides. We believed that these breakthroughs will lead to the creation of a new generation of peptide reagents which should have enormous potential in biological, therapeutic and basic applications. This anticipated explosion has in fact since occurred and many applications of these peptides are now being realized.

The sixth symposium was devoted to four major areas: **Human immunodeficiency virus, antibody immunoconjugates, bacterial vaccines and immunomodulators**. In this volume, many important papers will deal with various aspects of structure and biology of HIV and SIV, the expression and regulation of their genes and the immunology of their envelope proteins. Antibody immunoconjugates have become an important tool for specific targeting of drugs and radioisotopes in chemotherapy and radioimmunotherapy of certain malignancies. Papers by several leading investigators deal here with recent advances in this important field. Manipulation of the immune system by immunodulators, or by other strategies, is perhaps one of the most promising applications of the advances in immunology to disease therapy. Many important papers deal with designs and applications of vaccines against selected bacterial agents. To achieve an intelligent effective design of a vaccine, it is crucial to know details of the humoral and cellular immune responses against the infectious organism. How to maximize a required antibody or T-cell response or to reduce its magnitude is often desired in the design of an immunological defense strategy. What is the best means of delivery? How can tolerance be achieved in certain cases? These and other important questions that need to be appreciated in the design of vaccines are discussed in this volume.

Finally, I should like to express, on behalf of the organization, our gratitude to our sponsors whose generous support made this conference possible.

M. Zouhair Atassi

CONTENTS

STRUCTURE AND FUNCTION IN RECOMBINANT HIV-1 gp120 AND SPECULATION ABOUT THE DISULFIDE BONDING IN THE gp120 HOMOLOGS OF HIV-2 AND SIV

Timothy Gregory*[1], James Hoxie[#], Colin Watanabe*[2] and Michael Spellman*[3]

*Depts. of [1] Process Sciences, [2] Scientific Computing, [3] Medicinal and Analytical Chemistry, Genentech, Inc. 460 Pt. San Bruno Blvd., So. San Francisco, CA 94080

[#]Hematology-Oncology Section, Hospital of the University of Pennsylvania, 3400 Spruce St. Philadelphia, PA 19104

INTRODUCTION

The envelope glyco-proteins of the primate immunodeficiency viruses (HIV-1, HIV-2 and SIV) have been the objects of intense study since their discovery. The major envelope glycoprotein (gp120 in HIV-1) is of particular interest because it mediates the attachment of the virus to susceptible cells via the CD4 molecule[1,2], it contains most of the important epitopes for neutralization of the virus by antibodies[3,4,5], it plays an important role in the process by which the viral and host cell membranes fuse and the viral capsid gains access to the cytoplasm[6,7], and its sequence variability is central to the ability of the virus to adapt to and escape the protective immune response of the host organism[8]. Complete understanding of these processes requires an understanding of the molecular structure of gp120 in detail. Such structural information has proven to be difficult to obtain because of the large size of gp120 (approximately 480 amino acids), its high degree of glycosylation (approximately 50% by weight), the high degree of heterogeneity of the oligosaccharides on the molecule, and the scarcity of material available for analysis.

The scarcity of gp120 protein for structural analysis has largely been overcome by its production in recombinant mammalian cells[9]. Because recombinant gp120 (rgp120) is secreted from these cells with functional properties identical to those of gp120 produced by virally infected cells[10], it presumably has structural properties very similar to those of the viral protein. Mammalian cells, such as the CHO cells used for the production of rgp120, also glycosylate proteins in a manner similar to that expected for gp120 produced by virally infected cells and offer a degree of confidence that the structural attributes of rgp120 produced in them are also representative of the viral protein.

The ultimate goal of x-ray chrystalographic analysis of rgp120 has proven to be very dificult to achieve at least in part because of the extreme heterogeneity of the oligosaccharide moieties on the protein. A determination of the disulfide bonding pattern and the type of oligosaccharide at each of the potential N-linked glycosylation sites in rgp120 fron HIV-1 IIIB has however recently been reported[11]. In this paper we expand on the discussion of that analysis and extend its interpretations to some of the structural variants among the primate immunodefficiency virus gp120 homologs. We also use the HIV-1 gp120 data to predict the disulfide bonding pattern for the gp120 homolog of the HIV-2 isolate, ROD, and the SIV isolate, SIV-MM142.

THE PRIMARY STRUCTURE OF rgp120-IIIB

The structural analyses were performed on two forms of rgp120 from the IIIB isolate of HIV-1 produced in recombinant CHO cells[11]. For ease of expression and purification both of these proteins were constructed as fusion proteins consisting of a portion of the Herpes Simplex Virus type 1 glycoprotein D (gD) fused to the truncated N-terminus of gp120. One of the forms of rgp120 (CL44) was composed of the N-terminal 27 amino acids of mature gD fused to amino acid 31 of mature gp120 and the other form (9AA) was composed of the N-terminal 9 amino acids of gD fused to amino acid 4 of gp120. As a consequence of the construction the rgp120 CL44 protein is missing the first cysteine residue (C24) encoded by the mature gp120 sequence, and therefore has one unpaired cysteine. Because of this the determination of the disulfide bonding pattern was done first with the 9AA protein and then partially confirmed with the CL44 protein. Characterization of the 24 potential N-linked glycosylation sites was performed with the CL44 protein. The purified proteins (non-reduced for the disulfide determinations or reduced and carboxymethylated for the oligosaccharide analyses) were analyzed by tryptic digestion, rpHPLC purification of the resulting peptide fragments and identification of the peptides by quantitative amino acid analysis and N-terminal sequencing. Further proteolytic and/or glycosidic digestion of specific peptides and re-purification by rpHPLC were required to assign all of the disulfide bonds in the peptide fragments. The type of oligosaccharide present (i.e., complex or high mannose) at each potential N-linked site was determined on the basis of susceptability to endo glycosidase H.

The results of the analyses, summarized in Figure 1, indicate that the 18 cysteines of the rgp120-IIIB protein are all disulfide bonded to form a series of five domains: two domains, each with one disulfide bond, and three domains each containing a more complex pattern of two or more disulfide bonds. No heterogeneity of the disulfide bonding pattern was detected other than the unpaired cysteine (C44) in the CL44 form of rgp120-IIIB. All of the 24 potential N-linked glycosylation sites were utilized, with mostly complex type

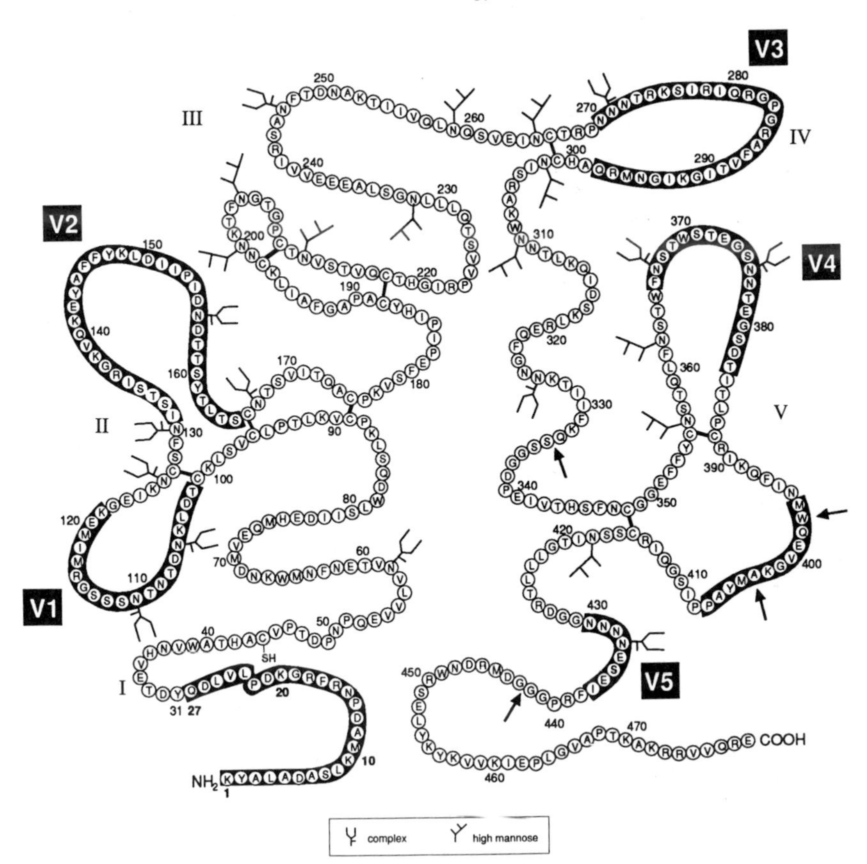

Figure 1. Model Of The Primary Structure Of Recombinant gp120 Of HIV-1 IIIB. This is a summary of the experimental work of Leonard, et al.[11] on the recombinant fusion protein CL44 expressed in CHO cells. The protein is composed of the N-terminal 27 amino acids of herpes symplex virus glycoprotein D (shaded) fused to amino acid 31 of gp120. The type of oligosaccharide structure present at each N-linked glycosylation site is indicated. The Roman Numerals refer to the five disulfide bonded domains and the boxed sequences accompanied by a boxed number refer to the hypervariable regions described by Modrow, et al.[8]. The arrows and the box around residues 396 to 407 designate the sites implicated in CD4 binding by mutagenesis studies[10, 13].

oligosaccharide structures at 13 of them and mostly high mannose or hybrid type structures at the remaining 11 sites. In general, the oligosaccharides at individual glycosylation sites were found, within the limits of detection, to be either completely resistant or completely susceptible to endo H. The only heterogeneity of this type detected at any of the N-linked sites was the presence of a trace amount of high mannose or hybrid type oligosaccharide at N246 instead of the predominant complex type structure. No O-linked oligosaccharides were detected.

REGIONS INVOLVED WITH CD4 BINDING

One of the most studied functional attributes of gp120 is its binding to the HIV cellular receptor CD4. This interaction has been well characterized using rgp120 and a soluble recombinant form of CD4[12]. The majority of information about the portions of gp120 that interact with CD4 has come from mutagenesis studies and the mapping of the epitope of an anti-rgp120 monoclonal antibody that blocks CD4 binding[10,13]. This information is summarized in Figure 1. Deletion of residues 396 to 407 was reported by Lasky et al[10] to abolish CD4 binding by the resulting recombinant mutant gp120 protein. More specific mutagenesis within this region of A403 to D[10] and W397 to S, G, V or R[14] abolished CD4 binding by the mutant proteins and implied that these residues were critically important for the interaction. Short insertions between residues 333-334, 388-390 and 442-443 by Kowalski et al[13] similarly abolished CD4 binding. The epitope of the murine monoclonal antibody 5C2-E5, which blocks the binding of rgp120 to CD4, was mapped to residues 392-402[10]. Further evidence comes from a report that a proteolytic fragment of gp120 from residue 322 to near the C-terminus retains the ability to bind to CD4[15]. In summation, the currently available data suggest that the portions of gp120 that are involved with its interaction with the CD4 molecule are found at various sites between residues 320 and 450.

On a linear map of the gp120 sequence the regions implicated in CD4 binding show little relation to each other. However, the disulfide bonding pattern shows that all of these sites are associated with a discrete disulfide bonded domain, domain V in Figure 1. The five sites identified above are located in conserved sequences[16] either between C388 and C415 or along the "neck" of domain V upstream from C348 and downstream from C415.

THE OLIGOSACCHARIDES OF rgp120

The structures of the carbohydrate moieties of gp120 have been determined for glyco-protein produced by virally infected H9 cells[17,18] and CL44 rgp120 produced in CHO cells[19]. A summary of the results of these analyses is presented in Table 1. A large proportion of the structures are of the high-mannose type which is sometimes associated with premature release of protein (e.g., by cell lysis) that has not completed the final stages of oligosaccharide processing. For CL44 rgp120 this is probably not the case as the high mannose (i.e., endo H susceptible) structures were found to be localized preferentially at 11 of the 24 sites. In the event of significant cell lysis, a mixture of completely processed (complex type) structure and incompletely processed (high-mannose type) structure would be expected at all glycosylation sites. In the analyses of the oligosaccharides on rgp120 and secreted viral gp120 summarized in Table 1[17,19], 39% and 54% of the total released structures were of the high-mannose or hybrid type, respectively. In the study in which the type of structure at each individual site was determined[11] 11 of the 24 total sites, 46%, had high-mannose or hybrid type structures. In contrast, and as expected, the gp120 isolated from virally infected cells had a higher proportion, 62% and 83%, of these structures, probably due to release of incompletely processed protein. The sum of this data suggests that high-mannose and/or hybrid type oligosaccharide moieties are normal components of mature gp120, produced either by recombinant CHO cells or virally infected cells, and that they are not an artefact of cell lysis.

In CL44 rgp120 7% of the oligosaccharides have a hybrid type structure[19]. The 7% hybrid structures in CL44 could represent the summation of trace amounts of such structures at some or all of the glycosylation sites in the final product. Also, it is known that glyco-proteins expressed in CHO cells can be secreted with a mixture of high mannose and complex type oligosacharides at a particular glycosylation site[20]. However 4% of the total oligosaccharides present on rgp120 corresponds to 1 of the 24 total N-linked sites and could suggest that at one particular site on gp120 the final processed oligosaccharide is of the hybrid type.

Table 1. Types of Oligosaccharides On gp120

	High Mannose	Hybrid	Complex
CHO rgp120 [19a]	32%	7%	61%
H9 sgp120 [17]	54%[b]	-	46%
H9 cgp120 [17]	83%[b]	-	17%
H9 cgp120 [18]	60%	2%	37%

[a]Mole % data summarized from the references in parentheses; CHO rgp120 is soluble recombinant gp120 expressed in CHO cells; H9 sgp120 and H9 cgp120 are soluble or cell-associated gp120, respectively, produced by HIV-1 IIIB infected H9 cells. [b]Numbers represent total Endo H released oligosaccharide and, as such, are the sum of the high-mannose and hybrid mole percentages.

DISULFIDE BOND VARIANTS

The pattern of cysteines in gp120 IIIB is highly conserved among the HIV-1, HIV-2 and SIV isolate sequences compiled in the Los Alamos data base[16]. Of the published envelope sequences each contains all of the 18 cysteines present in gp120 IIIB. In the HIV-1 Z3 sequence (Figure 2) and in all of the published HIV-2 and SIV sequences (compare Figure 1 with Figures 3 and 5) there are, in addition to the conserved 18, one or more extra pairs of cysteines. The gp120 of the Z3 isolate of HIV-1 has the only additional pair of cysteines reported thus far for an HIV-1 isolate. The extra pair occurs in the fourth hypervariable region (Modrow, et al.[8] and in the fifth disulfide bonded domain. Also shown in Figure 2 are the same regions from the HIV-1 isolates SF2 and IIIB which illustrate the variation in length and sequence that naturally occur within this region. Despite its proximity to that portion of gp120 involved in CD4 binding, this hypervariable loop region can be deleted without affecting CD4 binding[13]. Because the glypo-protein can tolerate extensive modification in hypervariable region 4 without major effect on one of its primary functional attributes, this suggests that one of the functions of the sequence variability in the hypervariable regions, like region 4, is to produce antigenic variability to facilitate escape from host immune surveillance.

In HIV-2 and SIV the gp120 homologs generally have extra cysteine residues in the region analogous to disulfide bonded domain II in HIV-1 gp120 (Figures 3 and 5). These additional cysteines are usually in pairs and are, with one exception discussed below, associated with regions that can be considered analogous to hypervariable regions one and two of HIV-1 gp120 (Figure 1). The gp120 proteins of two of the HIV-2 isolates (NIHZ and GH1) have unpaired cysteines. In both of these isolates apparently one cysteine from an additional pair in the region analogous to the second disulfide bonded domain of HIV-1 gp120 is mutated. Thus these cysteine residues present in addition to the conserved 18 usually occur within otherwise highly variable regions where they might be assumed to have minimal effect on the conserved structural features of the molecule. These regions are also not currently associated with cellular or species tropism, or CD4 binding. The apparent conservation of the 18 cysteines of gp120 IIIB among the other primate immunodefficiency virus major envelope glyco-proteins and the observation that, when present, additional cysteines are confined to regions of hypervariable sequence where structural variability can be tolerated suggests that the pattern of disulfide bonds among the conserved cysteines determined experimentally for rgp120 IIIB is similarly conserved in the HIV-2 and SIV gp120 homologs.

MODELS OF THE DISULFIDE BONDING IN THE HIV-2 AND SIV gp120 HOMOLOGS

A proposed model for the disulfide bonding in the gp120 homolog of the ROD HIV-2 isolate, based on the disulfide bonding pattern experimentally determined for HIV-1 gp120 IIIB, is presented in Figure 3. The major envelope glyco-protein of the ROD isolate was chosen for illustration because it contains the two additional pairs of cysteines that are characteristic of HIV-2 and SIV and because it is one of the most widely studied HIV-2 isolates. There is

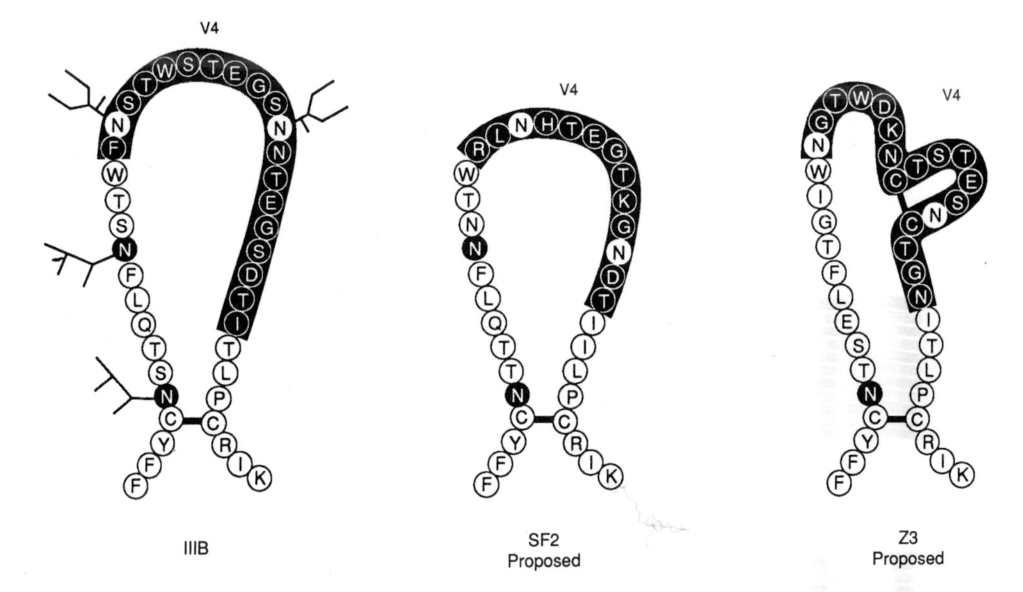

Figure 2. Variants Of The V4 Region of gp120. The primary structure of the loop containing the fourth hypervariable region of HIV-1 gp120[8] is presented for the HIV-1 isolates IIIB, SF2 and Z3. The IIIB structure was experimentally determined[11] and those for SF2 and Z3 are proposed based on that of IIIB. The hypervariable sequence is outlined in black and aspargines that are potential N-linked glycosylation sites are indicated by contrasting color.

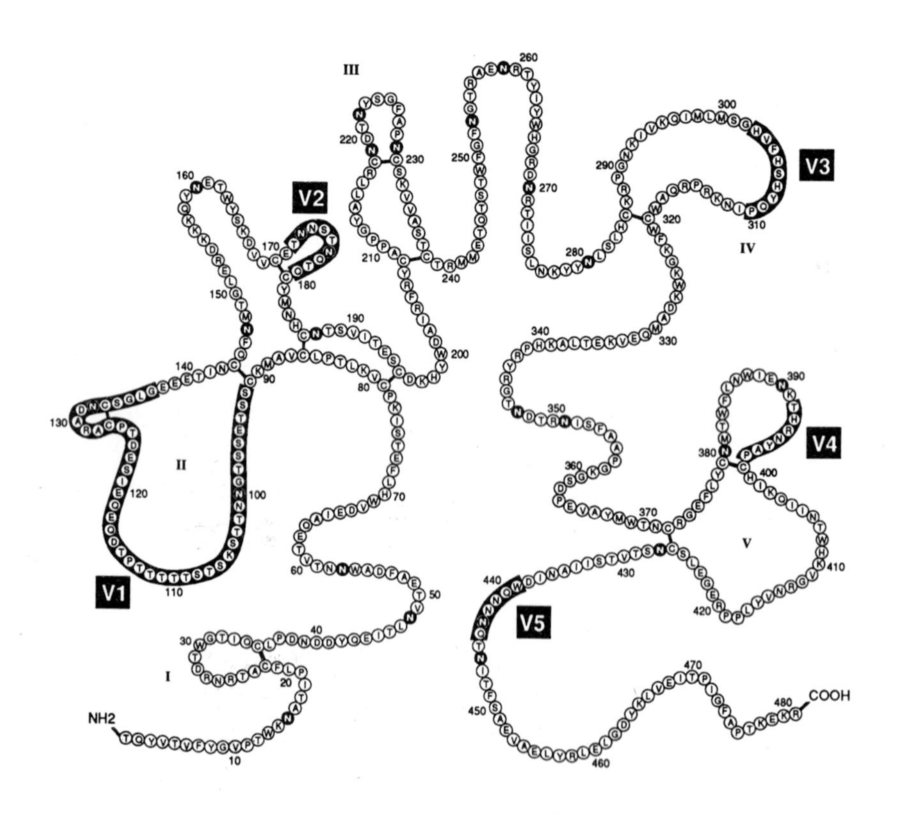

Figure 3. Proposed Model Of The Disulfide Bonding In The gp120 Homolog Of HIV-2 ROD. Roman Numerals refer to the five disulfide bonded domains analogous to those of HIV-1 gp120 (Figure 1) and the boxed sequences accompanied by a boxed number refer to the hypervariable regions described in the text. Potential N-linked glycosylation sites are indicated by boxes around the respective asparagine residues. The amino and carboxy termini are placed by analogy to HIV-1 gp120 and have not been experimentally determined.

considerable sequence homology between HIV-1 IIIB and HIV-2 ROD around the cysteines that form disulfide bonded domains I, III, IV, V and the "neck" of II in HIV-1 IIIB[16]. The five hypervariable regions in HIV-1 gp120 were originally defined as regions in which the amino acid sequence not only varied but also where there were insertions and/or deletions of sequence[8]. Following this definition there are five similar regions in the HIV-2 gp120 homolog that occur in analogous positions (16 and Figure 3). Functionally there are also similarities between HIV-1 and HIV-2 that are mediated by their respective major envelope glyco-proteins. Both HIV-1 and HIV-2 infect the same types of cells, and infection by both is inhibited in vitro by a soluble form of CD4[21,22]. The sequence homology and functional similarities suggest that the gp120 homolog of HIV-2 is structurally similar to gp120 of HIV-1 in its disulfide bonding pattern and allow postulation of the disulfide bonding pattern of ROD (Figure 3) based on that experimentally determined for IIIB. The pattern proposed in Figure 3 is only a postulation and is not based on any experimental evidence.

The arrangement of the disulfide bonds in the distal portion of the ROD sequence analogous to domain II are even more problematic than those in the other domains because of the presence of regions of high sequence variability. Figure 4 illustrates that there is some homology of the sequence around the eighth cysteine in HIV-2 (C-144 in ROD) with that around the sixth cysteine in HIV-1 (C-127 in IIIB). If we postulate a bond in ROD between C-91 and C-144 based on this homology (analogous to that between C-101 and C-127 in IIIB) then the additional pairs of cysteines and the first and second hypervariable regions of ROD can be positioned as proposed in Figure 3. This allows the ROD sequence to be molded into a domain analogous to domain II of IIIB; despite the appealing symmetry of this model it cannot be over-emphasized that it has been produced as an intellectual exercise and is not based on experimental evidence.

As for HIV-2, a proposed model for the disulfide bonding in the gp120 homolog of the MM142 SIV isolate is presented in Figure 5. This isolate was chosen for illustration because its gp120 homolog has the largest number of cysteine residues (24) reported to date and, as such, is the most divergent form relative to HIV-1 IIIB. It possesses the four additional cysteine residues present in the region analogous to domain II of HIV-1 gp120 that are characteristic of the gp120 homologs of HIV-2 and SIV, and it has an additional pair of cysteines in the region that corresponds to the sequence between domains II and III of IIIB gp120. If one applies to SIV the same criteria that were used to define hypervariable regions in HIV-1 gp120 (see above) then only the two regions noted in Figure 5 are defined. These are analogous to hypervariable regions one and four of HIV-1 gp120. There are however regions of sequence variability without concomitant insertions or deletions between residues 177 - 184, 325 - 333, and 453 - 461 of MM142[16]. These are analogous to the other hypervariable regions of HIV-1 gp120 except that the third region (325 - 333) is displaced out of the proposed disulfide loop of domain IV in the gp120 homolog

```
hivbru     150 G E M M M E K G E I K N C S F N I S T S I R G K V Q K E Y - A F F Y K L D I I P I
hivhxb2    145 G R M I M E K G E I K N C S F N I S T S I R G K V Q K E Y - A F F Y K L D I I P I
hivmn      150 S E G T I K G G E M K N C S F N I T T S I R D K M Q K E Y - A L L Y K L D I V S I
hivbrva    143 S G K M M E G G E M K N C S F N I T T S I R D K M Q K E Y - A L F Y K L D I V P I
hivsc      148 N R G K M E G G E M T N C S F N I T T S I R S K V Q K E Y - A L F Y K L D V V P I
hivjh32    148 G G E K M E K G E M K N C S F N I T T S I R D K V Q K E H - A L F Y K H D V V P I
hivcdc42   151 V W E Q R G K G E M R N C S F N I T T S I R D K V Q R E Y - A L F Y K L D V E P I
hivoyi     150 S W E T M E K G E L K N C S F N T T T S I R D K M Q E Q Y - A L F Y K L D V L P I
hivsf2     143 N W K E E I K G E I K N C S F N I T T S I R D K I Q K E N - A L F R N L D V V P I
hivhan     140 S W G R M E K G E I Q N C S F K V T T N I R D K V Q K E S - A L F Y K T D V V P I
hivjfl     142 S W G K M E E G E I K N C S F N T T T S I K N K M Q R E Y - A L F Y K L D V V P I
hivwmj22   140 N T T I I G G G E V K N C S F N I T T S R R D K V H K E Y - A L F Y K L D V V P I
hivrf      145 G G T M M E N G E I K N C S F Q V T T S R R D K T Q K K Y - A L F Y K L D V V P I
hiveli     142 N N V T T E E K G M K N C S F N V T T V L K D K K Q Q V Y - A L F Y R L D I V P I
hivz2z6    142 N N N V T E E I R M K N C S F N I T T V V R D K - T K Q V H A L F Y R L D I V P I
hivndk     140 N G K V E E E E K R K N C S F N - - - - V R D K R E Q V Y - A L F Y K L D I V P I
hivjy1     148 E E Q M M E K G E M K N C S F N I T T V I S D K - K K Q V H A L F Y R L D V V P I
hivmal     150 A E L K M E I G E V K N C S F N I T P V G S D K - R Q E Y - A T F Y N L D L V Q I
hivz321    141 N V D T E M K E E I K N C S Y N M T T E L R D K Q R K I Y - S L F Y R L D I V P I
hiv2ben    154 T C A G L G Y E E M M Q C E F N M K G L E Q D K - K R R Y K D T W Y L E D V V C D
hiv2gh1    141 N C T G L G K E E I V N C Q F Y M T G L E R D K - K K Q Y N E T W Y S K D V V C E
hiv2isy    142 N C S G L R E E D M V E C Q F N M T G L E L D K - K K Q Y S E T W Y S K D V V C E
hiv2nihz   142 N C T G L K E E E M I D C Q F S M T G L E R D K - R K Q Y T E A W Y S K D V V C D
hiv2rod    151 N C S G L G E E E T I N C Q F N M T G L E R D K - K K Q Y N E T W Y S K D V V C E
hiv2st     145 N C T G L G E E E M V D C Q F N M T G L E R D K - K K L Y N E T W Y S K D V V C E
sivmm251   158 N C T G L E Q E Q M I S C K F T M T G L K R D K - T K E Y N E T W Y S T D L V C E
sivmm142   159 N C T G L E Q E P M I S C K F N M T G L K R D K - K K E Y N E T W Y S A D L V C E
sivmmh4    158 S C A G L E Q E P M I G C K F N M T G L N R D K - K K E Y N E T W Y S R D L I C E
sivmm239   156 N C T G L E Q E Q M I S C K F N M T G L K R D K - K K E Y N E T W Y S A D L V C E
sivmne     156 N C T G L E Q E P M I S C K F N M T G L K R D K - R R E Y N E T W Y S A D L V C E
```

Figure 4. Sequence Homology In The Distal Portion Of Domain II Of gp120. Sequence data is from the Los Alamos data base[16] and numbering is from the first amino acid of the signal sequence. Alignment of the sequences around the sixth cysteine residue in mature HIV-1 gp120 (C162 in BRU) is compared to that around the eighth cysteine residue in the mature HIV-2 and SIV gp120 homologs (C163 in ROD and C171 in MM142). The amino acids that are conserved among the three virus envelope proteins are enclosed in boxes.

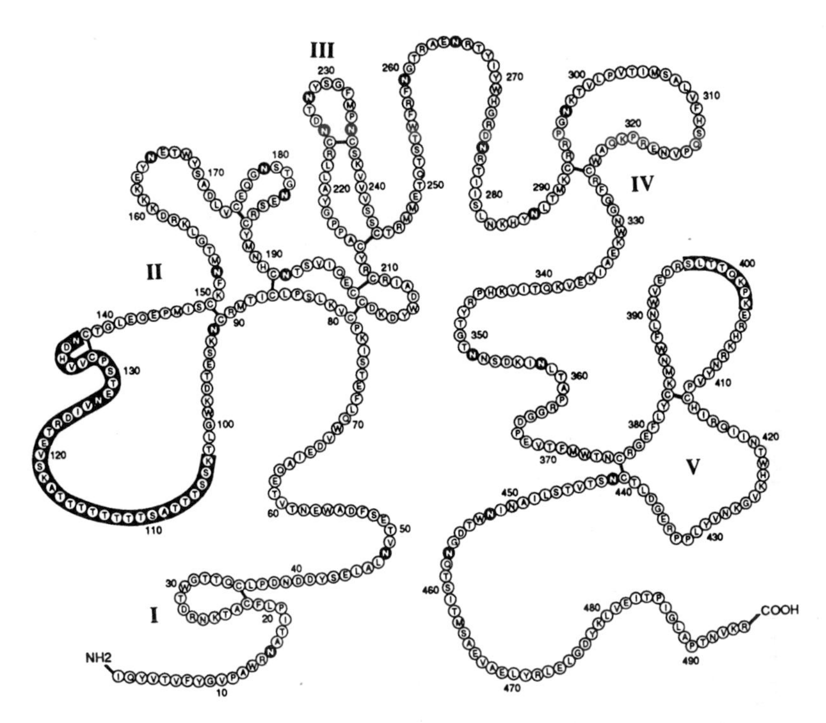

potential glycosylation site

Figure 5. Proposed Model Of The Disulfide Bonding In The gp120 Homolog Of SIV MM142. Roman Numerals refer to the five disulfide bonded domains analogous to those of HIV-1 gp120 (Figure 1) and the boxed sequences refer to the hypervariable regions described in the text. Potential N-linked glycosylation sites are indicated. The amino and carboxyl termini are placed by analogy to HIV-1 gp120 and have not been experimentally determined.

of SIV. The disulfide bonding pattern presented in Figure 5, as was stated for Figure 3, is purely speculative and not based on any experimental evidence. It is presented solely to suggest possible structural similarities between the major envelope proteins of the primate immunodefficiency viruses.

The illustrations in Figures 3 and 5 show that the sequences of the HIV-2 and SIV gp120 homologs can readily be molded into a pattern that is very similar to that determined for the HIV-1 primary structure. The sequence homology and inferred disulfide bonding suggest that domains I,III, IV, V and the proximal part of domain II are structurally conserved and that the positioning of regions of high sequence variability are similarly positioned in the three viral envelope glyco-proteins. This is not surprising for domain V, which has been associated with CD4 binding in HIV-1, because HIV-2 and SIV appear to use the same receptor or its simian counterpart[21,22]. The other domains have not yet been associated with any particular function so their apparent conservation only suggests that they have some other functional or structural significance. Indeed they may simply be conserved structural motifs. The suggested conservation of the structurally more complex domains II and III, which have been largely ignored in past mutagenesis studies, invites further investigation.

CONCLUSION

The complete primary structure of recombinant gp120 from HIV-1 IIIB has been reported. Secreted rgp120 contains a large proportion of high mannose and/or hybrid type oligosaccharide structures. Although these structures are frequently assumed to be intermediates in the maturation process for complex type oligosaccharides, they appear to be present on the completely processed mature gp120 protein. From the disulfide bonding pattern determined experimentally for HIV-1 gp120 and sequence homology between the HIV-1, HIV-2 and SIV major envelope glyco-proteins, models have been proposed for the disulfide bonding in the gp120 homologs of HIV-2 ROD and SIV MM142. These models are presented as the basis for speculations on the conservation of structure and function in these proteins.

REFERENCES

1. A. G. Dalgleish, P. C. L. Beverely, P. R. Clapham, D. H. Crawford, M. F. Greaves, and R. A. Weiss, The CD4 (T4) antigen is an essential component of the receptor for the AIDS retrovirus, Nature. 312:763 (1984).

2. D. H. Smith, R. A. Byrne, S. A. Marsters, T. J. Gregory, J. E. Groopman, and D. J. Capon, Blocking of human immunodeficiency virus infectivity by a soluble, secreted form of the CD4 antigen, Science. 238:1704 (1987).

3. K. Javaherian, A. J. Langlois, C. McDanal, K. L. Ross, L. I. Eckler, C. L. Jellis, A. T. Profy, J. R. Rusche, D. P. Bolognesi, S. D. Putney, and T. J. Matthews,

Principal neutralizing domain of the human immunodeficiency virus type 1 envelope protein, Proc. Natl. Acad. Sci. USA. 86: 6768 (1989).

4. P. W. Berman, T. J. Gregory, L. Riddle, G. R. Nakamura, M. A. Champe, J. P. Porter, F. M. Wurm, R. D. Hershberg, E. K. Cobb, and J. W. Eichberg, Protection of chimpanzees from infection by HIV-1 after vaccination with recombinant glycoprotein gp120 but not gp160, Nature. 345: 622 (1990).

5. I. Berkower, G. E. Smith, C. Giri, and D. Murphy, Human immunodeficiency virus type 1, predominance of a group-specific neutralizing epitope that persists despite genetic variation, J. Exp. Med. 170: 1681 (1989).

6. J. D. Lifson, M. B. Feinberg, G. R. Reyes, L. Rabins, B. Banapour, S. Chakrabati, B. Moss, F. Wong-Staal, K. S. Steimer, ans E. G. Engleman, Induction of CD4-dependent cell fusion by the HTLV-III/LAV envelope glycoprotein, Nature. 323: 725 (1986).

7. J. Sodroski, W. C. Goh, C. Rosen, K. Campbell, and W. A. Haseltine, Role of the HTLV-III/LAV envelope in syncitium formation and cytopathicity, Nature. 322:470 (1986).

8. S. Modrow, B. H. Hahn, G. M. Shaw, R. C. Gallo, F. Wong-Staal, and H. Wolf, Computer-assisted analysis of envelope protein sequences of seven human immunodeficiency virus isolates: prediction of antigenic epitopes in conserved and variable regions, J. Virol. 61: 570 (1987).

9. L. A. Lasky, J. E. Groopman, C. W. Fennie, P. M. Benz, D. J. Capon, D. J. Dowbenko, G. R. Nakamura, W. M. Nunes, E. M. Renz, and P. W. Berman, Neutralization of the AIDS retrovirus by antibodies to a recombinant envelope glycoprotein, Science. 237: 209 (1986).

10. L. A. Lasky, G. Nakamura, D. H. Smith, C. Fennie, C. Shimasaki, E. Patzer, P. Berman, T. Gregory, and D. J. Capon, Delineation of a region of the human immunodeficiency virus type 1 gp120 glycoprotein critical for interaction with the CD4 receptor, Cell. 50:975 (1987).

11. C. K. Leonard, M. W. Spellman, L. Riddle, R. J. Harris, J. N. Thomas, and T. J. Gregory, Assignment of intrachain disulfide bonds and characterization of potential glycosylation sites of the type 1 recombinant human immunodeficiency virus envelope glycoprotein (gp120) expressed in chinese hamster ovary cells, J. Biol. Chem. 265: 10373 (1990).

12. D. C. Capon, S. M. Chamow, J. Mordenti, S. A. Marsters, T. Gregory, H. Mitsuya, R. A. Byrn, C. Lucas, F. M. Wurm, J. E. Groopman, S. Broder, and D. H. Smith,

Designing CD4 immunoadhesins for AIDS therapy, Nature. 337: 525 (1989).

13. M. Kowalski, J. Potz, L. Basiripour, T. Dorfman, W. C. Goh, E. Terwilliger, A. Dayton, C. Rosen, W. Haseltine, and J. Sodroski, Functional regions of the envelope glycoprotein og human immunodeficiency virus type 1, Science. 237: 1351 (1987).

14. A. Cordonnier, L. Montagnier, and M. Emerman, Single amino acid changes in HIV envelope affect viral tropism and receptor binding, Nature. 340: 571 (1989).

15. A. Nygren, T. Bergman, T. Matthews, H. Jornvall, and H. Wigzell, 95 and 25 kDa fragments of the human immunodeficiency virus envelope glycoprotein gp120 bind to the CD4 receptor, Proc.Natl. Acad. Sci. USA. 85: 6543 (1988).

16. G. Meyers, A. Rabson, S. Josephs, T. Smith, J. Berzofsky, and F. Wong-Staal, eds., "Human Retroviruses and AIDS, LA-UR, 89-743, Los Alamos National Laboratory, Los Alamos (1989).

17. H. Geyer, C. Holschbach, G. Hunsmann, and J. Schneider, Carbohydrates of human immunodeficiency virus, J. Biol. Chem. 263: 11760 (1988).

18. T. Mizuochi, T. J. Matthews, M. Kato, J. Hamako, K. Titani, J. Solomon, and T. Feizi, Diversity of oligosaccharide structures on the envelope glycoprotein gp120 of human immunodeficiency virus 1 from the lymphoblastoid cell line H9, J. Biol. Chem. 265: 8519 (1990).

19. T. Mizuochi, M. W. Spellman, M. Larkin, J. Solomon, L. J. Basa, and T. Feizi, Carbohydrate structures of the human immunodeficiency virus (HIV) recombinant envelope glycoprotein gp120 produced in Chinese hamster ovary cells, Biochem. J. 254: 599 (1988).

20. R. J. Harris, S. M. Chamow, T. J. Gregory, and M. W. Spellman, Characterization of a soluble form of human CD4, Eur. J.Biochem. 188:291 (1990).

21 P. R. Clapham, J. N. Weber, D. Whitby, K. McIntosh, A. G. Dalgleish, P. J. Maddon, K. C. Deen, R. W. Sweet, and R. A. Weiss, Soluble CD4 blocks the infectivity of diverse strains of HIV and SIV for T cells and moncytes but not for brain and muscle cells, Nature. 337: 368 (1989).

22. R. A. Byrn, I. Sekigawa, S. M. Chamow, J. S. Johnson, T. J. Gregory, D. J. Capon, and J. E. Groopman, Characterization of in vitro inhibition of human immunodeficiency virus by purified CD4, J. Virol. 63: 4370 (1989).

THE LENTIVIRUS REGULATORY PROTEINS REV AND REX ARE SITE SPECIFIC RNA BINDING PROTEINS

G. King Farrington[1], Paul Lynch[1], Amy Jensen[1], Ernst Böhnlein[2], Reed Doten[1], Theodore Maione[1], Thomas Daly[1], James Rusche[1]

[1]Repligen Corporation, One Kendall Square, Bldg 700, Cambridge, MA 02139

[2]Sandoz Research Institute, Brunnerstrasse 59, A-1235 Vienna, Austria

Introduction

Lentiviruses encode proteins that regulate viral gene expression. One class of regulatory proteins are the HIV-1 REV and HTLV-1 REX proteins which are essential gene products for virus replication (Feinberg *et al.*, 1986; Sodroski *et al.*, 1986; Seiki *et al.*, 1988). Mutations within these genes ablate the expression of viral structural proteins without affecting the total level of viral RNA in the nucleus (Felber *et al.*, 1989; Terwilliger *et al.*, 1988; Knight *et al.*, 1987; Emerman *et al.*, 1989; Hammarskjöld *et al.*, 1989; Nosaka *et al.*, 1989). Transient expression and virus rescue experiments have led to the identification of a cis-acting sequence in the transcribed RNA of HIV-1 and HTLV-1 termed the REV or REX responsive element (RRE and RxRE respectively) that is essential for REV or REX function (Inoue *et al.*, 1987; Seiki, *et al.*, 1988; Malim *et al.*, 1989; Hadzopoulou-Cladaras *et al.*, 1989). Genetic studies of these genes and their cis-acting elements have led to the following hypothesis: the REV and REX gene products complete viral gene expression and replication by facilitating the transport from the nucleus of RNA transcripts encoding structural proteins and new viral genomes. One proposed step in this process involves the direct interaction of REV and REX with viral RNA transcripts in the nucleus.

To test the RNA binding prediction and further refine the RNA transport hypothesis we set about biochemically characterizing the REV and REX proteins. Preliminary studies have demonstrated that the purified REV protein does possess RRE specific RNA binding activity (Daly *et al.*, 1989; Heaphy *et al.*, 1990). This report describes similar studies with HTLV-1 REX protein and a further characterization of HIV-1 REV. Through these studies we hope to illuminate how gene expression can be regulated by selective RNA transport.

Immunobiology of Proteins and Peptides VI
Edited by M.Z. Atassi, Plenum Press, New York, 1991

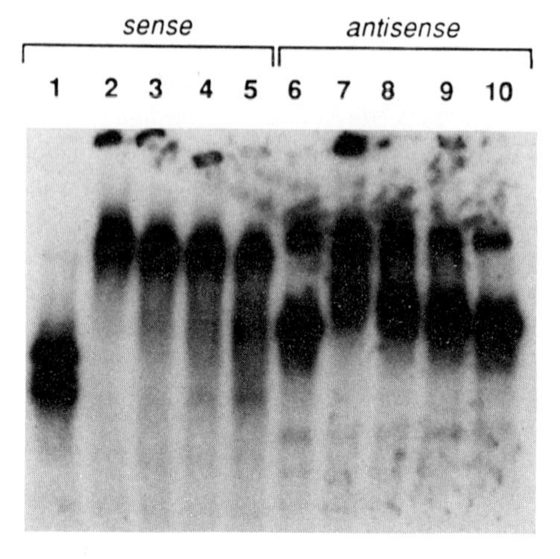

Fig. 1. Titration of RxRE RNA lanes 1-5 and antisense RxRE RNA lanes 6-7. The concentration of REX was varied from 0.62 to 2.5 μMolar as indicated, in the presence of phosphate buffered saline pH 7.2, 1.5mM DTT, 0.8M urea and 100μg/well tRNA.

HTLV-1 REX is a Sequence Specific RNA Binding Protein

This report briefly describes the expression of REX in *E. coli*, the purification of REX and demonstrates REX specifically binds to the putative RxRE under *in vitro* conditions. REX was expressed in *E. coli* by translationally coupling the REX gene to the *E. coli* β-glucuronidase gene on a high copy number plasmid. The bacterial cells were harvested, lysed, centrifuged and the pellet extracted with 8M urea in 50mM Mes, pH 6.5. The solubilized proteins were then subjected to sequential chromatography on S-sepharose and gel filtration, resulting in a near homogeneous preparation of REX (Farrington *et al.*, manuscript submitted).

The specific HTLV-I RxRE sequence used extends from nucleotides 8600 to 8879 (Malik *et al.*, 1988). The secondary solution structure of this construct has been reported (Toyoshima *et al.*, 1990).

The ability of REX to specifically bind this RxRE construct *in vitro* was tested in a gel retardation assay. The concentration of REX was varied in the presence of fixed concentrations of tRNA (Figure 1). A relatively narrow

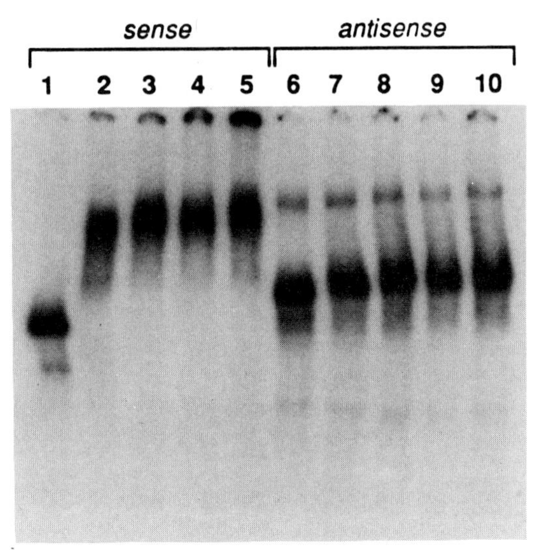

Fig.2. tRNA titration. The tRNA concentration was varied, as indicated below each lane, in the presence of phosphate buffered saline pH 7.2, 0.62 μMolar REX, 1.5mM DTT, 0.8M urea RxRE RNA in lanes 1-5 and antisense RxRE RNA in lanes 6-10.

concentration range of REX from about 0.5μM to 2.5μM demonstrated specificity for the RxRE in the gel retardation assay. To assess the level of specificity REX exhibits for binding to the RxRE versus antisense RxRE, fixed concentrations of REX were titrated with varied concentrations of tRNA (Figure 2). Lanes 1 and 6 are the RxRE and the antisense RxRE respectively with no REX, lanes 2-5 and 7-10 are in the presence of 0.62μM REX and contain varied concentrations of tRNA. The RxRE RNA (lanes 2-5) is retarded even in the presence of excess tRNA, while antisense RNA (lanes 7-10) mobility is unaffected. This result clearly demonstrates REX specifically binds the RxRE not the antisense RxRE. This result is in accord with genetic studies which demonstrated the loss of REX function if the RxRE orientation was reversed (Hanly *et al.*). In a further demonstration of specificity, the binding of REX to the RxRE could be competed with excess cold RxRE transcript (data not shown).

In additional studies, Ballaun *et al.* (manuscript submitted) have demonstrated that specific REX binding of RxRE mutants in the gel retardation assay correlated with *in vivo* activity of the same transcripts. These results confirm the hypothesis that REX

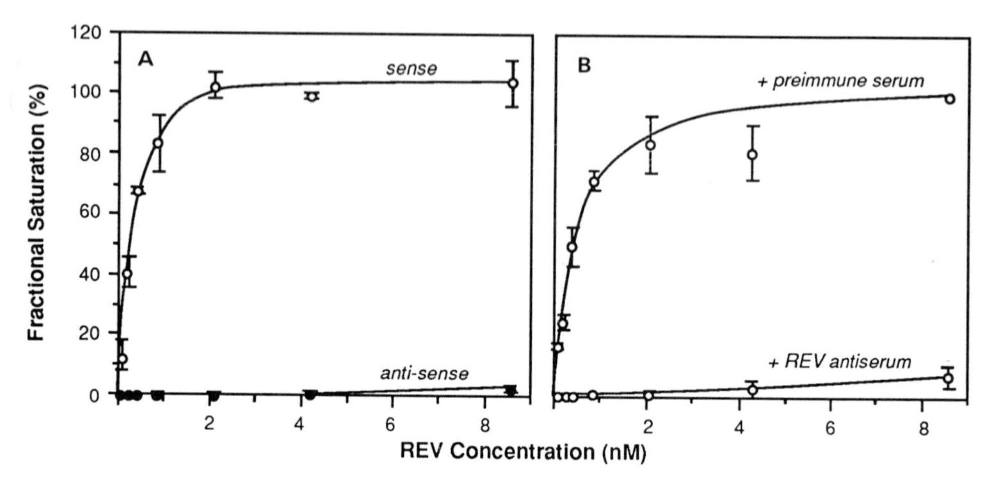

Fig. 3A. Direct binding of REV (0.1-9nM) to the RRE and antisense RRE by nitrocellulose filter binding assay. Assays typically contained 10 units RNAsin, 50µg/mL bovine serum albumin in PBS with 0.25M NaCl and no competitor RNA.

3B. REV antiserum or preimmune serum was added to each assay followed by an aliquot of Protein A Sepharose. The fractions were centrifugued filtered and subject to nitrocellulose filter binding.

binds the RxRE specifically and this binding is essential to REX function. Further, these results are in agreement with the biological results suggesting that REX and REV act in a mechanistically similar manner to alter viral gene expression.

Characterizing the HIV-1 REV Protein and Binding of the REV Responsive Element

The goal of this work was to define the region of the REV protein that interacts with RNA. Genetic studies have previously identified mutations within REV that result in the loss of RNA transport function (Malim *et al.*, 1989; Sadaie *et al.*, 1988). Deletion analysis of RRE has indicated a subregion of the RRE termed stem II which is essential for REV function (Malim *et al.*, 1990). Previously, we had purified recombinant REV protein from *E. coli* and characterized the binding of pure protein to the RRE (Daly *et al.*, 1989). Using similar techniques, we purified REV proteins with specific mutations and characterized the RNA binding properties of these proteins.

Protein dependent retention of radiolabeled RNA on nitro cellulose filters provides a quantitative measure of protein complexed with RRE-RNA of either the sense or antisense mRNA orientation. While both species are bound in low ionic strength buffers, at 0.25M NaCl only the sense RRE-RNA is efficiently retained on the filter (Figure 3A). This supports the genetic observation that the cis-acting RRE is orientation specific (Malim *et al.*, 1989). REV dependent filter retention of RNA is further demonstrated by observing that REV antiserum blocks the formation of the RNA-protein complex (Figure 3B). However, if the protein-RNA

SP

1 M A G R S G D S D E D L L K A V R L I K F L Y Q S N P P P N 30

M4

31 P E G T R Q A R R N R R R R W R E R Q R Q I H S I S E R I L 60

M5 M7

61 S T Y L G R S A E P V P L Q L P P L E R L T L D C N E D C G 90

M10

91 T S G T Q G V G S P Q I L V E S P T I L E S G A K E 116

Fig. 4. The 116 amino acid sequence of REV. Each boxed pair of amino acids were substituted by an aspartic acid and a leucine. SP indicates the border between the two REV coding exons.

complex is allowed to form before addition of the antiserum, the complex remains intact (data not shown).

REV proteins with missense mutations were identified that had apparently lost the ability to transport RRE-containing RNA (Malim *et al.*, 1989). One of the mutations, M10, demonstrated a dominant negative phenotype in that it could suppress the activity of wild type REV. We chose three of these mutations (Figure 4) for biochemical study to identify which domain affects RNA binding. M4, M5 and M10 were expressed and purified from *E. coli*. Quantitative binding studies by gel retardation and filter retention of nucleic acids (Figure 5) identified differences between these three proteins. The two mutant proteins from the poly-arg region showed impaired RNA binding characteristics. The M4 mutant was reduced 10 fold in the specificity for RRE RNA as compared to non-specific RNA. The M5 mutant was unable to distinguish between specific and non-specific RNA. However, the M10 mutation bound RNA as the wild-type REV protein. These results indicate that mutations in the poly-arg region affect the affinity and specificity of RNA binding while mutations in the second domain exemplified by the M10 mutant affects some other function of the protein, most likely its interaction with other factors within the cell.

Summary

These studies demonstrate that HIV-1 REV and HTLV-1 REX are site specific RNA binding proteins. In addition, the REV protein studies indicate a protein domain essential for biological function that is not involved in RRE binding. Lentiviruses are unique among retroviruses in providing gene products to switch expression from early to late genes. Rather than transcriptional control, this is accomplished by regulating RNA transport. A further understanding of these lentiviral genes may identify means of inhibiting viruses responsible for insidious human disease.

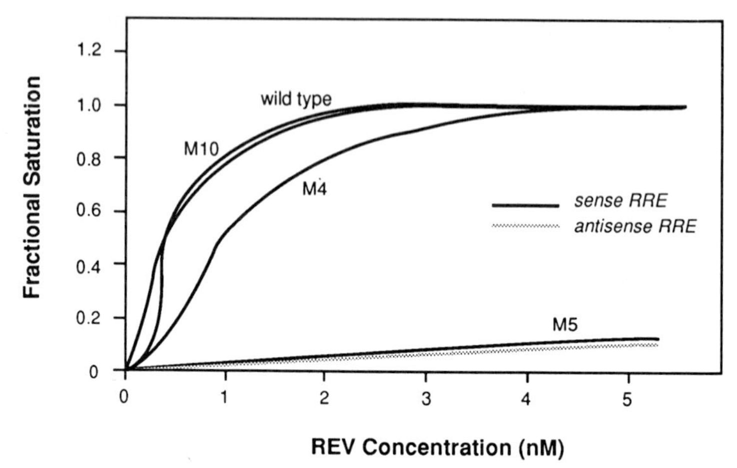

Fig.5. Binding of varied concentrations of REV and REV mutants M-10, M-5 and M-4 to the RRE or anti-sense RRE in the presence of 1mM DTT and excess MS-2 nonspecific RNA.

Acknowledgements

The authors wish to thank Greg Fisk and Mike White for technical contributions, Bryan Cullen for providing gene sequence of REV mutants, and Sue Cook, Joachim Hauber and Helmut Bachmayer for helpful discussions. Manuscript preparation was kindly provided by Marylou Cronin.

References

1) Feinberg, M.B., Jarrett, R.F., Aldovini, A., Gallo, R.C., and Wong-Staal, F., 1986, HTLV-III Expression and Production Involve Complex Regulation at the Levels of Splicing and Translation of Viral RNA, Cell, 46:807-817.

2) Sodroski, J., Goh, W.C., Rosen, C., Dayton, A., Terwilliger, E., and Haseltine, W., 1986, A second post-transcriptional *trans*-activator gene required for HTLV-III replication, Nature, 321:412-417.

3) Seiki, M., Inoue, J.-I., Hidaka, M., and Yoshida, M., 1988, Two cis-acting elements responsible for posttranscriptional trans-regulation of gene expression of human T-cell leukemia virus type I, Proc. Natl. Acad. Sci. USA, 85: 7124-7128.

4) Felber, B.K., Hadzopoulou-Cladaras, M., Cladaras, C., Copeland, T., and Pavlakis, G.N., 1989, rev protein of human immunodeficiency virus type 1 affects the stability and transport of the viral mRNA, Proc. Natl. Acad. Sci. USA, 86: 1495-1499.

5) Terwilliger, E., Burghoff, R., Sia, R., Sodraski, J., Haseltine, W., and Rosen, C., 1988, The Art gene product of human immunodeficiency virus is required for replication, J.of Virology, 62:655-658.

6) Knight, D.M., Flomerfelt, F.A., and Ghrayeb, J., 1987, Expression of the art/trs Protein of HIV and Study of Its Role in Viral Envelope Synthesis, Science, 236:837-840.

7) Emerman, M., Vazeux, R., and Peden, K., 1989, The *rev* Gene Product of the Human Immunodeficiency Virus Affects Envelope-Specific RNA Localization, Cell, 57:1155-1165.

8) Hammarskjöld, M.L., Heimer, J., Hammarskjöld, B., Sangwan, I., Albert, L., and Rekosk, D., 1989, Regulation of Human Immunodeficiency Virus *env* Expression by the *rev* Gene Product, J. of Virology, 63:1959-1966.

9) Nosaka, T., Siomi, H., Adachi, Y., Ishibashi, M., Kubota, S., Maki, M., and Hatanaka, M., 1989, Nucleolar targeting signal of human T-cell leukemia virus type I *rex*-encoded protein is essential for cytoplasmic accumulation of unspliced viral mRNA, Proc. Natl. Acad. Sci. USA, 86:9798-9802.

10) Inoue, J.-I., Yoshida, M., and Seiki, M., 1987, Transcriptional (p40x) and post-transcriptional ($p27^{x-III}$) regulators are required for the expression and replications of Human T-cell leukemia virus type I genes, Proc. Natl. Acad. Sci. USA, 84:3653-3657.

11) Malim, M.H., Hauber, J., Le, S.-Y., Maizel, J.V., and Cullen, B.R., 1989, The HIV-1 *rev trans*-activator acts through a structured target sequence to activate nuclear export of unspliced viral mRNA, Nature, 338:254-257.

12) Hadzopoulou-Cladaras, M., Felber, B.K., Cladaras, C., Athanassopoulos, A., Tse, A., and Pavlakis, G.N., 1989, The *rev* (*trs/art*) Protein of Human Immunodeficiency Virus Type 1 Affects Viral mRNA and Protein Expression via a *cis*-Acting Sequence in the *env* Region, J. of Virology, 63:1265-1274.

13) Daly, T.J., Cook, K.S., Gray, G.S., Maione, T.E., and Rusche, J.R., 1989, Specific binding of HIV-1 recombinant Rev protein to the Rev-responsive element *in vitro*, Nature, 342:816-819.

14) Heaphy, S., Dingwall, C., Ernberg, I., Gait, M.J., Green, S.M., Karn, J., Lowe, A.D., Singh, M., and Skinner, M.A., 1990, HIV-1 Regulator of Virion Expression (Rev) Protein Binds to an RNA Stem-Loop Structure Located within the Rev Response Element Region, Cell, 60:685-693.

15) Malik, K.T.A., Even, J., Karpas, A., 1988, Molecular cloning and complete nucleotide sequence of an adult T-cell leukemia virus/human T-cell leukemia virus type I (ATLV/HTLV-I) isolate of caribbean origin: relationship to other members of the ATLV/HTLV-I subgroup, J. Gen. Virology, 69, 1694-1710.

16) Toyoshima, H., Itoh, M., Inoue, J.-I, Seiki, M., Takaku, F., Yoshida, M., 1990, Secondary Structure of the Human T-cell Leukemia Virus Type I rex-Responsive Element is Essential for rex Regulation of RNA processing and Transport of Unspliced RNAs, J. of Virology, 64, 2825-2832.

17) Hanly, S.M., Rimsky, L.T., Malim, M.H., Kim, J.H., Hauber,

J., DucDodon, M., Le, S.-Y., Maizel, J.V., Cullen, B.R., Greene, W.C., 1989, Comparative analysis of the HTLV-I Rex and HIV-1 Rev trans-regulatory proteins and their RNA responsive elements. Genes and Development, 3, 1534-1544.

18) Malim, M.H., Böhnlein, S., Hauber, J., and Cullen, B.R., 1989, Functional Dissection of the HIV-1 Rev *Trans*-Activator- Derivation of a *Trans*-Dominant Repressor of Rev Function, Cell, 58:205-214.

19) Sadaie, M.R., Benter, T., and Wong-Staal, F., 1988, Site-Directed Mutagenesis of Two Trans-Regulatory Genes (*tat*-III, *trs*) of HIV-1, Science, 239:910-913.

20) Malim, M.H., Tiley, L.S., McCarn, D.F., Rusche, J.R., Hauber, J., and Cullen, B.R., 1990, HIV-1 Strucutral Gene Expression Requires Binding of the Rev *Trans*-Activator to Its RNA Target Sequence, Cell, 60, 675-683.

HIV-1 NEUTRALIZING ANTIBODY AND APPROACHES TO THE ENVELOPE DIVERSITY PROBLEM

Thomas J. Matthews[1], Alphonse J. Langlois[1], Stephano Butto'[1], Dani Bolognesi[1], and Kashi Javaherian[2]

[1] Department of Surgery, Duke University Medical Center Durham, North Carolina
[2] Repligen Corporation, Cambridge, Massachusetts

Since the discovery of human immunodeficiency virus (HIV) as the causative agent of AIDS in 1983-84, there has been considerable effort to design an efficacious vaccine. Principally for safety reasons, defined subunit polypeptides and proteins devoid of genomic material have been considered as alternatives to attenuated and/or killed virus. Much of the effort to develope such a component vaccine against HIV has been directed at the envelope gene products. These have included recombinant gp160 (1,2,3), recombinant gp120 (4,5,6), native viral gp120 (7,8,9), bacterially expressed envelope fragments (10), and a large number of synthetic peptides (11-19). In these early studies each of the various reagents were tested for their ability to raise neutralizing antibodies in experimental animals. When successful, resultant sera were usually found to be active only against homologous virus, i.e. were isolate restricted. The results suggested that variable sites on the envelope represented the major target of neutralizing antibody raised by the various experimental immunogens.

It is now apparent that a single region of the envelope gp120 is principally involved in this isolate restricted antibody response (15-18). The site falls in the third variable domain of gp120 and exists as a loop structure bounded by two disulfide bridge cysteine residues at positions 303 and 338 (20). It is often referred to as either the principal neutralizing determinant (PND) or the V3 loop. The PND can differ by as much as 50% among certain HIV-1 isolates and that variation appears to account for the isolate specificity usually observed in experimental sera. Efforts to uncover more conserved sites which serve as targets of neutralizing antibody have been less successful. For this reason as well as implication of the PND as an important determinant in recent successful chimp experiments (21,22,23) we have continued studies of this site in an effort to understand the extent of its variation and devise V3 based immunogens which will induce immunity to a broader spectrum of isolates.

During the course of these studies several experimental sera raised to synthetic peptides containing V3 sequences exhibited cross neutralizing activity to divergent isolates. That sort of cross reactivity contrasted sharply with the specificity more commonly observed in the study of a large number of experimental sera. For example in Table 1, sera taken from three guinea pigs immunized with a synthetic peptide representing the V3 sequence of HTLVIII$_B$ are shown to block cell/cell fusion events mediated by the III$_B$ virus but not the MN virus. In the same Table is shown similar studies of sera taken from another six animals immunized with a peptide based on the HIV-1 MN V3 sequence. Three of these (D4, 282, and 283) exhibited the expected specificity and blocked MN induced fusion but not III$_B$. Sera from the other three

Immunobiology of Proteins and Peptides VI
Edited by M.Z. Atassi, Plenum Press, New York, 1991

Table 1

Isolate Specificity of V3 Peptide Antisera

Guinea Pig	V3 Peptide Immunogen	No. Syncitia/Well 3B	MN
279	3B (135)	0	90
280	3B (135)	0	93
281	3B (135)	14	83
D4	MN (142)	43	0
282	MN (142)	47	0
283	MN (142)	51	0
284	MN (142)	2	0
89	MN (142)	0	0
D3	MN (142)	5	5
Normal Serum	none	49	60

Table 2

Cross Neutralizing Antibodies Directed At
Central Conserved Sequence

Animal	Immunogen	Competing Peptide	No. Syncitia/Well 3B	MN
gp 89	MN (142)	none	0	0
		135	52	0
		142	54	39
		111	8	0
		113	0	0
		116	60	0
		114	7	0

MN (142) Y N K R K R I H I G P G R A F Y T T K N I I G C

114 $R(G\ P\ G)_4C$

113 $(Q\ R\ G\ P\ G\ R)_3C$

111 $(I\ Q\ R\ G\ P\ G)_2$

116 $(G\ P\ G\ R\ A\ F)_3C$

IIIB135 N N T R K S I R I Q R G P G R A F V T I G K I G C

animals (D3, 89, and 284) in contrast were inhibitory to both the III_B and MN isolates. It thus was apparent that the latter animals recognized cross reactive epitopes carried by these two divergent isolates.

In efforts to further understand amino acid residues involved in the cross reacting sera, peptide competition studies were performed as indicated in Table 2 for the serum taken from gp89. Sequences of the peptides utilized in these studies are also indicated on Table 2. In this experiment uninhibited cultures gave 60 multinucleated syncytia for the III_B isolate and 44 for the MN isolate whereas serum from gp89 completely blocked cell fusion induced by both virus isolates. The same assay performed in the presence of the immunizing MN peptide (142) abrogated the blocking activity against both isolates. The anologous peptide, 135, based on the III_B sequence competed only for the IIIB activity. Comparison of the 142 MN sequence peptide to the corresponding III_B peptide as shown in Table 2 suggested that the central portion of these immunogens contained the most likely candidates for a cross reactive epitope. To test that possibility trimers of short peptides representing this region of the V3 loop were prepared and tested in similar competition analysis with gp89 sera. One of these, RP116, was found to compete effectively for the III_B but not the MN activity in a similar fashion as the larger III_B related peptide RP135. The results suggest that all of the III_B cross reactive antibodies induced by the MN based RP142 peptide were directed at the central GPGRAF sequence. Competition studies with the other cross reacting sera shown in Table 1 gave similar results.

We interpret these experiments to suggest the presence of multiple epitopes in the V3 domain which can serve as sensitive targets for neutralizing antibody. Following immunization with peptides or even full length gp160 the most common sites recognized by resultant sera usually involve the more variable residues. Some animals, however, as for example guinea pigs D3, 284, and 89, also made antibody to the central GPGRAF sequence which is conserved in isolates as diverse as III_B and MN. This latter set of antibodies then is apparently responsible for the cross-neutralization of the divergent isolates as outlined in Table 3. Note that although the population B antibodies shown in Table 3 are directed outside the GPGRAF sequence, residues from this latter site probably also contribute to their specificity.

Table 3. Proposed rationale for peptide competition studies shown in Table 2

Conclusion: Sera raised by the MN peptides contain several populations of antibody

Population A is directed at the conserved sequence GPGRAF and is the only population in these sera that neutralize the 3B isolate

Population B is probably more heterogenous and directed at sequences specific for the MN virus

Both populations neutralize the MN virus

Only population A neutralizes the 3B virus

MN	Y N K . R K R I H I . .	G P G R A F	Y T T K N I I G
3B	N N . T R K S I R I Q R	G P G R A F	V T I G K . I G
	B	A	B

In other studies, we have recently published the V3 sequences of 245 HIV-1 isolates (24). Most of the sequences were taken from replication competent field isolates after PCR amplification. The results indicated that in spite of the sequence diversity there is apparently a conserved structural motif based on computer models predictive of secondary structure. In addition residues at certain sites in the V3 domain are relatively invariant. For example, the strong turn predicting GPG sequence in the middle of the V3 loop was apparent in 237 of the 245 sequences. About 60% of the isolates contained the GPGRAF sequence as studied above. This suggests that immunity directed at this relatively conserved sequence might neutralize a large fraction of virus isolates. The results shown in Table 2 indicate that this short sequence can indeed be recognized as a subset of epitopes recognized by antisera raised by larger immunogens. But again when such peptides or full length glycoproteins are utilized more variable residues are more commonly involved in antibody recognition. We thus tested if peptides containing only the GPGRAF sequence would more reproducibly direct the response to this conserved region. Results of that study were encouraging in that two of two immunized rabbits developed antibodies that neutralized GPGRAF containing prototypic and field HIV-1 isolates (25). Yet immunization in several other species with the same peptide have yielded mixed results. We are thus in the process of studies aimed at the design of immunogens which will reproducibly direct the antibody response to this relatively conserved 6 residue sequence. Success to that end would increase the likelihood that this otherwise hypervariable domain of the envelope might ultimately play a role in vaccine design.

REFERENCES

1. Rusche, J., et al. Proc. Natl. Acad. Sci. USA 84, 6924 (1987).
2. Hu, S., et al. Nature 328, 721 (1987).
3. Earl, P. et al., AIDS Res. and Human Retroviruses 5, 23 (1989).
5. Barr, P. et.al., Vaccines 5, 90 (1987).
6. Berman, P. et al., Proc. Natl. Acad. Sci. USA 85, 7023 (1986).
7. Robey, W. et al., Proc. Natl. Acad. Sci. USA 83, 7023 (1986).
8. Matthews, T. et al., Proc. Natl. Acad. Sci. USA 83, 9709 (1986).
9. Arthur, L. et al., Proc. Natl. Acad. Sci. USA 84, 8583 (1987).
10. Putney, S. et al., Science 234, 1392 (1986).
11. Chanch, et al., European J. Mol. Biol. 5, 3065 (1986).
12. Dalgleish, A. et al., Virol. 165, 209 (1988).
13. Ho, D., et al., J. Virol. 61, 2024 (1987).
14. Ho, D. et al., Science 239, 1021 (1988).
15. Rusche, J. et al., Proc. Natl. Acad. Sci. USA 85, 3198 (1988).
16. Palker, T. et al., Proc. Natl. Acad. Sci. USA 85, 1932 (1988).
17. Goudsmit, J. et al., Proc. Natl. Acad. Sci. USA 85, 4478 (1988).
18. Kenealy, W. et al., AIDS Res. Human Retroviruses 5, 173 (1989).
19. Berzofsky, J. et al., Nature 334, 706 (1988).
20. Leonard, C. et al., J. Biol. Chem. 265, 10373 (1990).
21. Berman, P. et al., Nature 345, 622 (1990).
22. Emini, E. et al., J. Virol. 64, 3674 (1990).
23. Girard, M. AIDS Conference, Keystone, CO (1990).
24. LaRosa, G. et al., Science 249, 932 (1990).
25. Javaherian, K. et al., Science In press.

MOLECULAR MECHANISMS IN THE PATHOGENESIS OF AIDS-ASSOCIATED KAPOSI'S SARCOMA

Barbara Ensoli, Giovanni Barillari, Luigi Buonaguro and Robert C. Gallo

Laboratory of Tumor Cell Biology
National Cancer Institute
National Institutes of Health
Bethesda, MD 20892 USA

SUMMARY

Kaposi's Sarcoma (KS) is a tumor of mesenchymal origin of unclear etiology and pathogenesis. The epidemic form of KS (AIDS-associated) occurs in up to 30% of HIV-1 infected individuals with lesions characterized by mixed cellularity, spindle cells proliferation and neoangiogenesis.

The establishment of in vitro and in vivo model systems (AIDS-KS cell cultures and nude mouse) have allowed studies toward the understanding of the pathogenesis of KS.

The data presented here support the hypothesis that KS is a cytokine mediated disease and that interactions between mesenchymal cell types and HIV-1 gene products might lead to a composite lesion such as KS. In fact, in vitro and in vivo studies indicate that the HIV-1 Tat protein acts as a growth factor for cells derived from AIDS-KS lesions, thus establishing an experimental link between HIV-1 infection and the development of KS in humans.

Human immunodeficiency virus (HIV-1) is implicated in various clinical manifestations associated with AIDS, including KS. KS represents the most frequent tumor arising in infected individuals, particularly homosexual and bisexual men.[1] This form of KS (epidemic or AIDS-KS) is aggressive and often results in dissemination and invasion of lymph nodes and viscera. Histologically, KS is characterized by the proliferation of spindle-shaped cells ("KS cells"), considered to be the tumor element of the lesions, associated with endothelial cells, fibroblasts, inflammatory cells and new blood vessel formation (early stage lesions). In a later stage, the spindle cells tend to coalesce in larger tumor masses, although the slit-like spaces, which are characteristic of the lesion, usually remain evident. The histogenesis of the KS spindle cells, however, is still controversial and both types of mesenchymal cells, endothelial and smooth muscle cells, have been proposed as potential cell progenitors.

Although KS is clearly associated with HIV-1 infection, little is known about the molecular events underlying its pathogenesis.[1,2] Recently, however, two experimental advances (the establishment of long-term cell cultures derived from KS lesions of AIDS patients and the development of animal models) have made the study of the pathogenesis of AIDS-KS possible.[3-5]

Here we discuss results obtained from these new systems suggesting that the induction of the AIDS-KS lesions involves a pathway of events mediated by specific cytokines and that the HIV-1 _tat_ gene product may play a crucial role in the development and/or progression of KS in HIV-1 infected individuals.

THE ROLE OF CYTOKINES

The study of the biology and pathogenesis of KS has been hampered by the inability to maintain long-term cultures of KS cells. Recently, however, we have shown that spindle cells from KS lesions of AIDS patients (AIDS-KS cells) (Fig. 1) can be grown in long-term culture by initiating and supporting their growth with conditioned media (CM) derived from activated human CD4+ T cells, particularly when infected with either HTLV-I or HTLV-II,[3] but also phytohemagglutinin (PHA) stimulated T cells (Ensoli et al., in preparation). Although HIV infection of CD4+ T cells does not lead to their activation, growth stimulatory activity was also found in CM of HIV-1 infected CD4+ T cells.[3] With this cell culture system we could further characterize the phenotypic and functional characteristics of the AIDS-KS cells.

Morphological and immunohistochemical studies indicated that the AIDS-KS cells are of mesenchymal cell origin and have features in common with endothelial and smooth muscle cells,[4,6] both cell types are believed to be the potential progenitors of the tumor spindle cells of KS.

Furthermore, we determined that the AIDS-KS-derived cells produce potent biological activities (Table 1) which promote self growth, growth of normal endothelial cells, fibroblasts and other cell types,

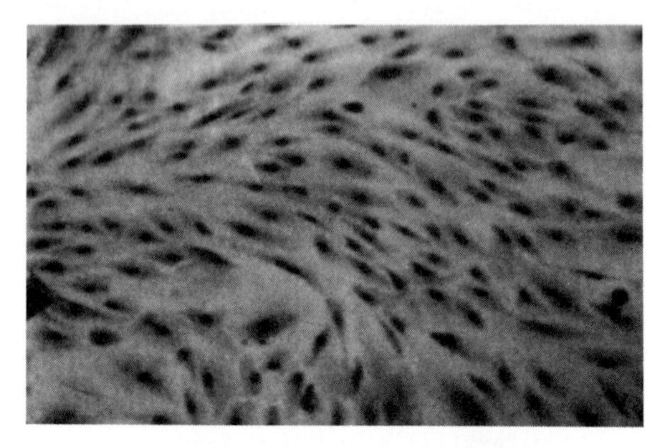

Fig. 1. Long-term cultured confluent AIDS-KS cells stained with Wright-Giemsa. Cells were grown in a growth medium (RPMI 1640, 15% fetal calf serum) supplemented with HTLV-II CM.

Table 1. Biological activities of conditioned medium and/or cellular extracts from AIDS-KS cells. The induction of KS-like lesions in nude mice resulted from subcutaneous inoculation of AIDS-KS cells.[4,5]

Biological Activities
- Growth-promoting activity for AIDS-KS cells
- Growth-promoting activity for normal endothelial cells and fibroblasts
- IL-1-like activity
- Granulocyte-monocyte colony stimulating-like activity
- Chemotactic activities for AIDS-KS, normal endothelial cells, fibroblasts and inflammatory cells
- Neoangiogenic activity (CAM assay)
- Induction of KS-like lesions in nude mice

chemotactic and chemoinvasive factors for AIDS-KS cells and normal cells, and new blood vessel formation[4,5] as monitored by the chorioallantoic membrane assay (CAM assay) (Fig. 2A and 2B). Contemporary with the in vitro studies, an animal model was also developed in which the induction of mouse lesions closely resembling KS were obtained by transplantation of AIDS-KS cells in nude mice[4] (Table 1, Fig. 2C and 2D). The lesions were transient, disappearing with the death of the transplanted AIDS-KS cells and were not induced by fixed cells or normal cell types (Fig. 2C). The histological features resembled early KS lesions in humans with evidence of neoangiogenesis, extravasated red blood cells and spindle-shaped cells (Fig. 2D).

These results suggest that the AIDS-KS cells produce factors which are capable of inducing histological changes similar to those observed in vivo in the KS lesions. To identify the factors responsible for these effects, we analyzed the mRNA expression by AIDS-KS and control cell types for those cytokines known to induce similar biological effects on target systems[5] (Table 2).

Acidic and basic fibroblast growth factor (aFGF and bFGF) are the prototypic angiogenic factors,[7] promoting growth of endothelial cells, fibroblasts and other cells derived from the mesoderm. These factors also stimulate chemotaxis and proliferation of endothelial cells with neovascularization.

Interleukin-1 (IL-1) may induce angiogenesis by interacting with cells of the immune system, and via induction of platelet-derived growth factor-AA (PDGF-AA), stimulates fibroblasts to produce granulocyte/ macrophage-colony stimulating factor (GM-CSF)[8] and IL-6.[9] GM-CSF induces growth and activation of responsive progenitor cells and, together with IL-1,[8] may induce infiltration by inflammatory cells. By inducing endothelial cell growth,[10] GM-CSF and PDGF can also enhance the effects of other angiogenic factors. IL-6 is a multifunctional cytokine capable of effects on several systems (vascular, immune and hemopoietic).[9] Transforming growth factor beta (TGFβ) is a multifunctional cytokine that activates fibroblasts to migrate and proliferate, induces chemotaxis and activation of monocytes-macrophages

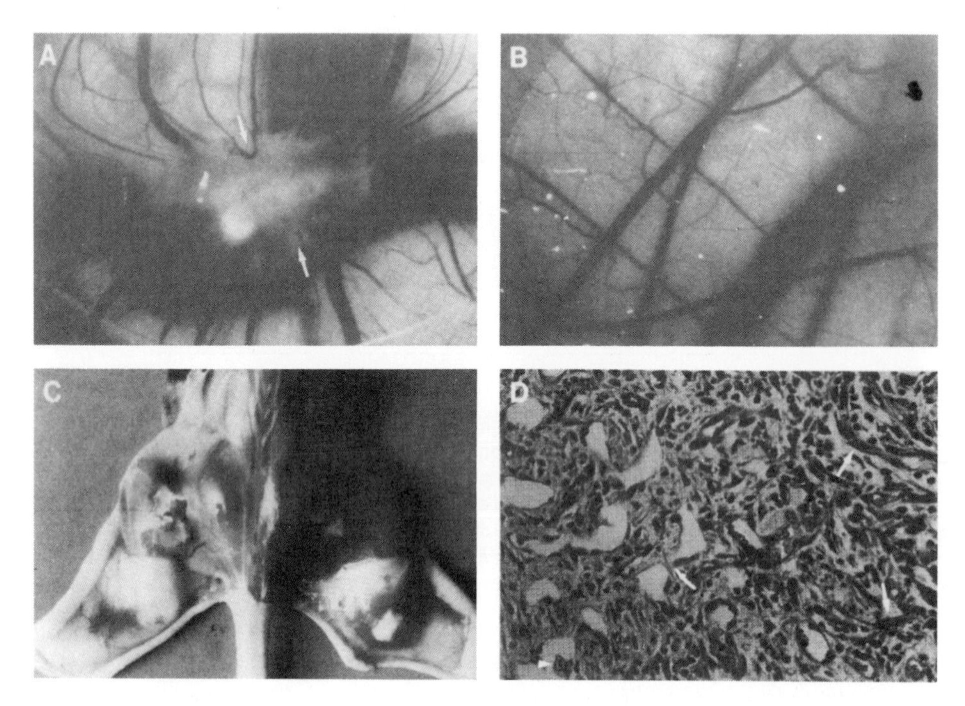

Fig. 2. A) Induction of angiogenic response in a CAM fertilized for 12 days by $2x10^4$ AIDS-KS cells. Numerous newly formed small vessels (→).

B) Control CAM showing no angiogenic response.

C) NCr-nude mouse injected subcutaneously in lower back with AIDS-KS cells [$4x10^6$ live (right) or fixed (left) cells per site] Note the angiogenesis on the site injected with the live AIDS-KS cells and absence of such reaction when fixed (0.00125% glutaraldehyde) cells were injected.

D) Histological appearance of the subcutaneous lesion in the NCr-nude mouse injected with AIDS-KS cells. Marked angiogenic response was evident 5 days post-injection. Note the presence of numerous slits and small vessels with and without red blood cells and spindle-like cells (→). Evidence of endothelial cell division (▲) (mitosis) is shown.

Table 2. The cytokine mRNA expression by AIDS-KS and control cells was analyzed by Northern blot analysis using antisense oligonucleotides or full-length probes.[5]

Cytokines mRNA Analysis
- Basic and acidic fibroblast growth factor
- Interleukin-1 α and β
- Monocyte and granulocyte-monocyte colony stimulation factor
- Transforming growth factor α and β
- Tumor necrosis factor α and β
- Angiogenin
- Platelet-derived growth factor A and B
- Interleukin-6

and promotes angiogenesis *in vivo*.[11] PDGF is another molecule with a key role in cell activation, proliferation and tissue repair.[12]

We have identified the specific cytokines responsible for most of these activities (Table 3).[5] By comparison with normal endothelial cells and fibroblasts, AIDS-KS cells constitutively express very high levels of mRNAs for bFGF and IL-1β; moderate levels of mRNAs for GM-CSF, TGFβ and PDGF-B; and low levels of mRNAs for aFGF and IL-1α. Further studies indicated expression of IL-6 mRNA and PDGF-A (Ensoli et al., unpublished data). There was no detectable expression of mRNA for other cytokines, such as TGFα, angiogenin, monocyte-colony stimulating factor (M-CSF), tumor necrosis factor α and β (TNFα and β), IL-2 and gamma interferon (γIFN).[5] Nor was there any detectable expression of the "KS oncogene," also called *hst*, which has been isolated from a KS lesion of one patient and belongs to the FGF family of angiogenic factors.[13,14] Moreover, this and other evidence [lack of expression in AIDS-KS tissues[15] and isolation from different tumors and normal tissues][16] indicate that the "KS oncogene" is unlikely to be involved in the development of KS in AIDS patients.

Table 3. Cytokine mRNA expression by AIDS-KS and H-UVE cells. No RNA expression was detected for γ IFN, IL-2, TGFα, M-CSF, TNFα, TNFβ and angiogenin, while further studies indicated that AIDS-KS cells also express IL-6 mRNA.

RNA source	Cytokine mRNA							
	bFGF	aFGF	IL-1α	IL-1β	IL-6	GM-CSF	TGFβ	PDGF
AIDS-KS cells	++++	+	+	+++	+++	++	+(+)	+
H-UVE cells	+/-	-	-	+/-	-	-	++	++

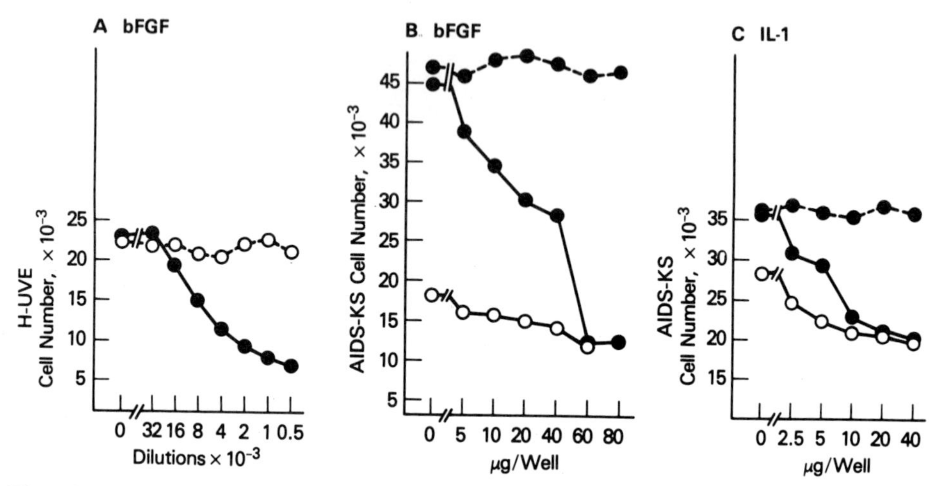

Fig. 3. A) Inhibition of the AIDS-KS CM-induced growth of normal vascular endothelial cells by antiserum to bFGF. H-UVE cells were grown with AIDS-KS-derived CM (•) (dilution 1:2) or with ECGS (30 μm/ml) (o). The CM were treated with several dilutions of antiserum to bFGF.

B) Inhibition of AIDS-KS cell growth by antibodies to bFGF. AIDS-KS-derived CM dilution 1:2 was treated with several concentrations of antibodies to bFGF (•, -) or with nonimmune serum (•, --), control medium (10% FCS in RPMI 1640) (o) was treated with antibodies to bFGF.

C) Inhibition of AIDS-KS cell growth by antibodies to IL-1. (•, -), AIDS-KS CM treated with antibodies to IL-1; (•, --), AIDS-KS CM treated with preimmune antibodies; (o, -), control medium treated with antibodies to IL-1.

Drawing on experimental data and previous information about the cellular source and behavior of these cytokines, we have assigned them the biological activities produced by the AIDS-KS cells that are consistent with the pathological features of KS lesions[5] (Table 4).

Radioimmunoprecipitation analysis of cell extracts and supernatants from metabolically labeled AIDS-KS cells confirmed the elevated protein expression of bFGF and IL-1. Blocking experiments with specific affinity purified antibodies further showed that the AIDS-KS cells produce and release biologically active bFGF and IL-1 capable of inducing proliferation of AIDS-KS and normal mesenchymal cell types in autocrine and paracrine fashion (Fig. 3).

The elevated expression and biological function of particular cytokines in the AIDS-KS cells suggest that these molecules may have important roles in the development of the KS lesion. In fact, the AIDS-KS cells have morphological and biological properties corresponding to and possibly explaining the typical histological features of the KS lesion (mixed cellularity of spindle cells, leukocytes, normal endothelial cells, fibroblasts and neoangiogenesis). Thus, AIDS-KS may be initiated by signals inducing the growth of particular mesenchymal cell types (spindle cells) and the expression of

Table 4. Contribution of the single cytokines to the biological activities released or induced by AIDS-KS cells.

Biological Activities	Cytokines
Growth promoting activity for AIDS-KS cells	bFGF, IL-1 (α, β) IL-6, PDGF
- Growth promoting activity for normal mesenchymal cell types:	
- endothelial cells	bFGF, aFGF, GM-CSF
- fibroblasts	bFGF, aFGF, PDGF, IL-1
- smooth muscle cells	bFGF, IL-1, PDGF
- IL-1-like activity	IL-1
- Granulocyte-monocyte colony stimulation-like activity	GM-CSF
- Chemotactic activities for AIDS-KS, normal mesenchymal and inflammatory cells	All above factors
- Neoangiogenic activity (CAM assay)	bFGF, aFGF
- Induction of KS-like lesions in mice	All above factors

specific cytokines with autocrine and paracrine activities that may play a central role in the pathogenesis of the histological changes observed in human KS.[2,6]

THE ROLE OF THE HIV-1 tat GENE PRODUCT

The previous data suggest that inflammatory and angiogenic cytokines may have a central role in the development and possibly in maintenance of the KS lesions. However, they do not explain the high frequency of KS in HIV-1 infected individuals, nor the mechanism by which HIV-1 may participate in the induction of KS in these patients. The ability of the AIDS-KS cells to grow with CM-derived from HIV-1 infected cells, to induce KS-lesions in the mice, and the high incidence of KS in HIV-1 infected individuals, strongly suggests that a tight link exists between KS development and HIV-1 infection. However, because HIV-1 sequences are not found in cells isolated from the tumor,[4] the role of HIV-1 in KS must be indirect and probably due to some kind of paracrine effect exerted by the infected cells.[4,5]

Concomitant with our studies, Vogel et al. showed that the tat gene of HIV-1 introduced into the germ line of mice induces skin lesions closely resembling KS.[17] Although the development of the tumors is correlated with the expression of Tat in the skin of the animals, the tumor cells do not express the trans-gene, supporting the hypothesis of a paracrine mechanism.[17] Because the tat gene alone could induce KS-like lesions, we hypothesize that the KS growth promoting activity present in CM from HIV-1 infected cells[3] might be Tat itself, and that the Tat protein could be released by infected cells and might promote AIDS-KS cell growth directly or indirectly by activating target cells to express growth promoting gene(s).

To prove this hypothesis, we analyzed whether the growth effect of CM from HIV-1 acutely infected cells on the AIDS-KS spindle cells was due to Tat.[18] CM from H9 infected (H9-HIV-1 CM) or uninfected (H9 CM) cells were harvested every 4 days after cell-free infection with low amounts of the virus (IIIB isolate, corresponding to 100,000 cpm of RT per 20 millions cells), virus cleared by high speed centrifugation, and

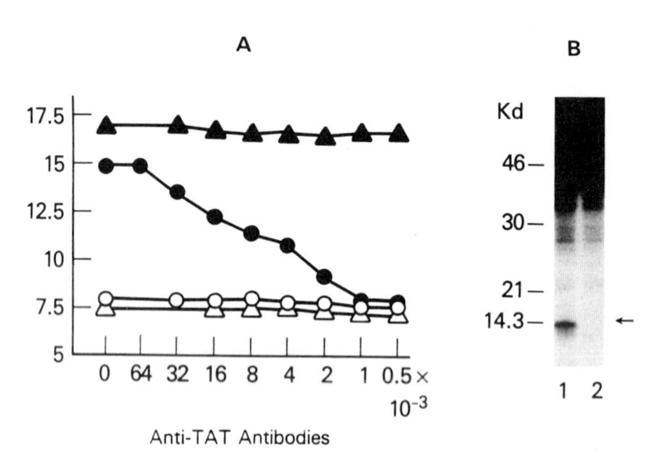

Fig. 4. A) Specific block of AIDS-KS cell growth response to H9-HIV-1-derived CM by anti-Tat antibodies. (•-•) H9-HIV-1-derived CM (1:16); (o-o) H9-derived CM (1:16); (Δ-Δ) medium without CM; (▲-▲) HTLV-II-derived CM (1:16). The CM or medium alone were preincubated with or without serial dilutions of affinity purified rabbit anti-Tat polyclonal antibodies. This figure shows the effects of CM 12 days post-infection.

B) Radioimmunoprecipitation assay of the Tat protein in CM from HIV-1 infected (12 days post-infection) or uninfected H9 cells. Lanes 1 and 2 show CM from HIV-1 infected or uninfected H9 cells, respectively. Arrows indicate Tat in CM from infected cells.

separately tested for proliferative activity on AIDS-KS cells and on normal control cells (endothelial and smooth muscle cells). The results with CM preparations from repeated infection experiments indicated that H9-HIV-1 CM was capable of inducing AIDS-KS cell proliferation, but not the growth of normal control cells, and that this proliferation was specifically blocked by several affinity purified anti-Tat antibodies (Fig. 4A).[18] Furthermore, the tat gene product was present in supernatant of metabolically labeled acutely infected cells from the same experiments (Fig. 4B).[18]

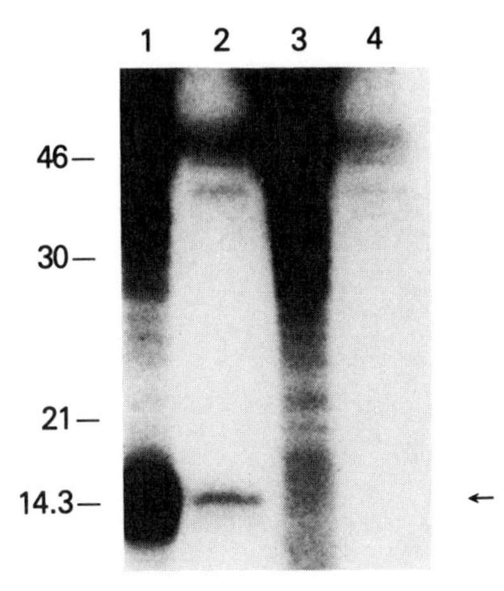

Fig. 5. Radioimmunoprecipitation of Tat in COS-TAT-transfected and metabolically labelled cells. Lanes 1 and 3, cell extracts; lanes 2 and 4, supernatants from tat-transfected COS-1 cells (COS-TAT) and COS-1 cells transfected with the vector alone (COS-CR), respectively, precipitated with affinity purified anti-Tat antibodies. The arrow indicates the Tat product.

Extracellular Tat was detectable 8-16 days after cell-free infection of T cells, and its presence correlated with the time of optimal growth promoting effect of the CM and with positive blocking effects by anti-Tat antibodies. Similar results were also obtained with CM from COS-1 cells transfected with a _tat_-expressing plasmid. In this system 5-15% of the transfected cells express Tat, as evaluated by immunostaining with anti-Tat purified monoclonal antibodies, and extracellular Tat is detected in the absence of cell death (Fig. 5).

Finally, partially purified recombinant Tat proteins from different sources (not gp160 or Nef) were capable of inducing specific proliferation of the AIDS-KS cells at very low doses (1-10 ng/ml) (Fig. 6A) and again the growth induction was blocked by specific antibodies (Fig. 6B and 6C).[18] Further studies indicated that the effective dose of purified Tat protein capable of inducing the growth effect is on the order of picograms of protein per ml of media (Ensoli et al., in preparation).

This is the first demonstration that a virus encoded regulatory gene product can be released in a biologically active form and acts as a growth factor for cells of mesenchymal origin. Interestingly, Tat induces cell growth at concentrations similar to several other cytokines with activity on the vascular-mesenchymal system and which, as for Tat, are released into the extracellular fluid in the absence of a leader sequence for transport (bFGF, aFGF, IL-1, etc.).

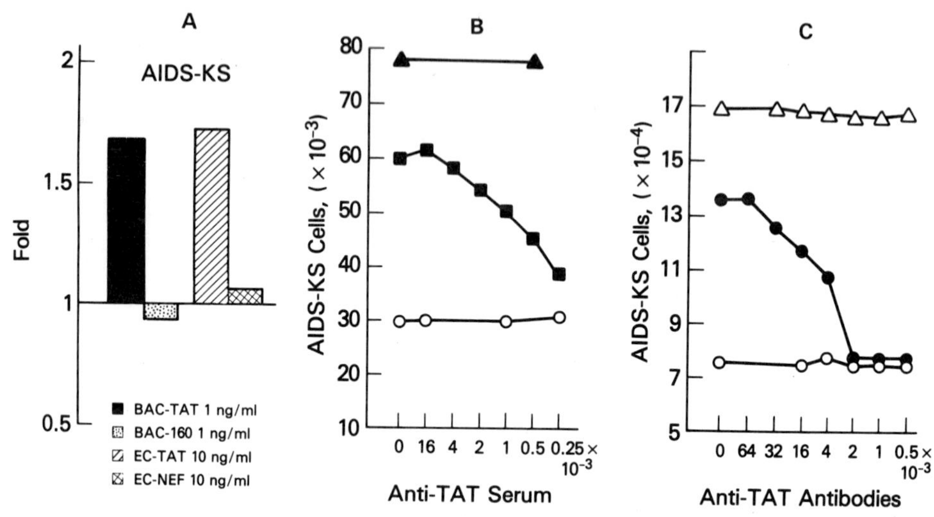

Fig. 6. A) Growth response of AIDS-KS-3 to recombinant Tat proteins or to recombinant control proteins (gp160 and Nef). BAC-TAT and BAC-160, baculovirus-derived Tat and gp160 proteins at 1 ng/ml. EC-TAT and EC-NEF, *E. coli*-derived Tat and Nef proteins at 10 ng/ml B,C) block of AIDS-KS cell growth response to EC-TAT (Fig. 6B) or BAC-TAT proteins (Fig. 6C) by anti-Tat antibodies. (■-■) EC-TAT at the final concentration of 10 ng/ml; (•-•) BAC-TAT at the final concentration of 1 ng/ml; (o-o) medium alone without CM; (Δ-Δ) and (▲-▲) HTLV-II-derived CM (1:16). Anti-Tat serum raised against a Tat synthetic peptide was used for EC-TAT, while anti-Tat affinity purified antibodies were preincubated with BAC-TAT.

Our results suggest that Tat, released during acute infection of CD4+ T cells by HIV-1, may contribute to the induction and/or progression of KS in HIV-1 infected individuals. However, these data do not rule out the possibility that at different stages of infection and disease development other factor(s) (of cellular and/or viral origin) also play a role in KS pathogenesis. This is further suggested by the observation that the normal mesenchymal cell types used in this study did not proliferate in response to Tat. Studies are ongoing to determine which other factors are needed by normal cell types, which may represent the cell progenitors of the KS spindle cells, to become responsive to Tat. Preliminary results suggest that cytokines released during immunoactivation, which is particularly frequent in homosexual and bisexual males, induce Tat-responsiveness of normal mesenchymal cells, and thus may represent a cofactor in AIDS-KS pathogenesis.

Our data support a model for AIDS-KS pathogenesis (Fig. 7) in which factor(s) released by HIV-1 infected cells (Tat), and activated T cells (cytokines), initiate and maintain cellular events leading to the production of angiogenic cytokines, such as IL-1, aFGF, bFGF, GM-CSF, IL-6, PDGF and TGFβ, that support the activation and the proliferation of normal cells (endothelial, smooth muscle cells, fibroblasts, monocytes and lymphocytes) with neoangiogenesis and inflammatory cell infiltration.

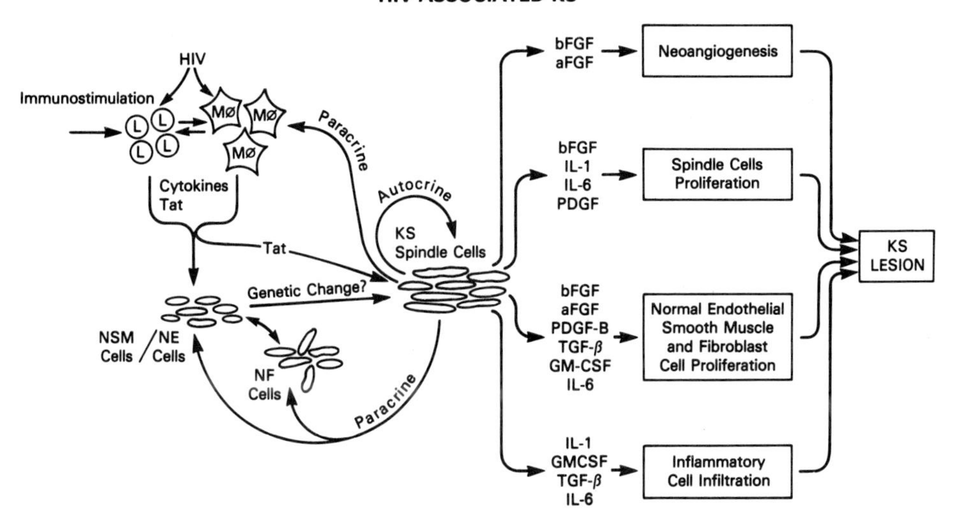

Fig. 7. A model for the pathogenesis of the KS lesion in AIDS patients. (NE) Normal endothelial cells; (NSM) normal smooth muscle cells; (NF) normal fibroblasts; (L) lymphocytes; (M) monocytes-macrophages. HIV-infected and activated cells (L and M) release Tat and/or other viral and cellular factor(s) capable of stimulating activation and proliferation of a particular cell type of mesenchymal origin. These cells (smooth muscle cells, endothelial cells?) then acquire the peculiar spindle-shaped morphology characteristic of KS cells. The KS cells, in turn, begin to produce and release several cytokines that maintain and amplify the cellular response via autocrine and paracrine pathways. Paracrine activation of normal cells (NE, NSMC, NF, L and M) leads to endothelial, smooth muscle cell and fibroblast proliferation, neoangiogenesis and inflammatory cell infiltration. These phenomena, together with the spindle cell proliferation, could underlie the typical histological changes observed in early KS lesions. Later, interactions between cells of the immune system and mesenchymal cells could amplify cell activation and cytokine production. If the initial stimulus persists, a vicious cycle could be established which, under certain circumstances (e.g. specific genetic changes), would lead to tumor transformation.

These results help to explain the biological mechanisms that underline the high incidence and progression of KS seen in HIV-1 infected individuals and support the hypothesis that KS is a cytokine-mediated disease, and that HIV-1 infection, via Tat release, could be directly involved in its onset and progression.

BIBLIOGRAPHY

1. H. W. Haverkos, D. P. Drotman, and M. Morgan, Prevalence of Kaposi's sarcoma among patients with AIDS, New Engl. J. Med. 312:1518 (1985).

2. B. Ensoli, G. Barillari, and R. C. Gallo, Pathogenesis of AIDS-associated Kaposi's sarcoma, Hematol. Oncol. Clin. North Am. 5:281 (1991).
3. S. Nakamura, S. Z. Salahuddin, P. Biberfeld, B. Ensoli, P. D. Markham, F. Wong-Staal, and R. C. Gallo, Kaposi's sarcoma cells: Long-term culture with growth factor from retrovirus-infected $CD4^+$ T cells, Science 242:426 (1988).
4. S. Z. Salahuddin, S. Nakamura, P. Biberfeld, M. H. Kaplan, P. D. Markham, L. Larsson, and R. C. Gallo, Angiogenic properties of Kaposi's sarcoma-derived cells after long-term culture in vitro, Science 242:430 (1988).
5. B. Ensoli, S. Nakamura, S. Z. Salahuddin, P. Biberfeld, L. Larsson, B. Beaver, F. Wong-Staal, and R. C. Gallo, AIDS-Kaposi's sarcoma-derived cells express cytokines with autocrine and paracrine growth effects, Science 243:223 (1989).
6. B. Ensoli, S. Z. Salahuddin, and R. C. Gallo, AIDS-associated Kaposi's sarcoma: A molecular model for its pathogenesis, Cancer Cells 1:93 (1989).
7. W. H. Burgess and T. Maciag, The heparin binding (fibroblast) growth factor family of proteins, Annu. Rev. Biochem 58:575 (1989).
8. S. C. Clark and R. Kamen, The human hematopoietic colony-stimulating factors, Science 236:1229 (1987).
9. I. Tamn, IL-6 current research and new questions, Ann. NY Acad. Sci. 557:478 (1989).
10. F. Bussolino, J. M. Wang, P. Defilippi, F. Turrini, F. Sanavio, C.-J. S. Edgell, M. Aglietta, P. Arese, A. Mantovani, Granulocyte- and granulocyte-macrophage-colony stimulating factors induce human endothelial cells to migrate and proliferate, Nature 337:471 (1989).
11. A. B. Roberts, and M. B. Sporn, Transforming growth factor, Adv. Cancer Res. 51:107 (1988).
12. R. Ross, E. W. Raines, and D. F. Bowen-Pope, The biology of platelet-derived growth factor, Cell 46:155 (1986).
13. P. Delli Bovi and C. Basilico, Isolation of a rearranged human transforming gene following transfection of Kaposi's sarcoma DNA, Proc. Natl. Acad. Sci. USA 84:5660 (1987).
14. P. Delli Bovi, A. M. Curatola, F. G. Kern, A. Greco, M. Ittmann, C. Basilico, An oncogene isolated by transfection of Kaposi's sarcoma DNA encodes a growth factor that is a member of the FGF family, Cell 50:729 (1987).
15. P. Delli Bovi, E. Donti, D. M. Knowles, A. Friedman-Kien, P. A. Luciw, D. Dina, R. Dalla-Favera, C. Basilico, Presence of chromosomal abnormalities and lack of AIDS-retrovirus DNA sequences in AIDS-associated Kaposi's sarcoma, Cancer Res. 46:6333 (1986).
16. H. Sakamoto, M. Mori, M. Taira, T. Yoshida, S. Matsukawa, K. Shimizu, M. Sekiguchi, M. Terada, T. Sugimura, Transforming gene from human stomach cancer and a noncancerous portion of stomach mucosa, Proc. Natl. Acad. Sci. USA 83:3997 (1986).
17. J. Vogel, S. H. Hinrichs, R. K. Reynolds, P. A. Luciw, G. Jay, The HIV-1 tat gene induces dermal lesions resembling Kaposi's sarcoma in transgenic mice, Nature 335:606 (1988).
18. B. Ensoli, G. Barillari, S. Z. Salahuddin, R. C. Gallo, and F. Wong-Staal, Tat protein of HIV-1 stimulates growth of cells derived from Kaposi's sarcoma lesions of AIDS patients, Nature 345:84 1990.

STRUCTURE-FUNCTION ANALYSIS OF THE HIV GLYCOPROTEIN

John W. Dubay, Hae-ja Shin, Jian-yun Dong, Susan Roberts and Eric Hunter

Department of Microbiology
University of Alabama at Birmingham
UAB Station, Birmingham, AL 35294

INTRODUCTION

The envelope glycoprotein complex of replication competent retroviruses is comprised of two polypeptides, an external, glycosylated, hydrophilic polypeptide (SU) and a membrane-spanning protein (TM), that form a knob or knobbed spike on the surface of the virion. Both polypeptides are encoded in the *env* gene of the virus and are synthesized in the form of a polyprotein precursor that is proteolytically cleaved during its transport to the surface of the cell. While these proteins are not required for the assembly of enveloped virus particles, they do play a critical role in the virus replication cycle by recognizing and binding to specific receptors, via the SU domain, and by mediating the fusion of viral and cell membranes, via the TM domain: virus particles lacking envelope glycoproteins are thus non-infectious.

Although the sizes and amino acid composition of the SU and TM proteins from different retroviruses vary, there is an overall structural similarity between the different glycoproteins. The basic organization is defined by three hydrophobic or apolar regions that are arranged along the envelope molecule (Figure 1). At the amino-terminus is the signal peptide, a short stretch of hydrophobic amino acids that directs the nascent *env* gene product into the secretory pathway of the infected cell. It is removed from the nascent *env* gene product during translation.

Near the carboxy-terminus is a longer hydrophobic region that stops the translocation process, anchors the molecule in the membrane and causes

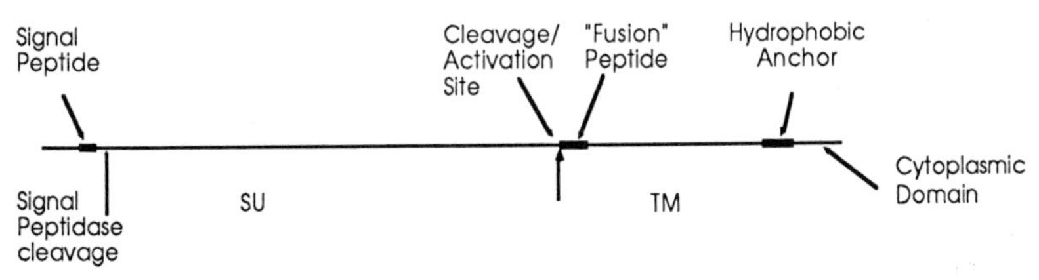

Figure 1. Schematic of the *env* gene product of a typical retrovirus.

it to span the lipid bilayer. The molecule is thus oriented with its N-terminus outside and its C-terminus in the cytoplasm - a type 1 glycoprotein - and is therefore similar to that of several membrane-spanning virus or cell-encoded glycoproteins. For most retroviruses, the cytoplasmic domain is relatively short (approximately 20-40 amino acids long), and in some cases this region may not be essential for assembly of an infectious virus (Perez et al., 1987). In HIV and the other primate lentiviruses, however, the cytoplasmic domain is over 150 amino acids long. The length of this region suggests that it might play a functional role in HIV infection, although previous mutational analyses have led to contradicting results (Terwilliger et al., 1986; Lee et al., 1989). We report here on the effect of single point mutations within this region that through the insertion of stop codons result in a progressive truncation of this domain.

A third apolar domain, postulated to be involved in the process of membrane fusion during virus entry, is located at the N-terminus of the TM protein, and C-terminal to a conserved stretch of basic amino acids (Arg-X-Lys-Arg) at which proteolytic cleavage of the precursor molecule occurs. So that it does not induce membrane fusion during transport out of the cell, this "fusion peptide" is probably sequestered in the center of a multimeric structure, most likely comprised of three or four SU-TM heterodimers that assemble prior to transport out of the rough endoplasmic reticulum (RER). This oligomeric structure must thus dissociate after binding its receptor molecule in order to allow the fusion sequence that is freed by precursor cleavage to function.

ANALYSIS OF GLYCOPROTEIN PRECURSOR CLEAVAGE

The precursor glycoprotein molecule is cleaved by a host cell protease that is located in the Golgi. Cleavage occurs after a tetra-peptide sequence (Arg-X-Lys-Arg) that is highly conserved among the different retroviruses. Through mutational studies of the tetra-peptide cleavage site it has become clear that by blocking cleavage of the precursor molecule infectivity of resulting virions can be abrogated (McCune et al., 1988; Bosch and Pawlita, 1990). This ability to block infectivity by blocking cleavage of the precursor molecule makes the cellular protease responsible for cleavage a potential target for inhibitors of retroviral infection.

In an effort to better understand the sequence requirements for cleavage of the glycoprotein precursor, and thus better understand the proteases responsible, a series of mutations have been engineered into the cleavage site of both Rous sarcoma virus (RSV) and the human immunodeficiency virus type 1 (HIV-1) that are designed to answer questions regarding the specificities of the enzymes responsible for cleavage.

Previous studies with Rous sarcoma virus, where cleavage occurs after the sequence Arg-Arg-Lys-Arg, showed that mutations of this sequence could dramatically affect cleavage. Deletion of the entire tetrapeptide cleavage region completely abrogated precursor cleavage without affecting intracellular transport. Similarly, we have demonstrated the importance of dibasic sequences within this region since mutation (S19) which leave only unpaired arginines within the cleavage site (-Ser-Arg-Glu-Arg-) also completely block cleavage of the glycoprotein precursor.

Furthermore, as shown below, when the mutations are engineered back into a proviral genome these mutations which completely block cleavage, render the virus non-infectious. This confirms the requirement of glycoprotein cleavage for virus infectivity. Mutagenesis of the RSV *env* gene has led to the conclusion that at least two enzymes are responsible for cleavage. One that preferentially cleaves at the Lys-Arg and another that cleaves less efficiently at the dibasic Arg-Arg residue.

Table 1. Summary of Rous sarcoma virus cleavage site mutant phenotypes

Mutant	Cleavage Sequence	% Cleavage	Surface Expression	Trypsin Sensitivity	Infectivity
Wt	-Arg-Arg-Lys-Arg-	>95	+	NA	+
Dr1	.... Δ	0	+	-	-
G1	-Arg-Arg-**Glu**-Arg-	10	+	+	+
S19	-**Ser**-Arg-**Glu**-Arg-	<5	+	+	-

It should be noted that the mutations do not appear to disrupt the overall structure of the glycoprotein or the accessibility of the cleavage site since arginines that remain in the mutated cleavage sites can be cleaved by extracellular trypsin added to the culture medium (Table 1).

The immunodeficiency viruses present a different situation. While in the case of HIV-1, HIV-2 and SIV, the precursor molecule is cleaved to SU and TM proteins in a manner similar to RSV, there is a significant difference between these immunodeficiency viruses and other retroviruses in the presence of two potential cleavage sites near the N-terminus of the TM polypeptide. This is shown for HIV-1 in Figure 2. Both potential cleavage sites contain the Lys-Arg pair that is highly conserved at retroviral glycoprotein cleavage sites. While it has been shown experimentally that cleavage can occur at the downstream site (designated the primary site in Figure 3), it is possible that the upstream or secondary site is used either alternatively or in addition to the primary site.

-**Lys-Ala-Lys-Arg-Arg**-Val-Val-Gln-**Arg-Glu-Lys-Arg**-↓-gp41
2° Cleavage Site 1° Cleavage Site

Figure 2. Amino acid sequences at the HIV gp120/gp41 junction.

In order to investigate the roles of these two potential cleavage sites, the mutations shown in Figure 3 were created using site directed mutagenesis.

The mutated *env* genes were cloned into pSRH, an SV40-based expression vector that has been shown to synthesize and process the HIV glycoprotein normally (Data not shown). Transfection of Cos-1 cells with the wild type pSRH plasmid results in the normal synthesis and processing of HIV-1 glycoprotein, as can be seen in lanes 1 and 2 of Figure 4.

Initially a mutant (CS1) was made that changed the lysine at amino acid 510 to a glutamic acid so that no dibasic lysine-arginine pair was present at the primary cleavage site. When this mutation was inserted into the pSRH vector a number of observations were made. The most striking was that the CS1 precursor protein, which lacks a dibasic pair at the primary cleavage site, appeared to be cleaved in a manner similar to that of the wild type protein.

	↓
Wt	T-**K**-**A**-**K**-**R**-**R**-V-V-Q-**R**-**E**-**K**-**R**--A-V-G
CS1	T-**K**-**A**-**K**-**R**-**R**-V-V-Q-**R**-**E**-E-**R**--A-V-G
CS2	T-**K**-**A**-E-**R**-**R**-V-V-Q-**R**-**E**-**K**-**R**--A-V-G
CS3	T-**K**-**A**-E-**R**-**R**-V-V-Q-**R**-**E**-E-**R**--A-V-G
CS4	T-**K**-**A**-E-**R**-S-V-V-Q-**R**-**E**-**K**-**R**--A-V-G
CS5	T-**K**-**A**-E-**R**-S-V-V-Q-**R**-**E**-E-**R**--A-V-G
A.A.	**500** **505** **510**

Figure 3. Mutations within the HIV glycoprotein cleavage sequence.

This can be clearly seen in the CS-1 lanes in Figure 4. This result was unexpected since a similar mutation in the RSV *env* gene blocks cleavage of the substituted precursor and suggested that either a novel protease lacking the requirement for dibasic residues at the cleavage site was present or that the secondary cleavage site was being utilized. Since expression of both the RSV and HIV mutated *env* genes was carried out in COS cells, the latter possibility seemed most likely. In order to test this experimentally, mutations were made that changed residues in the secondary site so that either the lysine-arginine pair or any dibasic sequences in the secondary site were removed.

Expression of mutant CS2, which contains a glutamic acid substitution for the lysine within the lysine-arginine pair of the secondary cleavage

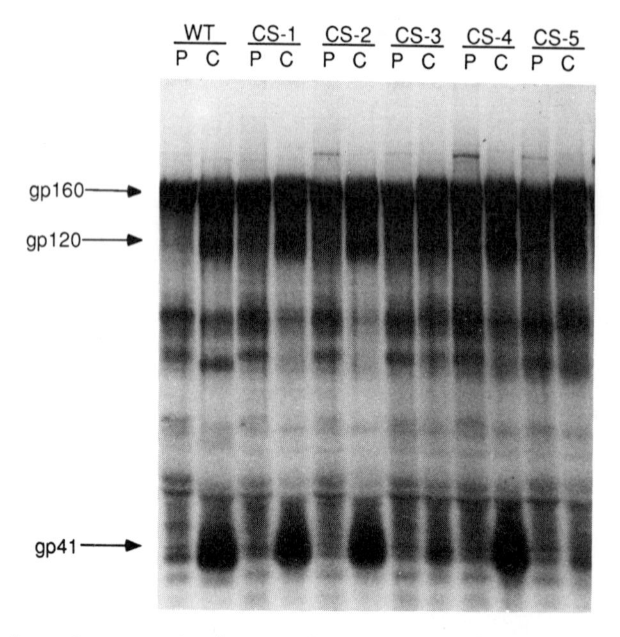

Figure 4. Pulse-chase analysis of HIV glycoprotein cleavage site mutants. Cos-1 cells were pulse-labeled (15 min) with ^{3}H-leucine 48h. post-transfection. Chases were for 3h. after which cells were lysed and immunoprecipitated with patient sera.

site, and CS4, in which all dibasic pairs within this site have ben removed, showed that both mutant precursors were cleaved normally (CS2 and CS4 Lanes, Figure 4). However, when either of these mutations (CS2 or CS4) was combined with the mutation at the primary site, cleavage is completely blocked (CS3 and CS5 Lanes, Figure 4). These results suggested that cleavage can occur with high efficiency at either the primary or secondary cleavage site and that the reason for cleavage of the CS1 precursor protein was due to cleavage at the secondary site.

Using microsequence analysis of the mutant gp41 product, we have confirmed that cleavage of CS1 does indeed occur at the secondary cleavage site and have shown that the N-terminus of the this mutant protein is heterogeneous. It appears likely that amino-peptidases and serum proteases can both remove residues from the mutant N-terminus. However, a majority of the molecules retain at least the -Arg-Glu-Glu-Arg- sequence at the N-terminus (Data not shown).

In order to determine the effects of these mutations on the biological activity of the glycoprotein the mutants were assayed 1). For their ability to fuse cells expressing the CD4 molecule, the receptor for HIV-1; 2). Transport to and expression on the plasma membrane and 3). The ability to confer infectivity on virus after reconstruction into the infectious HXB provirus.

Table 2. Summary of HIV Cleavage Mutant Phenotypes

Mutant	Cleavage	Fusion	Surface IF	Infectivity
W T	+	+	++	+
CS1	+	+	++	-
CS2	+	+	++	+
CS3	-	-	+++	-
CS4	+	+	++	+
CS5	-	-	++++	-

In the fusion assay HeLa cells expressing CD4 were transfected with the pSRH expression vector. After 48 hours the cells were stained and examined for the presence of multinucleate cell (syncytia). Transfection of HeLa-T4 cells with the wild type construct resulted in many large syncytia that contained 50-100 nuclei per giant cell. When the same assay was done using pSRH expressing the mutant envelope proteins the results corresponded with the level of cleavage detected in the the pulse-chase experiment. Those proteins that can be cleaved and presumably expose a functional fusion peptide thus have the ability to fuse the HeLa-T4 cells.

One possible explanation for the inability of the double mutants to form syncytia was that the mutations altered the tertiary structure of the protein to such an extent that they were no longer able to be transported and processed normally. This might prevent the molecule from being directed to the cell surface. However, using a monoclonal antibody to gp120, surface immunofluorescence staining of transfected Cos-1 cells clearly showed that all of the mutant glycoproteins were transported to the cell surface. Interestingly, the double mutants showed a more intense level of immunofluorescent staining than that observed with wild-type transfected cells, presumably because the uncleaved gp120 could not dissociate from gp41.

The results from this mutational analysis of the sequence requirements for cleavage of the HIV-1 glycoprotein clearly demonstrate that both cleavage sites can be utilized. Pulse-chase experiments showed that when the dibasic residues at the primary cleavage site were disrupted, cleavage of the precursor glycoprotein could still occur at the secondary site. However, when dibasic residues at both sites were removed, cleavage was completely blocked despite the fact that the glycoprotein could be detected at higher levels on the cell surface. These results support our previous observations which indicated that a protease with a stringent requirement for dibasic residues at the cleavage site is responsible for precursor cleavage. Interestingly, when the CS1 mutation was engineered into the infectious HXB2 proviral genome it rendered it non-infectious, even though cleavage of the precursor was observed. The reason for the discrepancy between HeLa-T4 cell fusion and infectivity for H9 cells remains to be determined, but may reflect the difference in density of mutant glycoprotein (with a modified gp41 N-terminus) on HeLa cells and virions. We have confirmed other reports that mutations at arginine 511 in the primary site completely block cleavage (data not shown). However, mutations at this site probably result in a failure of the the glycoprotein to be transported to the site of cleavage or result in more drastic changes in protein structure that render the cleavage site inaccessible to the enzyme.

FUNCTIONAL ANALYSIS OF THE HIV CYTOPLASMIC DOMAIN

HIV-1 encodes a transmembrane protein that contains several common structural elements as we have described above. A notable characteristic of HIV-1 and other lentivruses is an extremely long cytoplasmic "tail" often comprising nearly half of the total amino acid in the transmembrane protein. In the case of HIV-1 this region is 150-160 amino acids long depending on the isolate. In order to define a role for the nearly 20 kD of protein that make up the cytoplasmic tail, we have begun a systematic series of mutations that insert premature termination codon and result in the deletion of 18 to 191 amino acids from the C-terminus of gp41. These mutations were inserted into the *env* gene of the BH10 clone using site-directed mutagenesis to change single nucleotide residues and were designed to avoid altering the *rev* and *tat* reading frames that overlap the coding region of gp41.

Table 3. Summary of cytoplasmic domain mutant phenotypes

Mutant (# of a.a. deleted)	Release of virions	Fusion	Infectivity
191	+	-	-
136	+	+	-
108	+	+	-
80	+	+	-
42	+	+	-
18	+	+	-
Wild Type (BH10)	+	+	+
Wild Type (HXB2)	+	+	+

The entire region mutagenized was sequenced to ensure no secondary site mutations and subsequently engineered into the infectious molecular HIV-1 clone pHXB2Dgpt and the SV40-based *env*-expression vector pSRH, in order to determine the effects of these mutations. In infectivity studies the mutant pHXB2Dgpt constructs were transfected into Cos-1 cells and the virus containing supernatants were used to infect the continuous T-cell line H9. Virions derived from Cos-1 cells transfected with the wild-type pHXB2Dgpt induced rapid infection of the H9 cells, as assayed by release of reverse transcriptase activity into the culture medium, the formation of large multinucleate cells and subsequent cell death. All of the mutant proviruses released virions from Cos-1 cells at levels similar to wild-type, but when equivalent amounts of mutant virus were assayed for their ability to infect H9 cells none showed any evidence of infectivity (Table 3). Titration of the wild-type, virus containing COS supernatants indicated that the mutants were 3-4 $\log_{10}$ less infectious than HXB2gpt.

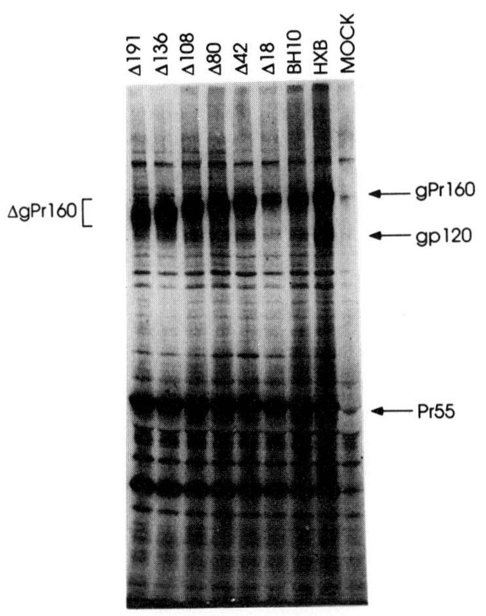

Figure 5. Metabolic labeling of mutant HIV glycoproteins. Cos-1 cells were labeled for 3 h at 48h. post-transfection with ^{35}S-methionine. Cells were lysed and immunoprecipitated using patient serum.

In order to determine if the mutations affected the synthesis of viral proteins, Cos-1 cells were transfected with the wild-type mutant constructs and metabolically labelled with ^{35}S-methionine. Cellular lysates immunoprecipitated with serum from an HIV-1 infected individual showed, as expected, that the glycoprotein precursor (gp160) was slightly smaller in the mutants. It gradually increased in size as the truncation in gp41 became smaller (Figure 5).

The viral *gag*-encoded proteins are synthesized and processed in a manner similar to that of wild type. Thus these mutations do not effect the synthesis or processing of viral proteins nor the release of virions.

As can be seen in Figure 5 while the glycoprotein precursor molecule gradually gets larger in size, gp120 remains the same size as wild-type and is synthesized in amounts similar to that observed with wild-type. The one exception is Δ191 in which little cell associated gp120 is observed due to the rapid secretion of the glycoprotein into the medium due to the lack of a membrane anchor domain.

In order to define the defect that results in a lack of infectivity in these mutants, the mutations were cloned into the SV40-based glycoprotein expression vector, pSRH. In pulse-chase experiments the glycoprotein was synthesized from this vector and processed in a manner indistinguishable from glycoproteins synthesized by the viral constructs. Similarly, the products of mutant genes introduced into pSRH were synthesized and processed in a similar manner to wild-type. In order to test the functionality of the glycoprotein, the HeLa-T4 fusion assay described above was used. Transfection of the wild type envelope gene in this assay results in the formation of large multi-nucleate induced by the HIV-1 glycoprotein. Interestingly, all of the mutants except 191 retain the ability to cause fusion of cells in this assay. Mutant 191 inserts a stop codon prior to the membrane anchor domain which results in a glycoprotein that is secreted into the medium.

While the defect in these truncated glycoprotein molecules is not yet fully understood, it is clear that, in contrast to some previous reports (Fisher et al., 1986; Terwilliger et al., 1986; Lee et al., 1989), small deletions from the C-terminus of gp41 can profoundly affect the ability of the virus to infect cells. The mutant glycoproteins appear to be synthesized, processed, and transported normally to the plasma membrane and retain the ability to bind to and fuse CD4-expressing cells. However, virions that encode the mutant glycoproteins lack the ability to infect cells, even when large amounts of virus are used,. These data indicate an important and not yet fully understood role for the TM protein in virus replication.

ACKNOWLEDGEMENTS

The work described in this manuscript was supported by grants CA-29884, AI-27290 and AI-25784. HIV virus culture was carried out in the UAB Center for AIDS Research Central Virus Culture Core Facility supported by the Center core grant AI-27767.

REFERENCES

Bosch, V. and Pawlita, M., 1990, Mutational analysis of the human immunodeficiency virus type 1 *env* gene product proteolytic cleavage site, J. Virol., 64, 2337.

Fisher, A. G., Ratner, L., Mitsuya, H., Marselle, L. M., Harper, M. E., Broder, S., Gallo, R. C. and Wong-Staal, F., 1986, Infectious mutants of HTLV-III with changes in the 3' region and markedly reduced cytopathic effects, Science., 233, 655.

Lee, S.-J., Hu, W., Fisher, A. G., Looney, D. J., Kao, V. F., Mitsuya, H., Ratner, L. and Wong-Staal, F., 1989, Role of the carboxy-terminal portion of the HIV-1 transmembrane protein in viral transmission and cytopathogenicity, AIDS Res. Human Retrovirus., 5, 441.

McCune, J., Rabin, L., Feinberg, M., Lieberman, M., Kosek, J., Reyes, G. and Weissman, I., 1988, Endoproteolytic cleavage of gp160 is required for activation of human immunodeficiency virus, Cell., 53, 55.

Perez, L. G., Davis, G. L. and Hunter, E., 1987, Mutants of the Rous sarcoma virus envelope glycoprotein that lack the transmembrane anchor and/or cytoplasmic domains: analysis of intracellular transport and assembly into virions, J. Virol., 61, 2981.

Terwilliger, E., Sodroski, J. G., Rosen, C. A. and Haseltine, W. A., 1986, Effects of mutations within the 3' orf open reading frame region of human T-cell lymphotropic virus type III (HTLV-III/LAV) on replication and cytopathogenicity, J Virol., 60, 754.

MATERNAL ANTIBODY EPITOPE MAPPING IN MOTHER-TO-CHILD TRANSMISSION OF HIV

Paolo Rossi, Viviana Moschese, Anita de Rossi, Britta Wahren, Marianne Jansson, Valter Lombardi and Hans Wigzell

Dept. of Immunology, Karolinska Inst., Dept. of Virology, SBL Stockholm, Sweden and Dept. of Pediatrics Univ. of Rome, Dept. of Oncology Univ. of Padua, Italy

The problem of pediatric HIV infection has recently been the object of great concern since the seriousness and urgency of the issue of pediatric AIDS threatens to grow in the coming years.[1,2,3] Increasing numbers of women of childbearing age become HIV infected through intravenous drug abuse or heterosexual transmission[4]. The seroprevalence in an unselected population of childbearing women has for instance been reported to be as high as 2% in an inner New York City hospital and, more strikingly, in Central and East Africa rates up to 10% have been documented. The epidemiology of pediatric HIV infection is, of course, different from that of its adults counterpart. At present, approximately 80% of the AIDS children have been infected through perinatal transmission, whereas only 20% through HIV-infected blood or blood products. In the developed countries, the decrease in prevalence of this latter route due to HIV screening of blood transfusions since 1985 will account for perinatal transmission as the only significant source of HIV infection. Yet, our knowledge about the route and the exact timing of vertical transmission is still incomplete. Based on classical seroepidemiological surveys, the likelihood of mother-to child transmission appears to range from 22 to 39%. Divergent rates have been accounted for by the length of follow-up, the particular cohort studied, the criteria used to define pediatric HIV infection . Nevertheless, is clear that only a minor proportion of infants born to HIV infected mothers will acquire the infection in the fetal life. Mother-to-fetus interaction is likely to play a major role on determining the final outcome of the pregnancy in relation to infection.As a matter of fact transmission in utero occurs under special circumstances such as the contemporary presence of a specific maternal immune response to HIV which includes possible tranfer of functional antibodies mediating viral neutralization and/or cellular cytotoxicity [5].The identifications of factors that influence this mode of transmission is critical to develop strategies of immune intervention and drug therapy in pregnant women.

Several epitopes of the HIV genome have been mapped that are potential target sites for the immune attacks[6]. Only the envelope glycoprotein complex gp120/gp41has been found to be target antigen for HIV-specific ADCC mediating antibodies whereas HIV-1 neutralizing antibodies can be produced against various subregions of gp120, gp 41 and maybe p17. One major site inducing neutralizing antibodies and syncytia-inhibition has been described in the gp 120 V3 domain, including 34-36 aa. depending on the viral strain, surrounded by two cysteines. In strain HTLV-IIIB, these cysteines are located at positions 296 and 331. Relatively conserved sequences include not only the cysteine residues but also the GPGRAF sequence located at the tip of the loop. Normally, isolate-specific neutralizing monoclonal antibodies have been shown to bind to this domain. Recently a mouse monoclonal antibody mediating both neutralization and ADCC has been produced which also showed a strong reactivity to a part of the hypervariable V3 loop region of gp120 represented by aa. 304-323.

To elucidate the potential protective and/or predictive role of antibodies directed to specific HIV epitopes we used the mother-to-child model in relation to transferred maternal anti-HIV antibody pattern and the clinical outcome of the at-risk pregnancy. An initial retrospective analysis of antibodies to synthetic peptides and recombinant proteins, representing structural gene products, was carried out . Sera from 33 children born to HIV-infected mothers and whose clinical outcome was known at the time of the analysis were studied. Sera from uninfected at-risk children before 6 months of age were found to selectively contain maternal antibodies reactive to certain peptides representing epitopes of the gp 120 envelope protein (Table 1).

TABLE 1. ANTIBODY REACTIVITY TO SPECIFIC PEPTIDES - PROTEINS IN NEWBORNS BELOW 6 MONTHS OF AGE

Antibody reactivity to	Uninfected (n=19)	Infected n=14)
	number of positive reactions (%)*	
peptide gp 120/C51	2 (11)	0
peptide gp 120/C53	2 (11)	0
peptide gp 120/C57	5 (26)**	0**
peptide gp 120/C58	2 (11)	0
SP22	4 (21)	2 (14)
peptide p17/9	1 (5)	2 (14)
peptide pol/B98	5 (26)	3 (21)
PB1	7 (37)	5 (35)
pENV9	15 (78)	14 (100)

* All controls (n=5) were negative ** Fisher's test, p<0,05.

In this group of sera, only samples from uninfected newborns reacted to peptides describing conserved sequences of HIV IIIb V3 loop region. In particular the peptide gp120C57 containing the COOH-terminal sequence of the loop reacted in a significant clustered fashion to sera of uninfected newborns.[7] Since the V3 loop region has been described to be the major neutraliziong domain of the envelope glicoprotein gp120 and its structural integrity seems to be essential for viral infectivity we envisaged that a maternal antibody response to such region might play a role in preventing mother-to child transmission of HIV. To test this hypothesis a prospective analysis was then performed on sera of HIV infected pregnant women [8,9]. In our studies we have analysed sera collected at the time of delivery from infected mothers who gave birth to uninfected or infected infants. Criteria for infection/non infection of the infants were those from CDC ,. in addition several DNA samples were assayed for HIV viral sequences by polymerase chain reaction to confirm at the molecular level the diagnosis of infection/non infection. Serum antibody reactivity was assayed in a peptide ELISA test using 15 mer peptides overlapping by ten aminoacids which represent most of the structural proteins of HIV (gag, env). To date 86 mother samples have been evaluated with a large panel of structural HIV peptides Although no significant difference was found in reactivity to the majority of the epitopes tested , sera of mothers of uninfected children (MUC) react at a significant level with the cysteine-containing peptides of the gp120 V3 hypervariable loop when the HIV IIIb peptides were, used, whereas only few sera from mothers of infected children (MIC) had such reactivity(Tab.2)

TABLE 2

HIV IIIb Peptides	MIC (30)	MUC (56)	P
C120-51(INCTRPNNNTRKSIR)	2	22	<0.01
C120-53(RKSIRIQRGPGRAFV)	12	34	NS
C120-57(GNMRQAHCNISRAKW)	2	19	<0.02

Statistical analysis performed by Statview software. Chi square evaluated with continuity correction.Peptide A.A. sequences according to Ratner et al.

These data confirmed our previous observation of a skewed distribution of the reactivity to conserved regions of the V3 loop among MIC and MUC and pointed out that a significant association exists between lack of vertical transmission and reactivity to certain V3 loop sequences.

Since the appearance of our preliminary observations, two other papers evaluated the correlation between HIV vertical transmission and maternal antibody response to HIV. Goedert et al [10] showed that transmitting mothers have lower titers of antibodies to gp120 than non-tranmitting mothers. In addition, Devash et al [11] in a limited number of HIV mothers showed that transmitting mothers lack high affinity antibodies to a conserved sequences of the V3 loop which is considered to be the principal neutralizing domain (PND) of HIV envelope. These antibodies were present in 3 out of 4 of the non-transmitting mothers. Alltogether these data are in keeping with the hypothesis that a good immuneresponse to conserved linear epitopes within a crucial structure of the HIV envelope correlates with a reduced rate of vertical transmission. A major difference between our studies and those of Devash et al is that they have used peptides prepared on the basis of the HIV MN strain sequence which recently has been shown to be the most prevalent strain in the American and European infected cohorts. This difference could account for the lack of correlation between antibody response to gp120-C53 which includes the PND and MUC status in our study. To evaluate this hypothesis we have recently synthesised a group of HIV MN-peptides and tested sera from a cohort of HIV infected mothers classified as MIC or MUC according to the previous defined criteria. When the PND containing sequence was evaluated 100% of the sera from MUC reacted to the gp120-C53 MN and/or IIIb peptide in contrast to only 50% of MIC The details of reactivity to single MN and IIIB peptides are shown in the following table (tab.3).

TABLE 3

Peptides	**MIC**	**MUC**
A) C120-53 IIIb (RKSIRIQRGPGRAFV)	1/12	3/18#
B) C120-53 MN (RKSIHIGPGRAFYTK)	2/12	9/18*
C) C120-53 IIIb & MN	3/12	6/18*

C*) Identifies cross reactivity to IIIB and MN peptides*
*Chi Square with contiguity correction: #=ns, *=p< 0.05. p value calculate as reactivity to C120-53 in MUC vs MIC = <0.01*

These data show that when an appropriate combination of peptides representative of the most prevalent HIV strins is used there is a significant different distribution of the reactivity to PND between MIC and MUC. Moreover the data point out that,at least with out peptide configuration, there exist a consistent cross reactivity between C53MN and C53IIIb which might have account for the high percentage of IIIb reactive sera in HIV infected population.Conversely, only 4 out of 30 sera reacted exclusively to C53IIIb.
Recent reports have shown that within the V3 loop at least three sequences are very conserved among more than 240 HIV isolates and these are the cysteine-containing flanking regions and the top sequence identifying the PND.Taken together all these observation, althogh they do not provide any direct evidence of protective immunity, they are strongly suggestive for a non casual cluster of the presence of anti V3 loop

antibodies and reduced rate of intrauterine vertical transmission.
Another set of experiments has aimed at evaluating the potential of epitope mapping by site directed serology for identifying early markers of infection in children born to HIV infected mothers. The analysis of sets of sequential sera from birth up to the third year of life with a panel of peptides including those of the V3 loop has revealed that in uninfected children the end detection point of HIV antibody measured be peptide ELISA is far longer than that measured by standard commercial Lysate ELISA and Western blot. Thus,we could detect antibodies to HIV envelope peptides in WB seroreverted sera of uninfected children up to 24 months of life.

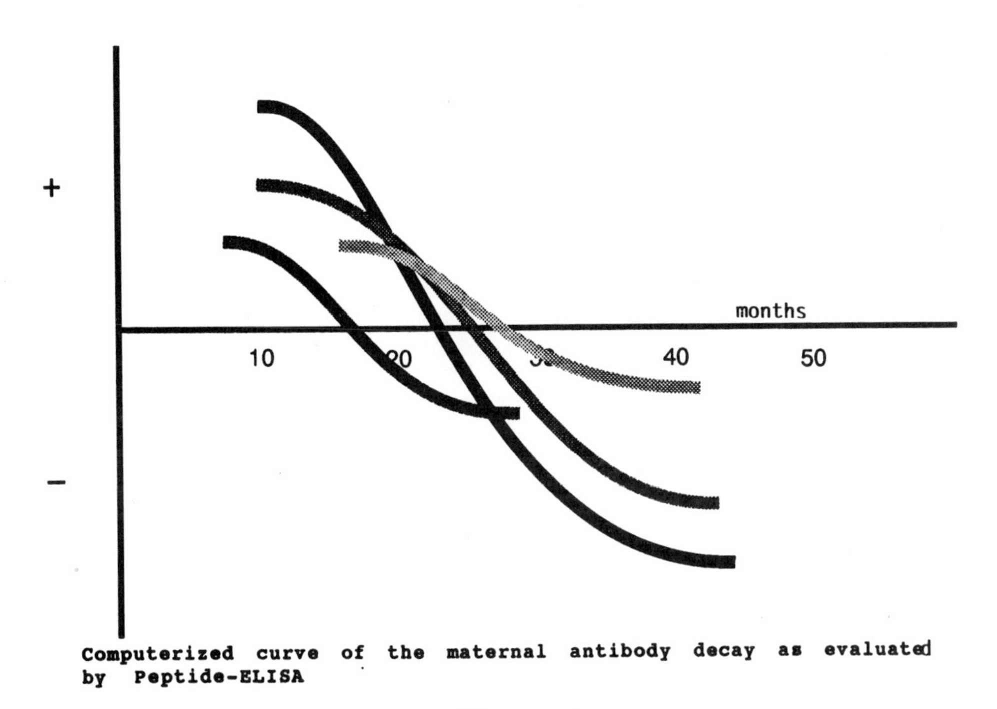

Computerized curve of the maternal antibody decay as evaluated by Peptide-ELISA

Figure 1

These results although preliminary show that the use of synthetic peptides on gp120 describing immunoreactive linear epitope sites of HIV may greatly improve the sensitivity of HIV antibody detection.
A V3-Peptide ELISA might be a crucial technique to evaluate the natural history of immune response and reactivities to HIV in mothers and in children born to infected mothers.

REFERENCES

1 J.Q.Mok, Giaquinto C.,De Rossi A et al Lancet i, 1164-1168.(1987)

2. P Piot, Plummer F.A., Mhalu F.S et al. Science 239:573-579 (1988)

3. Italian Multicenter Study. Lancet ii, 1043-1046. (1988).

4. C.A Hankins. J Acquired Immune deficiency Syndromes. 3:443-448 (1990)

5. D.P. Bolognesi. AIDS 3,s, 111-118 (1989).

6. D.P. Bolognesi. J Acquired Immune deficiency Syndromes. 3:390-340 (1990)

7. P.A Broliden., Moschese V., Ljunggren K et al. AIDS 9:577-582 (1989)

8. P Rossi, Moschese V., Broliden P.A., et al. Proc Natl Acad Sci. USA 86:8055-8058 (1989).

9. P Rossi, Moschese V., Lombardi V et al. Lancet 335:359-360 (1990)

10. J.J Goedert., Mendez H., Robert-Guroff M et al. Lancet ii:1351-1354 (1989).

11. Y Devash., Calvelli T., Wood D G et al. Proc Natl Acad Sci, USA 87:3445-3449 (1990)

IMMUNOGENICITY OF SYNTHETIC PEPTIDES CORRESPONDING TO VARIOUS EPITOPES OF THE HUMAN IMMUNODEFICIENCY VIRUS ENVELOPE PROTEIN

Habib Zaghouani, Brenda Hall, Himanshu Shah and Constantin Bona

Department of Microbiology
Mount Sinai School of Medicine
New York, NY 10029

INTRODUCTION

The envelope gene of the human immunodeficiency virus (HIV) encodes a 160 kd precursor protein which during virus maturation is cleaved into 120 kd and 41 kd proteins, respectively [1,2]. The outer (gp120) and the transmembrane (gp41) proteins are non-covalently associated and are involved in virus infectivity [3-6]. The gp120 has been suggested to bear the CD4 binding site mediating the attachment of the virus to the CD4 antigen [4,7]. Recently, using deletion mutants [8] and monoclonal antibodies [9] the CD4 binding site was mapped within the amino acid sequence 420-437 of the gp120 protein. The comparison of sequences among the fusogenic sites of various viruses with the sequence of HIV predicted that the sequence 526-535 of the amino terminal part of gp41 to be the fusogenic site of HIV-I [10]. Beside these sites playing an important role in infectivity, the gp160 protein bears B cell epitopes inducing neutralizing antibodies and T cell epitopes recognized by T helper and cytotoxic T cells (data reviewed in 11). To date, the best characterized neutralizing epitope is the so called principle neutralizing determinant or PND which maps within the cysteine loop of the gp120 protein [12,13]. It has been suggested that the hexapeptide, GPGRAF, which is conserved in the majority of known HIV-1 isolates is responsible for the induction of broadly neutralizing antibodies [14]. Recently, it was shown that the peptide corresponding to amino acid residues 254-274 of gp120 represents an immunodominant epitope capable of inducing antibodies in HIV-1 infected patients [15]. Also the C-terminal peptide 504-518 of gp120 reacted with sera of HIV-infected patients [16]. In the present work, we present data on the immunogenicity of 4 peptides corresponding to various epitopes of the gp160 protein in various animal species.

PEPTIDE SYNTHESIS AND STRUCTURE

Peptides corresponding to amino acid residues 254-274, 420-445, 503-535 and 526-535 of the gp160 envelope protein from the HIV-1 IIIB isolate (see fig. 1) were synthesized on a p-methyl-

Immunobiology of Proteins and Peptides VI
Edited by M.Z. Atassi, Plenum Press, New York, 1991

benz-hydrylamine resin (IAF Biochem International Inc. Laval, Quebec Canda). The structure of these peptides are as illutrated in figure 1. A glycine-cysteine dipeptide was added at the carboxyl-terminus of both 503-535 and 526-535 to provide a functional SH group facilitating the coupling of peptides to a carrier protein. Three lysine residues were added at the amino terminus of 526-535 to improve its solubility.

The peptides were coupled to either KLH or BSA through their cysteine residues using sulfosuccinimidyl 4-(P-maleimidophenyl) butyrate (sulfo-SMPB) as a coupling reagent [17]. The peptide BSA conjugates were coupled to Sepharose 4B.

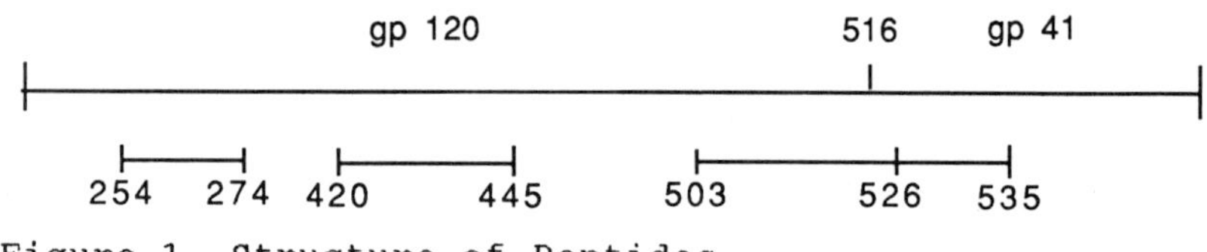

Figure 1. Structure of Peptides

254-274	CTHGIRPVVSTQLLLNGSLAE
420-445	GITLPCRIKQIINMWQEVGKAMYAPP
503-535	VAPTKAKRRVVQREKRAVGIGALFLGFLGAAGS
526-535	FLGFLGAAGS

It was reported that the peptide 254-274 is conserved among various isolates and represented an immunodominant epitope [15]. The peptide corresponding to amino acid residues 420-445 is relatively conserved and might encompass the CD4 binding site [8,9]. The sequence 503-535 encompasses a conserved amino acid stretch (504-518), which has been suggested to represent an immunodominant epitope [16], and a conserved amino terminal part (519-535) of gp41, which was predicted to bear the fusogenic site (526-535) [10].

REACTIVITY OF SERA FROM HIV POSITIVE PATIENTS WITH SYNTHETIC PEPTIDES

A panel of 122 HIV infected sera collected from HIV+ patients in various stages of disease according to the Walter Reed classification were tested for reactivity with synthetic peptides corresponding to various epitopes of the gp160 envelope protein. This was done using an ELISA technique in which microtiter plates were coated with free peptide (2µg/ml), saturated with goat serum than incubated with various dilutions of human serum. Bound antibodies were revealed with goat anti-human IgG antibodies coupled to peroxidase. The serum was considered reactive only when the optical density was three fold higher than the average of 10 sera collected from HIV negative subjects. As can be seen in table 1, the peptide 503-535 reacted significantly with 18 out of 122 sera tested. The remaining 3 peptides did not react with any of the sera tested. Most of the patients 16/18 who reacted with the peptide 503-535 are at stage 1 or 2 of the disease, and the remaining were at stage 4 according to the Walter Reed classification. The fact that the peptide 503-535 reacted with human serum indicates that this epitope of the envelope protein is seen as an immunogenic epitope. The data also indicates that this linear epitope is exposed like in the native gp120 protein. On the other hand, the peptides 254-274, 420-445 and 526-535 did not react with any of the 122 human sera. We would like to point out a few possibilities which might account for these observations:

1. Not all peptides are antigenic within the envelope protein.
2. A particular 3-dimensional configuration is required in order to interact with antibodies.
3. It is possible that the antibody titer is below the limit of detection by the technique used in this study.
4. Finally, one may imagine that the peptides become reactive with antibodies only during a particular interaction between virus and the target cell.

Table 1. Reactivity of HIV envelope synthetic peptides with human sera from HIV infected patients

Peptide	Number of reactive Sera
254-274	0/122
420-445	0/122
503-535	18/122
526-535	0/122

IMMUNOGENICITY OF SYNTHETIC PEPTIDES IN VARIOUS SPECIES

In order to evaluate the immunologic functions of the above peptides, we studied their immunogenicity in various animal species. For this purpose, mice, rabbits and baboons were immunized with peptides coupled to KLH and the anti-peptide antibody titer was determined (table 2). Five BALB/c mice and 2 New Zealand white rabbits were injected with 100 and 500μg of peptide coupled to KLH emulsified in CFA. Mice were boosted 3 weeks later with 500μg of peptide-KLH in IFA and rabbits 15 days later with 500μg of peptide in IFA. Mice were bled 7 days after boosting and rabbits 11 days after the completion of immunization. Also, 2 adult baboons were immunized as follows: 100μg of peptide coupled to KLH emulsified in stearyl-tyrosine (ST) adjuvant [18] on day 1, 7, 20; and 1000 and 700μg of peptides on day 69 and 178, respectively.

The baboons were bled on day 34, 82, 96, 123, 151, 192 and 206. The antibody titer was determined by RIA by using microtiter plates coated with 2μg/ml of free peptide. In the case of rabbits and baboons, several serum dilutions were added, incubated for 2 hours at room temperature and bound antibodies were revealed with ^{125}I-rat anti-mouse κ light chain monoclonal antibody, ^{125}I-goat anti-rabbit IgG, or ^{125}I-goat anti-human IgG for baboons. Table 2 shows the mean of the highest serum dilutions expressed in cpms ^{125}I bound. From these data it appears that the peptide 420-445 induced low antibody titer in all animals. The antibody titer against various peptides varied from species to species. A high titer was observed in rabbits as compared to mice, in spite of the immunization being carried out in FCA.

Table 2. Anti-peptide antibody titer in mice, baboons and rabbits

Immunogen	Mice	Baboons	Rabbits
254-274	4860	100	>36000
420-445	80	<100	2187
503-535	>4860	2700	>36000
526-535	80	300	36000

FINE SPECIFICITY OF AFFINITY PURIFIED RABBIT ANTI-PEPTIDE ANTIBODIES

For further analysis, the rabbit antibodies were purified on a peptide-BSA Sepharose 4B column, and their specificity was studied by incubating various amounts of antibodies on free peptide coated plates. Bound immunoglobulin was revealed using ^{125}I-labeled anti-isotypic antibodies. Figure 2 shows the binding specificty of affinity purified rabbit anti-peptide antibodies.

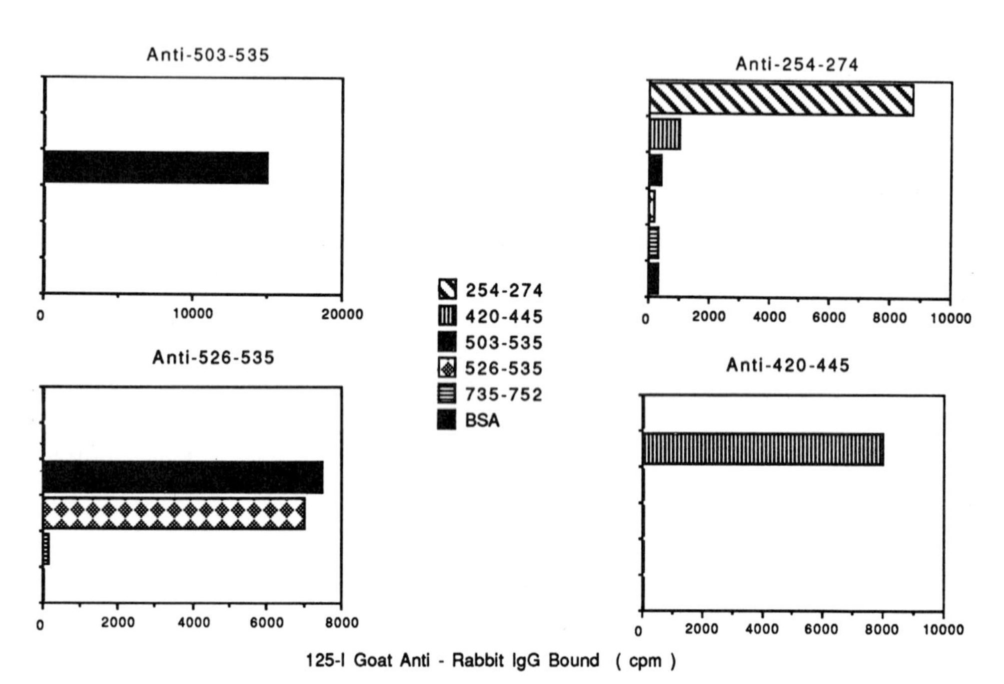

Figure 2. Specificity of anti-peptide antibodies.

In all cases a dose effect binding was observed. Only the binding obtained with 10ng of antibodies is illustrated. It is clear that all rabbit antibodies bind specifically to uncoupled peptide in spite of the fact that the animals were immunized with peptides coupled to KLH. While rabbit anti-254-274 shows a high binding to 254-274 and low insignificant binding to the remaining peptides, the anti-420-445 antibodies bound only the immunizing peptide. Immunization of rabbits with 526-535 peptide induced rabbit antibodies which bound to both the 526-535 and to the 503-535 peptides. However, immunization with the 503-535 peptide induced antibodies which bound to 503-535 but not to 526-535, which is part of the immunizing peptide. Considering the hydrophobic nature of the amino acid residues 526-535, one can speculate that this epitope is hidden within the 503-535-KLH peptide and is not recognized by antibodies. On the other hand, it is possible that the folding of this amino acid stretch within the 503-535 peptide in solution is different from the folding whenever the peptide is bound to plastic plates. In fact, immunization with the 526-535 peptide induced antibodies which bound to both 526-535 and 503-535. These data indicate that the hydrophilic amino part of the 503-535 peptide is recognized by rabbit lymphocytes, but the amino acid stretch 526-535 cannot be seen unless provided in its shortest form.

The binding specificity was then studied by competitive inhibition RIA, in which, ^{125}I-labeled rabbit (50,000 cpms) antibodies and various amounts of competitor were simultanously added to peptide coated plates and incubated overnight at 37°C. Table 3 shows that the binding of rabbit anti-254-274 antibody to the 254-274 peptide is inhibited by both 254-274 free peptide and 254-274 coupled to BSA. The irrelevant 735-752 peptide and BSA did not show any inhibition. In the case of the rabbit anti-420-445 antibody, only the corresponding free and BSA coupled peptide inhibited the binding to 420-445-BSA antibodies indicating the high specificity of these antibodies. The binding of anti-503-535 to 503-535-BSA coated plates is inhibited by 503-535, 503-535-BSA and to some extent by 526-535-BSA but not by 526-535 free peptide.

Table 3. Inhibition of binding of rabbit anti-peptide antibodies by various peptides

	Rabbit antibodies to			
Inhibitor	254-274	420-445	503-535	526-535
254-274	45*	ND	ND	ND
254-274-BSA	78	ND	ND	ND
420-445	ND+	53	-	-
420-445-BSA	ND	75	-	-
503-535	ND	-	90	75
503-535-BSA	ND	-	85	80
526-535	ND	-	-	30
526-535-BSA	ND	-	35	80
735-752-BSA	-*+	-	-	-
BSA	-	-	-	-

* % Inhibition obtained with 7200ng of peptide.
+ Not done.
*+ No inhibition obtained with 7200ng peptide

In a separate experiment, we found that the anti-503-535 antibody binds to 526-535-BSA but not to 526-535 free. This observation suggested that the inhibition seen by 526-535-BSA is due to neoantigens resulting from the coupling of the peptide to both BSA and KLH. Finally, the binding of the rabbit anti-526-535 antibody to 526-535-BSA was inhibited by 526-535-BSA as well as by 503-535 free and 503-535-BSA. Poor inhibition was obtained with the 526-535 peptide. This phenomenon might be due to the small size of the peptide consisting of only 9 amino acid residues.

Taken collectively, these data indicate that the rabbit antibodies are highly specific and that small peptides can be used for the production of specific antibodies. The lack of antibodies specific for the 526-535 epitope subsequent to immunization with the 503-535 peptide may be related to the high hydrophobicity of the epitope or that it is hidden within the structure of a longer peptide bearing this sequence.

INHIBITION OF SYNCYTIA FORMATION BY ANTI-PEPTIDE ANTI-SERA

Since it was predicted that these epitopes play a critical role either in viral infection or immunity, we examined the ability of these antibodies to inhibit syncytia formation mediated by two different virus isolates (MN and IIIB). This was done by incubating 10^4 H9 cells productively infected with either IIIB or MN isolates with 10^5 uninfected HUT78 cells in the presence or absence of a 1/10 dilution of anti-peptide anti-sera or preimmune control. Syncytia inhibition was evaluated by comparing the number of syncytia obtained in the presence or absence of anti-peptide anti-sera. Inhibition of syncytia formation was considered significant when higher than 90% of syncytia were inhibited by the presence of a given anti-sera. Table 4 summarizes the results.

Table 4. Inhibition of IIIB and MN mediated syncytia formation by anti-peptide anti-sera from various animal species

	Sera				
	Preimmune	254-274	420-445	503-535	526-535
		IIIB			
Mice	No	No	No	Yes	No
Rabbits	No	No	No	Yes	No
Baboons	No	No	No	No	No
		MN			
Mice	No	No	No	Yes	No
Rabbits	No	No	No	Yes	No
Baboons	No	No	No	No	No

In each case, a pool of 5 mice, 2 rabbits and 2 baboons sera were tested.
Yes: indicates that 90% or higher inhibtion was obtained.
No: indicates that there was no inhibition.

As illustrated in table 4, mouse and rabbit anti-503-535 peptide sera inhibtited syncytia formation mediated by both IIIB and MN isolates as does a human serum, from an HIV infected patient used as a control. The baboons anti-503-535 sera did not inhibit syncytia formation. This might be due to the low titer of anti-503-535 antibodies in baboon sera (see table 2). On the other hand, none of the remaining anti-peptide sera inhibited the formation of syncytia caused by either IIIB or MN isolates. Several explanations can be entertained:

1. The titer of the antibodies specific for a particular epitope involved in the syncytia formation event is low among the total anti-peptide antibodies.
2. The epitopes represented by those peptides are not linear and/or hidden within the native envelope protein.
3. These peptides are not involved in this particular event.

PRECIPITATION OF GP120 PROTEIN BY AFFINITY PURIFIED ANTI-PEPTIDE ANTIBODIES

In order to verify these possiblities, the anti-peptide antibodies were assayed by radioimmunoprecipatation for recognition of the envelope protein. Mature virus lysates were labeled with ^{125} Iodine and 1-2 X 10^6 cpm were mixed with either anti-peptide antibodies or purified preimmune Ig and incubated overnight at 4°C. Complexes were precipitated using pansorbian cells. Proteins were than denatured by boiling at 100°C for 5 minutes and separated on a 10% acrylamide gel. The gel was dried and autoradiographed.

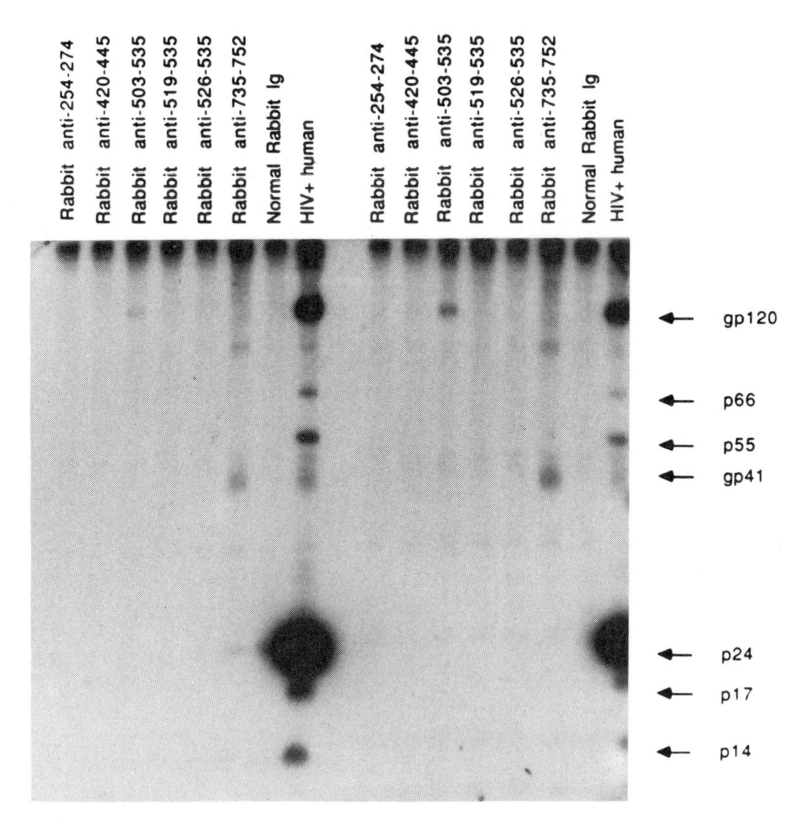

Figure 3. Radioimmunoprecipitation of HIV envelope proteins with purified rabbit anti-peptides

As can be seen in figure 3, 1μg of rabbit anti-503-535 antibodies precipitated a 120 kd protein as does a control human serum obtained from an HIV patient (left pannel). Using 5ug of antibodies (right panel), the rabbit anti-503-535 antibodies precipitated the 120 kd protein. The anti-503-535 antibodies did not precipitate a 41 kd protein as do the rabbit anti-735-752 antibodies used as a positive control. This observation confirmed the specificity of the anti-503-535 antibodies which did not react with the peptide 526-535, which belongs to the gp41 protein. Neither the rabbit antibodies specific for 254-274, 420-445, 526-535 peptides nor the normal rabbit and human sera precipitated any viral protein.

To be certin of these finding we repeated the radioimmunoprecipitation with 10ug of rabbit and baboon antibodies to peptides 254-274, 420-445 and 526-535 (fig 4A) and with 10ug of rabbit 3, rabbit 4, mouse, baboon and human anti 503-535 antibodies (fig 4B). Again the rabbit and baboon antibodies specific for 254-274, 420-445, and 536-535 peptides did not precipitate any viral proteins (fig. 4A). In contrast, rabbit 3, rabbit 4, mouse, baboon, and human anti-503-535 antibodies but not their respective preimmune Ig, precipitated the gp120 protein (fig. 4B).

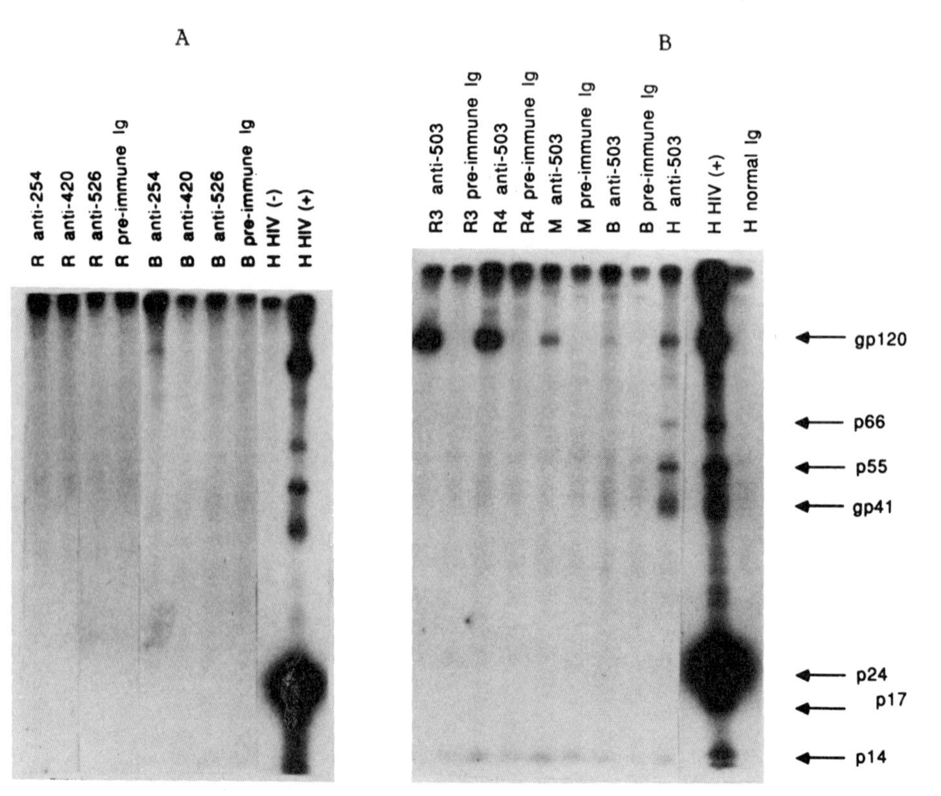

Figure 4. Radioimmunoprecipitation of HIV envelope protein with high amounts of anti-peptide antibodies. R, rabbit; B, baboon; H HIV(-), human serum from HIV negative patient; H HIV(+), human serum from HIV positive patient; H anti-503, anti-503-535 antibodies purified from HIV infected patient.

One exception is that the purified human anti 503-535 antibodies also precipitated gp41, gag (P55) and reverse transcriptase (p66) proteins. This may be due to incomplete absorption of non-anti-503-535 antibodies or to the contamination with sticky antibodies.

The fact that anti-503-535 antibodies from various animal species did not precipitate the gp41 protein indicates that the 526-535 epitope borne by the 503-535 peptide did not induce antibodies toward itself. Alternatively, this hydrophobic sequence requires a special folding or is hidden within the gp41 protein. Thus, one may speculate that the absence of anti-526-535 antibodies after immunization with the 503-535 peptide and the lack of binding to gp41 protein is related to the inaccessibility of this epitope in gp41 to the immune system. On the other hand, the peptides 254-274 and 420-445 induced specific antibodies in both baboons and rabbits but did not inhibit syncytia formation or precipitate the gp120 protein. Since a high amount of antibodies were used in the preciptation assay and no viral protein was recognized, one can suggest that this peptide is not involved in syncytia formation.

In conclusion, our study demonstrates that four peptides corresponding to various epitopes of gp160 protein are immunogenic in various species but only antibodies against 503-535 peptide inhibit syncytia formation.

REFERENCES

1. J.S. Allan, J.E. Coligan, F. Barin, M.F. McLane, J.G. Sodroski, C.A. Rosen, W.A. Haseltine, T-H. Lee, and M. Essex, Major glycoprotein antigens that induce antibodies in AIDS patients are encoded by HTLV-III, Sci. 228:1091-1093 (1985).
2. F.D. Veronese, A.L. DeVico, T.D. Copeland, S. Oroszlan, R.C. Gallo, and M.G. Sarngadharan, Characterization of gp41 as the transmembrane protein coded by the HTLV-III/LAV envelope gene, Sci. 229:1402-1405 (1985).
3. J.M. McCune, L.B. Rabin, M.B. Feinberg, M. Lieberman, J.C. Kosek, G.R. Reyes, and I.L. Weissman, Endoproteolytic cleavage of gp160 is required for the activation of human immunodeficiency virus, Cell 53:55-67 (1988).
4. J.S. McDougal, M.S. Kennedy, J.M. Sligh, S.P. Cort, A. Mawle, and J.K.A. Nicholson, Binding of HTLV-III/LAV to T4+ T cells by a complex of the 110K viral protein and the T4 molecule, Sci. 231:382-385 (1986).
5. B.S. Stein, S.D. Gowda, J.D. Lifson, R.C. Penhallow, K.G. Bensch, and E.G. Engleman, pH-independent HIV entry into CD4-positive T cells via virus envelope fusion to the plasma membrane, Cell 49:659-668 (1987).
6. M. Kowalski, J. Potz, L. Basiripour, T. Dorfman, W.C. Goh, E. Terwilliger, A. Dayton, C. Rosen, W. Haseltine, and J. Sodroski, Functional regions of the envelope glycoprotein of human immunodeficiency virus type 1, Sci. 237:1351-1355 (1987).
7. A.G. Dalgleish, P.C.L. Beverly, P.R. Clapham, D.H. Crawford, M.F. Graves, and R.A. Weiss, The CD4 (T4) antigen is an essential component of the receptor for the AIDS retrovirus, Nature (London) 312:763-767 (1984).
8. L.A. Lasky, G. Nakamura, D.H. Smith, C. Fennie, C. Shimasaki, E. Patzer, P. Berman, T. Gregory, and D. Capon, Delineation of a region of the human immunodeficiency virus type 1 gp120

glycoprotein critical for interaction with the CD4 receptor, Cell 50:975-985 (1987).
9. N. Sun, D.D. Ho, C.R.Y. Sun, R. Liou, W. Gordon, M.S.C. Fung, X. Li, R.C. Ting, T-H. Lee, N.T. Chang, and T-W. Chang, Generation and characterization of monoclonal antibodies to the putative CD4-binding domain of human immunodeficiency virus type 1 gp120, J. Virol. 63:3579-3585 (1989).
10. W.R. Gallaher, J.M. Ball, R.F. Garry, M.C. Griffin, and R.C. Montelaro, A general model for the transmembrane proteins of HIV and other retroviruses, AIDS Res. Hum. Retroviruses 5:431-440 (1989).
11. Q.J. Sattentau, HIV infection and the immune system, Biochimica et Biophysica Acta 989:255-268 (1989).
12. G. Goudsmit, C. Debouck, R.H. Meloen, L. Smit, M. Bakker, D.M. Asher, A.V. Wolff, C.J. Gibbs, and C. Gajdusek, Human immunodeficiency virus type 1 neutralization epitope with conserved architecture elicits early type-specific antibodies in experimentally infected chimpanzees, Proc. Natl. Acad. Sci. USA 85:4478-4482 (1988).
13. G.J. LaRosa, J.P. Davide, K. Weinhold, J.A. Waterbury, A.T. Profy, J.A. Lewis, A.J. Langlois, G.R. Dreesman, R.N. Boswell, P. Shadduck, L.H. Holley, M. Karplus, D.P. Bolognesi, T.J. Matthews, E.A. Emini, and S. D. Putney, Conserved sequence and structural elements in the HIV-1 principal neutralizing determinant, Sci. 249:932-935 (1990).
14. K. Javaherian, A.J. Langlois, G.J. LaRosa, A.T. Profy, D.P. Bolognesi, W.C. Herlihy, S.D. Putney, and T.J. Matthews, Broadly neutralizing antibodies elicited by the hypervariable neutralizing determinant of HIV-1, Sci. 250:1590-1593 (1990).
15. D.D. Ho, J.C. Kaplan, I.E. Rackauskas, and M.E. Gurney, Second conserved domain of gp120 is important for HIV infectivity and antibody neutralization, Sci. 239:1020-1023 (1988).
16. T.J. Palker, M.E. Clark, A.J. Langlois, T.J. Matthews, K.J. Weinhold, R.R. Randall, D.P. Bolognesi, and B.F. Haynes, Type-specific neutralization of the human immunodeficiency virus with antibodies to env-encoded synthetic peptides, Proc. Natl. Acad. Sci. USA 85:1932-1936 (1988).
17. F-T. Liu, M. Zinnecker, T. Hamaoka, and D.H. Katz, New procedures for preparation and isolation of conjugates of proteins and a synthetic copolymer of D-amino acids and immunochemical characterization of such conjugates, Biochemistry 18:690-697 (1979).
18. A. Nixon-George, T. Moran, G. Dionne, C.L. Penney, D. Lafleur, and C.A. Bona, The adjuvant effect of stearyl tyrosine on a recombinant subunit hepatitis B surface antigen, J. Immunol. 144:4798-4802 (1990).

COMMON SEQUENCE IN HIV 1 GP41 AND HLA CLASS II BETA CHAINS CAN GENERATE CROSSREACTIVE AUTOANTIBODIES WITH IMMUNOSUPPRESSIVE POTENTIAL EARLY IN THE COURSE OF HIV 1 INFECTION

Robert Blackburn[*], Mario Clerici[#], Dean Mann[@], Daniel R. Lucey[+], James Goedert[&], Basil Golding[^], Gene M. Shearer[#], and Hana Golding[*]

[*]Division of Virology, and [^]Division of Hematology, CBER, FDA. [#]Experimental Immunology Branch, [@]Viral Carcinogensis Branch, and [&]Viral Epidemiology Branch, NCI, NIH, Bethesda, Maryland 20892. [+]HIV Unit / SGHMMM, Lackland Air Force Base, Texas 78236

ABSTRACT

We have previously reported the identification of highly conserved homologous regions located in the carboxy terminus of the HIV 1 gp41 (aa 837-844), and the amino-terminal of the beta chain of all human HLA class II antigens (aa 19-25). Murine monoclonal antibodies raised against synthetic peptides from these homologous regions bound not only to the isolated peptides, but also to "native" HLA class II molecules on cells. Screening of sera from HIV 1 infected individuals revealed high frequency of sera (35%) containing anti-class II crossreactive antibodies (CRAb), not only in AIDS patients, but also in early, asymptomatic patients. The CRAb containing sera caused potent inhibition of normal CD4-bearing cells' proliferative responses to tetanus toxoid in vitro. They could also kill class II bearing cells by ADCC. The possible contribution of these antibodies to the establishment of immunodeficiency state in HIV 1 infected individuals and/or to disease progression, was examined in two clinical studies: I. Asymptomatic patients were tested in parallel for their PBL responses to flu/tetanus, HLA alloantigens, and PHA (proliferation and IL2 production), and for the presence of anti-class II CRAb. About 50% of these patients showed a selective loss of their in vitro responses to recall antigens (flu/tetanus), which depend on $CD4^+$ cells, while still responding to PHA and ALLO. Interestingly, positive correlation was found ($P<0.001$) between patients' lack of responsiveness to flu/tetanus and the presence in their sera of anti-class II CRAb. II. Retrospective study of HIV 1-infected hemophiliacs, suggest that patients with high titers of CRAb early in the disease progressed faster to full blown disease.

Immunobiology of Proteins and Peptides VI
Edited by M.Z. Atassi, Plenum Press, New York, 1991

INTRODUCTION

"Molecular mimicry" between an infectious agent and "self" molecules may lead to the breakdown of immunological self-tolerance and the generation of harmful autoreactive antibodies and/or effector T cells (Oldstone, 1987; Fujinami and Oldstone 1985). In case of the HIV 1 virus, several studies identified sequences with significant homology to known sequences of self proteins (Auffray and Novotny, 1986; Levy, 1989, Lewis, et al. 1990). However, most of these studies did not provide evidence that the molecular mimicry indeed resulted in the generation of autoantibodies in infected individuals. We have reported the identification of a sequence in the cytoplasmic tail of HIV 1 envelope (gp41), which is homologous to a conservered sequence in the N-terminal of all human HLA class II beta chains (Golding et al., 1988 and Figure 1). We also provided evidence that this "molecular mimicry" indeed leads to the generation, in HIV 1 infected individuals, of crossreactive antibodies which recognize both the virus env

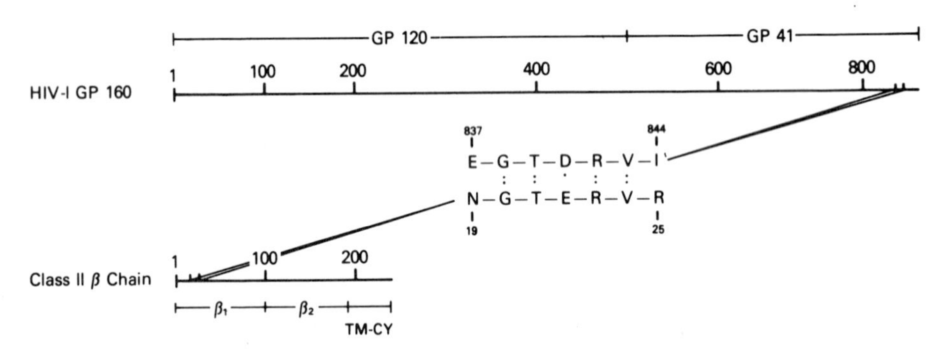

Figure 1. The homologous sequences found in HIV 1 gp41 and human HLA class II beta chains. Identical aa (:); conservative aa change (.).

product and self class II molecules in their native form, as expressed by B cells, T cell lines, and murine L cells transfected with the structural genes of HLA DR, DQ, and DP (Golding et al., 1988, and Figure 2). These CRAb were found in the sera of 30-40% patients irrespective of their disease state including asymptomatic patients. The potential biological function of such anti-class II CRAb was examined in vitro. It was found that the CRAb are IgG antibodies capable of blocking the proliferative responses of normal $CD4^+$ T cells to recall antigens such as tetanus toxoid, and allogeneic class II antigens. Furthermore, these antibodies could also kill class II bearing cells of various haplotypes by antibody dependent cell cytotoxicity (ADCC) (Golding et al., 1989). In the present study we asked whether the generation of CRAb early in the course of HIV 1 infection affect the immune status of the patients and/or their progression to a full blown AIDS.

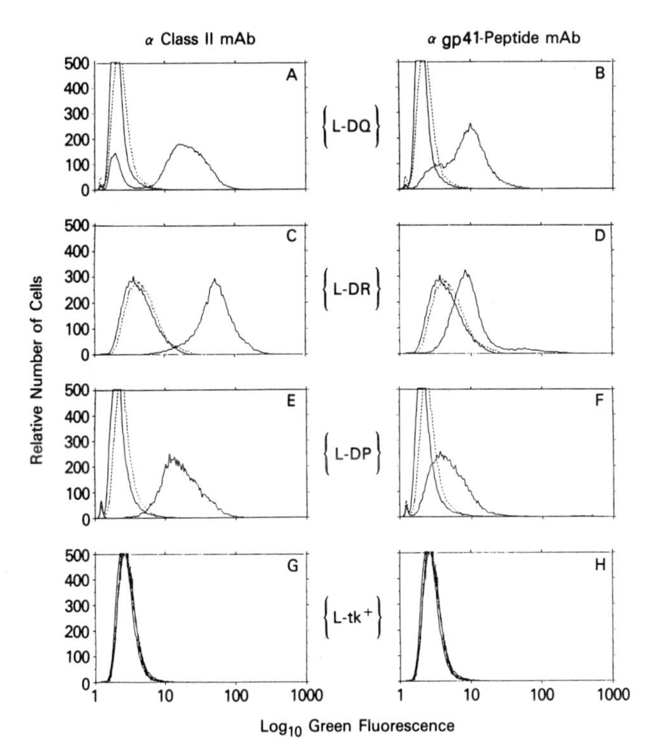

Figure 2. Monoclonal antibody specific for HIV 1 gp41 peptide binds to L cells transfected with human DQ (A-B), DR (C-D), and DP (E-F), but not with the herpes simplex tk gene (G-H). Transfectants were stained with Becton Dickinson class II specific mAb (A,C,E,G solid lines) or the murine anti-gp41 peptide mAb (B,D,F,H solid lines), or anti-TNP mAb (dotted lines).

MATERIAL AND METHODS

Patients and Clinical Evaluation: HIV 1^+ patients were obtained from Wilford Hall United States Air Force (USAF) Medical center, Lackland Air Force Base, TX. Individuals were diagnosed as being HIV 1 infected if they had anti-HIV antibodies demonstrated on two specimens tested by the HIV enzyme immunoassay (Abbott Laboratories, Irving TX) and confirmed by Western Blot analysis (Roche Biomedical Laboratories, Burlington, NC). Patients were classified according to the Walter Reed staging system (Redfield et al., 1986). The HIV 1^- control donors were from USAF. Lymphocytes counts and T cell subsets were determined using laser-based flow cytometry (Coulter Epics Profile; Coulter Electronics, Inc., Hialea, FL) and OKT4A (anti-CD4) and OKT8 (anti-CD8) mAb (Orthodiagnostics Systems, Raritan, NJ). Serial sera samples from HIV 1 infected hemophiliacs (and HIV 1^- controls) were obtained from the "multi-center study of AIDS in hemophiliacs, sponsored by NCI, NIH.

In Vitro tests for T_h function: The processing of the patients' bloods and their in vitro stimulations were described in detail before (Clerici et al., 1989).

Enzyme Linked Immunosorbent Assay (ELISA): Patients' sera were diluted 1:100-1:1600, and 100ul were added to ELISA plates coated with the homologous gp41 or class II derived peptides, or control peptides, as previously described (Golding et al. 1988, 1989). A particular serum dilution was considered positive if it gave an OD(405) reading of at least 3 SD above the mean of normal sera from HIV 1^- individuals (usually >0.4).

Statistical Analysis of Data: Row (R) X column (C) contingency table was set up as shown in Table II for testing the correlation of the different T_h functional categories with the presence or absence of anti-class II CRAb in patients' sera (Snedecor and Cochran, 1980). The sum of the X^2 value was calculated by the sums of the equation $X^2 = (f-F)^2/F$, where f is the observed frequency and F is the expected frequency. The degrees of freedom for this analysis are: df = (R-1)(C-1). Student's t test was performed for the comparison of two independent samples of unequal size, and P value was determined.

RESULTS AND DISCUSSION

In a previous study (Clerici et al. 1989), it was found that the in vitro proliferative responses and IL2 production of PBL from HIV 1 infected patients in stages WR 1 and WR 2, did not correlate with their $CD4^+$ cell numbers. The different stimuli used to evaluate the immunological status of the patients could either stimulate only $CD4^+$ T_h cells (flu/tetanus), or both $CD4^+$ and $CD8^+$ T cells (PHA, allogeneic cells,(ALLO)). Significant number of asymptomatic patients with $CD4^+$-T cells >800/ml demonstrated immune dysfunction in vitro. Three different pattern of T_h dysfunction were identified: (a) 54% showed a selective loss of T_h response to recall antigens such as flu and tetanus, while retaining their responsiveness to PHA and ALLO stimuli; (b) 16% did not respond to flu/tetanus as well as ALLO, but retained their response to PHA; and (c) 14% of patients showed lack of responses to all stimuli.

The underlining mechanism(s) responsible for the early selective loss of $CD4^+$ T_h-cells' function (but not numbers), are not clear yet. In the current study we asked in a double blind fashion whether the production of anti-class II CRAb in asymptomatic patients correlates with any particular pattern of T_h dysfunction pattern.

As can be seen in Table 1, of the 22 patients studied, 9 (41%) responded to all stimuli, 12 (54%) did not respond to flu/tetanus but did respond to ALLO and PHA, and 1 patient (.5%) did not respond to flu/tetanus and ALLO, while retaining his response to PHA. Interestingly, 10 of the 13 patients unresponsive to flu/tetanus produced CRAb to the nonpolymorphic class II sequence. Statistical analysis, depicted in Table 2, showed that this correlation is highly significant ($X^2 = 10.4$, P<.001).

This finding suggests that the presence of circulating antibodies capable of binding to all class II bearing cells, may result in T cell dysfunction. The mechanism of the suppression is not determined yet. It could be a simple blockade of antigen presentation to T cells, since the epitope recognized by the CRAb is very close to the antigen binding sites on the class II molecules. Alternatively, such anti-class II antibodies may be able to deliver a negative signal to class II expressing cells (as was previously demonstrated for B cells and monocytes, Chen et al., 1987 Santamaria et al., 1989), resulting in long term anergy which manifested itself in the in vitro cultures.

To determine the long term effects of such autoantibodies on disease progression, requires many years of follow up. Instead, we were able to obtain multiple bleeds from

Table 1. The correlation between IL2 production to various stimuli, and the presence of anti-class II CRAb in sera of asymptomatic HIV 1 patients

Patient	anti-class II CRAb*	IL2 production in response to: FLU/TETANUS	ALLO	PHA
257	+	-	+	+
718	++	-	+	+
724	+	-	+	+
318	+	-	+	+
300	+	-	+	+
516	++	-	+	+
575	++	-	+	+
176	+	-	+	+
234	+	-	+	+
473	++	-	-	+
380	-	-	+	+
717	-	-	+	+
716	-	-	+	+
820	-	+	+	+
382	-	+	+	+
108	-	+	+	+
756	-	+	+	+
425	-	+	+	+
160	-	+	+	+
363	-	+	+	+
149	-	+	+	+
328	-	+	+	+

*Titers: (-) negative, (+) 1:200, (++) 1:400 .

Table 2. Statitical analysis of the data in Table 1

		PRODUCTION OF ANTI CLASS II CRAb (+)	PRODUCTION OF ANTI CLASS II CRAb (-)
IL2 RESPONSE TO FLU/TET ($CD4^+$-T CELLS)	(+)	0	9
	(-)	10	3

$X^2 = 10.4$ $P < 0.001$

hemophiliacs with HIV 1 infection. Table II summarises the first retrospective study. The numbers in brackets represent the year of sero-conversion. As can be seen, patients with undetectable or low titers of anti-class II CRAb remained asymptomatic for a long period of time and showed a mild reduction in their % $CD4^+$ T cells (e.g. patient HER059, 8 years; COR183, 9 years). In contrast, patients with high serum titers of CRAb, in general deteriorated much faster to full blown AIDS, and showed a rapid decline of their $CD4^+$ T cells (e.g. patient HER025, 1 year; HER212, 2 years; HER003, 4 years; HER068, 5 years). This study is by no mean finished. But the two studies in concert suggest that the early appearance of anti-class II autoantibodies as a result of the molecular mimicry between HIV 1 gp41 and self HLA may be a factor in the prognosis of HIV 1 infected individuals.

Table 3. Correlation between anti-class II CRAb and disease progression in HIV 1 infected hemophiliacs

Patient	Year	HIV serol/disease	%CD4 cells	CRAb to class II
COR 115	1986	negative/negative	24%	--
	1987	negative/negative	27%	--
COR 029	1986	negative/negative	30%	--
COR 003	1986	positive/asymptom	42%	--
(1984)	1988	positive/asymptom	48%	--
HER 026	1982	positive/asymptom	38%	--
(1982)	1984	positive/asymptom	35%	±
	1985	positive/asymptom	30%	±
	1987	positive/asymptom	25%	ND
HER 059	1984	positive/asymptom	28%	--
(1980)	1986	positive/asymptom	20%	--
	1988	positive/asymptom	22%	ND
COR 183	1987	positive/asymptom	45%	--
(1980)	1988	positive/asymptom	25%	--
	1989	positive/asymptom	25%	--
HER 005	1982	positive/asymptom	>30%	±
(1982)	1986	positive/asymptom	>30%	±
	1988	positive/asymptom	20%	±
PHI 129	1985	positive/asymptom	30%	±
(1984)	1987	positive/asymptom	40%	--
HER 093	1980	positive/asymptom	>30%	--
(1980)	1986	positive/AIDS	15%	+++
HER 068	1980	positive/asymptom	>30%	+++
(1980)	1982	positive/asymptom	30%	+++
	1985	positive/AIDS	5%	+++
HER 003	1982	positive/asymptom	>30%	++
(1980)	1984	positive/AIDS	17%	+++
	1985	positive/AIDS	7%	+
HER 212	1985	positive/AIDS	20%	++
(1983)	1986	positive/AIDS	4%	+
HER 011	1986	positive/AIDS	7%	++
(1979)	1987	positive/AIDS	5%	+++
HER 025	1985	positive/AIDS	9%	+++
(1984)	1987	positive/AIDS	9%	+++

Titers: (--) negative; (±) 1:100; (+) 1:200; (++) 1:400; (+++) 1:800-1600. ND not done.
numbers in brackets represent the year of HIV 1 sero-conversion.

REFERENCES

Auffray, C., and J. Novotny. (1986), Speculatios on sequence homologies between the fibronectin cell-attachment site, Major Histocompatibility Antigens, and a putative AIDS virus polypeptide. Human Immunol. 15, 381-390.

Chen, Z. Z., J. C. McGuire, K. L. Leach, and J. C. Cambier. (1987), Transmembrane signaling through B cell MHC class II molecules: anti- Ia antibodies induce protein kinase C translocation to the nuclear fraction. J. Immunol. 138, 2345-2352

Clerici, M., N. I. Stocks, R. A. Zajac, R. N. Bothwell, D. R. Lucey, C. S. Via, and G. M. Shearer. (1989), Detection of three distinct patterns of T helper cell dysfunction in asymptomatic, Human Immunodeficiency Virus-seropositive patients. Independence of $CD4^+$ cell numbers and clinical staging. J. of Clinical Invest. 84, 1892-1899.

Fujinami, R. S. and M. B. A. Oldstone, (1985), Amino acid homology between the encephalitogenic site of myelin basic protein and virus: mechanism for autoimmunity. Science, 230, 1043-1045.

Golding, H., F. A. Robey, F. T. Gates III, W. Linder, P. R. Beining, T. Hoffman, and B. Golding. (1988), Identification of homologous regions in Human Immunodeficiency Virus 1 gp41 and human MHC class II beta 1 domain. I. Monoclonal antibodies against the gp41-derived peptide and patients' sera react with native HLA class II antigens, suggesting a role for autoimmunity in the pathogenesis of Acquired Immunodeficiency Syndrom. J. Exp. Med. 167, 914-923.

Golding, H., G. M. Shearer, K. Hillman, P. Lucas, J.Manischewitz R. A. Zajac, M. Clerici, R. E. Gress, R. N. Boswell, and B. Golding. (1989), Common epitope in Human Immunodeficiency Virus (HIV) 1 - gp41 and HLA class II elicits immunosuppressive autoantibodies capable of contributing to immune dysfunction in HIV 1 - infected individuals. J. Clinical Invest. 83, 1430-1435.

Levy, J. A. (1989), Human Immunodeficiency Viruses and the pathogenesis of AIDS. JAMA 261, 2997-3006.

Lewis, D. E., R. G. Ulrich, H. Atassi, and M. Z. Atassi. (1990) HLA DR peptide inhibits HIV-induced Syncytia. Immunol. Lett. 24, 127-132.

Oldstone, M. B. A. (1987), Molecular mimicry and autoimmune disease. Cell 50, 819-820.

Redfield, R. R., D. C. Wright, and E. C. Tramont. (1986), The Walter Reed staging classification for HTLV III/LAV infection. N. Engl. J. Med. 314, 131-132.

Snedecor, G. W. and W. G. Cochran. (1980), Statistical Methods. 7th Ed. The University of Iowa Press, Ames, IA.

Santamaria, P., R. C. Gehrz, M. K. Bryan, and J. J. Barbosa. (1989), Involvement of class II MHC molecules in the LPS-induction of IL1/TNF secretions by human monocytes. Quantitative differences at the polymorphic level. J. Immunol. 143, 913-922.

ISOLATION AND CHARACTERIZATION OF THE NEUTRALIZABLE EPITOPE OF SIMIAN RETROVIRUS-1 (SRV-1) AND OF THE CELL RECEPTOR FOR THE VIRUS

Eli Benjamini[1], Jose V. Torres[1], Linda L. Werner[1] and Arthur Malley[2]

Department of Medical Microbiology and Immunology, School of Medicine, University of California, Davis, California[1] and Oregon Regional Primate Research Center, Beaverton, Oregon[2]

ABSTRACT

An area encompassing residues 142-167 of the envelope protein of type D simian retrovirus (SRV-1) has been shown to contain the epitope to which neutralizing antibodies are directed. This area has been synthesized and shown to bind to monkey and mouse antiviral antibodies and to a virus neutralizing mouse monoclonal antibody. Protein conjugates of this peptide as well as the cross-linked or the free peptide induce antibodies capable of neutralizing, _in vitro_, viral infectivity.

The cell receptor to the virus was isolated following extraction of Raji cells with non-ionic detergents. The receptor was isolated and characterized following radioimmuno-precipitation of ^{125}I labeled cell extract bound to viral envelope protein. This immunoprecipitation could be inhibited by antiserum to peptide 142-167. Analysis in gels indicate that the receptor is of molecular weight of approximately 60 KDa.

These results indicate that the neutralizing antibodies and the receptor recognize the same area on the viral envelope protein and that neutralization is the result of blocking the virus-receptor interaction by antibodies.

INTRODUCTION

Simian acquired immunodeficiency syndrome (SAIDS), in many species of Asian macaques, is caused by a group of type D retroviruses. Infection rates have been estimated to be about 5% in zoos and as high as 25% in primate research centers. The clinical syndrome in monkeys infected with SRV shares many features in common with human AIDS which is caused by the genetically unrelated lentivirus, the human immunodeficiency virus (HIV) (Hendrickson et al, 1984; Maul et al, 1985, 1986). SAIDS, symptoms include lymphadenopathy, splenomegaly, diarrhea, hematologic abnormalities, immunologic deficiency, infections by opportunistic microorganisms and chronic wasting, resulting in 25-50% fatality rates. However, unlike HIV, protective immunity to SRV is achieved in some animals who recover from viral infection and in most animals immunized with a killed virus vaccine. In fact, some SRV infections result in recovery to a carrier state or to a completely nonviremic state (Marx et al. 1986; Maul et al., 1986). These findings make SRV an interesting model for HIV and for simian immunodeficiency virus (SIV) infection.

CHARACTERIZATION OF THE NEUTRALIZABLE EPITOPE

At least five serotypes of SRV have thus far been identified. The envelope proteins of SRV-1, SRV-2 and SRV-3 (Mason Pfizer Monkey Virus) have been molecularly cloned and sequenced. They are closely related but distinct from each other (Power et al., 1986; Sonigo et al., 1986; Thayer et al., 1987). The various SRV serotypes exhibit extensive immunological cross reactivity, but there is no cross-reactivity at the level of serum neutralizing antibodies. For example, although SRV-1 and SRV-2 exhibit serological cross reactivity (Bryant et al., 1986), the envelope protein sequences of these two serotypes show approximately 40% divergence (Thayer et al. 1987) and these two serotypes do not exhibit cross reactivity at the level of neutralizing antibodies. Based on these findings and on the comparison of amino acid sequences of the envelope proteins of SRV-1 and SRV-2 we identified unique areas on the envelope proteins which may serve as potential epitopes for the neutralizing antibodies specific to each of the serotypes. Two such areas consisted of residues 147-162 of SRV-1 and residues 92-106 of SRV-2 (Werner et al., 1990). The amino acid sequences of these areas are shown in Table 1.

Table 1. The Amino Acid Sequences of SRV-1 and SRV-2 Peptides

	142								150										160							167
SRV-1	L	T	A	T	M	I	R	D	K	S	P	S	S	G	D	G	N	V	P	T	I	L	C	N	N	Q
SRV-2	Y	-	-	I	L	A	S	N	R	A	-	T	I	-	T	S	-	-	-	-	V	-	-	-	T	H

The peptides SRV-1 147-162 and SRV-2 92-106 were synthesized; a portion of each of the synthetic peptides was conjugated to KLH and to BSA. Peptide SRV-1 147-162 conjugated to KLH exhibited specific binding with neutralizing mouse antisera raised by immunization with SRV-1 or with its recombinant envelope protein, rEP (Werner et al., 1990). Also, the free peptide SRV-1 147-162 cross-linked to itself with glutaraldehyde exhibited binding with an anti-SRV-1 neutralizing mouse monoclonal antibody (Werner et al., 1990). Moreover, the free SRV-1 peptide 147-162 exhibited the capacity to compete with the whole live virus for neutralizing antibodies since it inhibited the neutralization of SRV-1 infectivity, <u>in vitro</u>, by monkey anti-SRV-1 neutralizing serum but did not inhibit neutralization of SRV-2 infectivity by anti-SRV-2 serum. The specificity of the neutralizable epitope of SRV-1 and SRV-2 is further indicated by the findings that the SRV-2 peptide, 92-106 inhibited the neutralization of SRV-2 infectivity, <u>in vitro</u>, by anti-SRV-2 neutralizing serum but did not inhibit neutralization of SRV-1 infectivity by anti-SRV-1 serum. Finally, the immunoreactivity of SRV-1 and SRV-2 peptides with neutralizing antibodies for each strain was demonstrated by the findings that SRV-1 peptide 147-162-KLH conjugate could serve as a specific immunoadsorbent for rhesus anti-SRV-1 neutralizing antibodies (Werner et al., 1990). Similarly SRV-2 peptide 92-106-KLH conjugate served as an immunoadsorbent for anti-SRV-2 neutralizing antibodies. A summary of the immunoreactive properties of SRV-1 peptide 147-162 and of SRV-2 peptide 92-106 is given in Table 2.

Table 2. Immunological Reactivity of SRV-1 and SRV-2 Peptide

Reactivity	Peptide or Conjugate	
	SRV-1 147-162	SRV-2 92-106
Binding with anti-SRV-1	+	-
Binding with anti-SRV-2	-	N.D.[1]
Inhibition of neutralization of SRV-1 infectivity by anti-SRV-1	+	-
Inhibition of neutralization of SRV-2 infectivity by anti-SRV-2	-	+
Serving as immunoadsorbent for anti-SRV-1	+	-
Serving as immunoadsorbent for anti-SRV-2	-	+
Inducing, in mice, antibodies reacting with SRV-1 or rEP of SRV-1	+	N.D.

[1] ND - not done

Immunization of mice with SRV-1 peptide 147-162-KLH conjugate induced antibodies capable of reacting specifically with SRV-1 and with its recombinant envelope protein rEP (Werner et al., 1990). However, these antibodies were incapable of neutralizing in vitro the infectivity of SRV-1. To explain the inability to the conjugate to induce neutralizing antibodies in spite of the immuno-reactivity of SRV-1 peptide 147-162 with neutralizing antibodies we considered the possibility that this peptide represented only a portion of the neutralizable epitope. To test this hypothesis we synthesized a larger peptide comprising of residues 142-167 of SRV-1 envelope protein (Table 1). Mice were immunized with peptide 142-167 conjugated to KLH or to BSA, with the glutaraldehyde cross-linked peptide prepared as described (Werner et al, 1990) or with the free peptide. All immunogens were injected twice at a dose of 100 µg in Freund's complete adjuvant followed by an aqueous boost with 100 µg of the conjugate or peptide. Results summarized in Table 3 indicate that these immunogens induced antibodies capable of neutralizing in vitro, the infectivity of SRV-1 but not of SRV-2.

The results summarized so far indicate that peptide 142-167 constitutes an epitope to which neutralizing antibodies are directed. Moreover, the neutralizing antibodies produced by mice or by rhesus monkeys recognize the same area on the envelope protein of SRV-1.

Table 3. The Immunogenicity of Peptides and Peptide Conjugates

Immunogen	Average* Titer of Binding To		Neutralizing Titers** of Infectivity by	
	SRV-1	SRV-1 rEP	SRV-1	SRV-2
SRV-1 142-167-KLH	$1x10^4$	$1x10^4$	1:32	0
SRV-1 142-167-BSA	$1x10^4$	$1x10^4$	1:8	0
Cross linked SRV-1 142-167	$1x10^3$	$1x10^4$	1:32	0
Free SRV-1 142-167	$1x10^3$	$1x10^3$	1:8	0
Cross linked control***peptide	0	0	0	0
Free control*** peptide	0	0	0	0
Control*** peptide-KLH conjugate	0	0	0	0

*As assessed by ELISA.
**Highest titers of several antisera.
***Peptide representing residues 303-321 of simian immunodeficiency virus (SIV) having the sequence NKHYNLTMKCRRPGNKT.

CHARACTERIZATION OF THE VIRUS RECEPTOR

Unlike HIV and SIV which infect primarily $CD4^+$ cells, SRV infects a variety of cells which include B cells, T cells, macrophages, fibroblasts and epithelial cells (Maul et al., 1988). It has recently been reported that the receptor for SRV is encoded by genes present on human chromosome 19 (Sommerfelt et al., 1990). Also, it has been recently reported that all of the SRV serotypes utilize the same cell surface receptor (Sommerfelt and Weiss, 1990). However, the nature of the receptor for SRV has not been determined.

Our findings 1) that there is no cross reaction between SRV-1 and SRV-2 on the level of neutralizing antibodies 2) that the neutralizing antibodies are directed to different sequences of amino acids on SRV-1 and SRV-2 envelope proteins and 3) that SRV-1 rEP inhibits, *in vitro*, infectivity by SRV-1 but not by SRV-2 (Werner et al., 1991) are not compatible with the conclusions that all serotypes utilize the same cell receptor. The discrepancy may be reconciled by the possibility that although the various serotypes utilize the same receptor they recognize different areas on the receptor.

In an attempt to ascertain the mechanism of viral infectivity, the nature of cell receptor and the mechanism of neutralization by antibodies we proceeded to isolate the viral cell receptor. This was achieved by immunoprecipitation. Briefly, Raji cell surface proteins were labeled with ^{125}I using lactoperoxidase as described (Marchalonis 1969). The cells (approximately 10 x 10^6) were lysed in phosphate buffered saline (PBS) containing 0.5% Triton X 100 detergent and protease inhibitors; immunoprecipitation was performed basically as described (Urdal et al., 1988) with a preclearing stage prior to immunoprecipitation (Knowels, 1987). The immunoprecipitating ligand for the receptor consisted of SRV-1. The virus-receptor complex was reacted with mouse anti-rEP and the entire complex was precipitated using immunobeads coated with rabbit anti-mouse antibodies. For gel analysis, the beads bound to the virus receptor complex were boiled in the presence of buffer containing 4% SDS. Following centrifugation the supernatant was applied to a 12% SDS gel. The developed gel was dried and radioautographed, revealing a single radioactive spot of molecular weight of approximately 58-60 KDa. For radioactivity analysis, an aliquot of the supernatant was counted and the amount of radioactivity bound to the virus was determined.

Table 4. The immunoprecipitation of ^{125}I labeled Raji cell surface by SRV-1 proteins and its inhibition by antiserum to cross linked SRV-1 peptide 142-167 and by antiserum to cross linked SIV peptide 514-528 serving as control

Treatment of SRV-1 Prior to Immunoprecipitation	Total CPM in Cell Lysate	Total CPM Precipitated	Inhibition (%)
Reacted with anti-SIV 514-528	92,627	12,367	0
Reacted with anti-SRV 142-167	92,904	5,675	54

An experiment was performed in which the SRV-1 used as the ligand for binding the cell receptor was first incubated with mouse antibodies raised against glutaraldehyde cross-linked SRV-1 peptide 142-167. These antibodies have been shown in our laboratory to neutralize the _in vitro_ infectivity of Raji cells by SRV-1. Results in table 4 clearly indicate that the addition of these antibodies to SRV-1 prior to the addition of ^{125}I labelled cell lysate significantly inhibited the immuno-precipitation of ^{125}I protein(s) in comparison to the addition of mouse antibodies to a glutaraldehyde cross-linked control peptide (peptide 514-528 of SIV). The results indicate that the immunoprecipitated labeled protein(s) (or at least a portion thereof) contain the cell receptor for the virus. Moreover, the results of the inhibition by the anti-SRV-1 142-167 indicate that the portion of the viral envelope protein recognized by the receptor is in the area of residues 142-167 of the envelope protein. Work is in progress to map the precise area on the envelope protein of SRV-1 which is recognized by the viral cell receptor. Results of the above experiments strongly indicate that the neutralizing antibodies exert their effect by preventing the viral envelope attachment to the cell receptor.

ACKNOWLEDGEMENTS

This work was supported in part by Grants AI27027 and AI28570 from the National Institute of Allergy and Infectious Diseases.

REFERENCES

Bryant, M.L., Gardner, M.B., Marx, P.A., Maul, D.H., Lerche, N.W., Osborn, K.G., Lowenstine, L.J., Bogden, A., Arthur, L.O., and Hunter, E., Immunodeficiency in rhesus monkeys associated with the original Mason-Pfizer Monkey Virus, J. Natl. Cancer Inst., 77, 957-965 (1986).

Hendrickson, R.V., Maul, D.H., Lerche, N.W., Osborn, K.G., Lowenstein, L.J. Prahalada, S., Sever, J.L., Madden, D.L., and Gardner, M.B., Clinical features of simian acquired immunodeficiency syndrome (SAIDS) in rhesus monkeys, Lab. Anim. Sci., 34, 140-145 (1984).

Knowels, R.W., Two dimensional gel analysis of transmembrane proteins, In:Histocompatibility Testing (ed. B. Dupont), pp. 1-45, Springer Verlag, New York (1987).

Marchaloms, J.J., 1969, An enzyme method for the trade iodination of immunoglobulins and other proteins, Biochem. J., 113, 299-305 (1969).

Marx, P.A., Pedersen, N.C., Lerche, N.W., Osborn, K.G., Lowenstine, L., Lackner, A.A., Maul, D.H., Kluge, J.D., Zaiss, C., Sharpe, V., Spinner, A., and Gardner, M.D., Prevention of simian acquired immune deficiency syndrome with a formalin-inactivated type D retrovirus vaccine, J. Virol., 60, 431-435 (1986).

Maul, D.H., Lerche, N.W., Osborn, K.G., Marx, P.A., Zaiss, C., Spinner, A., Kluge, J.D., Mackewise, M.R., Lowenstine, L.J., Bryant, M.L., Blakeslee, J.R., Hendrickson, R.V., and Gardner, M.B., Pathogenesis of simian AIDS in rhesus macaques inoculated with type D retrovirus, Am. J. Vet. Res., 47, 863-868 (1986).

Maul, D.H., Miller, C.M., Marx, P.A., Bleviss, M.L., Madden, D.L., Hendrickson, R.V., and Gardner, M.B., Immune defects in simian acquired immunodeficiency syndrome, Vet. Immunol. Immunopathol., 8, 201-294 (1985).

Power, M.D., Marx, P.A., Bryant, M.L., Gardner, M.B., Barr, P.J., and Luciw, P.A., Nucleotide sequence of SRV-1, a type D simian acquired immune deficiency syndrome retrovirus, Science, 231, 1567-1572 (1986).

Sommerfelt, M.A., and Weiss, R.A., Receptor interference groups of 20 retroviruses plating on human cells, Virology, 176, 58-69 (1990).

Sommerfelt, M.A., Williams, B.P., McKnight, A., Goodfellow, P.N., and Weiss, R.A., Human chromosome 19 localization of the receptor gene for type D simian retroviruses, J. Virol, (in press) (1990).

Sonigo, P., Barker, P.C., Hunter, E., and Wain-Holeson, S., 1986, Nucleotide sequence of Mason-Pfizer monkey virus: an immune suppressive D type retrovirus, Cell, 45, 375-385 (1986).

Thayer, R.M., Power, M.D., Bryant, M.L., Gardner, M.B., Barr, P.J., and Luciw, P.A., Sequence relationship of type D retroviruses which cause simian acquired immunodeficiency syndrome, Virology, 157, 1317-329 (1987).

Urdal, D.L., Call, S.M., Jackson, J.L., and Dower, S.K., Affinity purification and chemical analysis of the Interleukin-1 receptor, J. Biol. Chem., 236, 2870-2877 (1988).
Werner, L.L., Malley, A., Torres, J.V., Leung, C.Y., Kwang, H.S., and Benjamini, E., Synthetic peptides of envelope proteins of two different strains of simian AIDS retroviruses (SRV-1 and SRV-2) represent unique antigenic determinants for serum neutralizing antibodies, Mol. Immunol., (in press 1990).
Werner, L.L., Torres, J.V., Leung, C.Y., Kwang, H.S., Malley, A., and Benjamini, E., Immunological properties of a recombinant simian retrovirus-1 envelope protein and a neutralizing monoclonal antibody directed against it, Mol. Immunol., (submitted 1990).

COMPLEXES AND CONJUGATES OF CIS-Pt

FOR IMMUNOTARGETED CHEMOTHERAPY

Ruth Arnon[1], Bilha Schechter[1] and Meir Wilchek[2]

Departments of Chemical Immunology
and Biophysics
The Weizmann Institute of Science
Rehovot, Israel

INTRODUCTION

A major problem in cancer chemotherapy is how to enhance the effectiveness of currently available anti cancer drugs. Due to the lack of selectivity of cytotoxic agents, the administration of single drug doses is restricted to sub-toxic levels. Drug clearance and excretion and/or metabolic conversion of the drug result in only transient increased levels at the tumor site which are generally insufficient to effect complete cure. The problem is therefore, how to attain sufficient amounts of drug at the tumor site for the required period of time, without using drug doses which are above the threshold of toxicity. One possible way is to use molecules with inherent or acquired ability to interact selectively with the target organ, e.g. anti-tumor antibodies, as carriers for the drug, thus leading to specific targeting to the tumor site. An alternative approach is to attach anti-cancer agents to polymeric carriers that will act as control release units by affecting the biodistribution and maintenance of the drug. These delivery devices may cause retardation of drug clearance or inactivation as well as prolonged or sustained release of the drug at non-toxic levels, thus overcoming the problem of harmful peak concentrations. Such polymeric drug systems, in addition to being advantageous to the free drug as such, may be also used for conjugation to antitumor antibodies for the purpose of immunotargeting.

This study describes the preparation and characterization of polymeric drug complexes comprising the chemotherapeutic agent cis-diamminedichloroplatinum(II) (cis-Pt), and their possible employment in a site-directed targeted system using antitumor antibodies.

Immunobiology of Proteins and Peptides VI
Edited by M.Z. Atassi, Plenum Press, New York, 1991

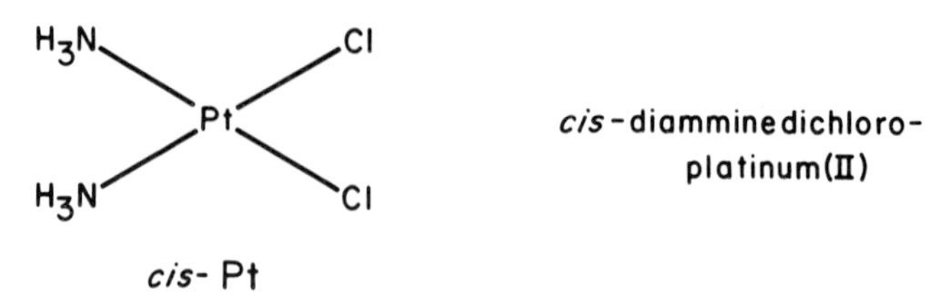

Fig. 1. Chemical structure of cis-Pt.

POLYMERIC-DRUG COMPLEXES OF CIS-PT

Cis-Pt is one of the most important chemotherapeutic agents used today in the treatment of solid tumors, both as a single agent and in combination regimens (1). The major toxic manifestation of cis-Pt is nephrotoxicity, which can be partially controlled by hydration or diuresis (2), but then other related toxicities such as myelosuppression, gastro-intestinal, oto- or neuro-toxicities become dose limiting (3). The chemistry of cis-Pt (Figure 1), which is described as a square planar metal complex, is dominated by the high affinity of the NH_3 groups to the Pt(II) center. The two coordinated chlorides act as leaving groups, which are open to nucleophilic substitution by incoming ligands exhibiting higher affinity towards the Pt(II) (4,5). Cis-Pt can thus react with a variety of ligands, present on low or high molecular weight substances, such as carboxyl or sulfhydryl groups in proteins, or certain nitrogen atoms in DNA. The antitumor activity of cis-Pt most likely results from its complexing to DNA or from cross-linking of DNA and proteins, which in turn inhibits further DNA replication. Interactions of cis-Pt with plasma membranes and intervention with membrane nutrient transport systems have also been reported (6,7).

Due to its chemical reactivity, cis-Pt could easily be employed in polymeric drug systems, since by replacing the chloride leaving groups it can form complexes of varying degrees of stability with a variety of reactive groups. The potential use of a macromolecule to serve as a carrier for cis-Pt is dependent on the presence of carrier-associated ligands that can form partially stable bonds with the platinum(II) compound. Such ligands will serve as more appropriate leaving groups for a better controlled release of the drug in favour of its target molecule in the tumor cell.

Cis-Pt was complexed to various macromolecular carriers in an attempt to identify (a) adequate carriers for the preparation of pharmacologically active platinum multicomplexes of decreased toxicity and increased therapeutic range and (b) complexes suitable for attachment to antitumor antibodies for the purpose of site directed immunotargeting.

Preparation and Characterization of Cis-Pt-Carrier Complexes

Several requirements must be considered in the design of a polymeric drug: the polymeric carrier should be soluble

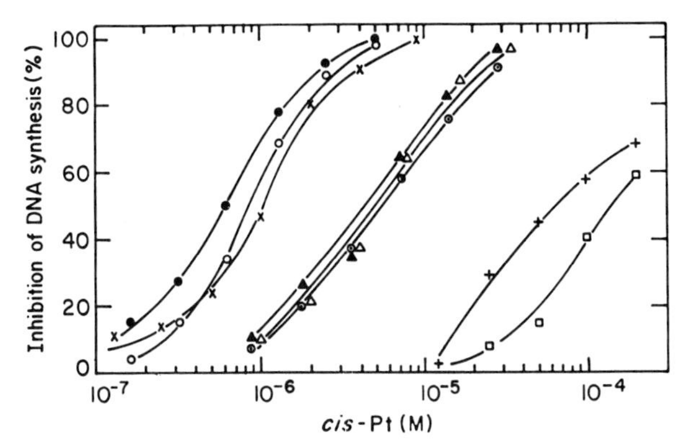

Fig. 2. Inhibition of DNA synthesis in 38C-13 lymphoma cells by free cis-pt (●) or by cis-Pt complexed to HA (o), CM-dex (x), p-L-Glu (▲), p-D-Glu (⊙), p-L-Asp (△), DNA (+) and BSA (□). The free drug or drug complexes were incubated with the cells for 21 hr followed by a 3 hr pulse of [H^3]-methylthymidine.

in the conjugated or complexed form, it should carry appropriate functional groups for attachment to the drug, and the linkage between the carrier and the drug should be cleavable or dissociable in a manner that will release the drug in an active form. With these criteria in mind we have prepared complexes between cis-Pt and several potential polymeric carriers and analyzed them for drug dissociability and activity. Reaction mixtures containing cis-Pt and the carrier at different drug/carrier molar ratios (8-10) were kept in double distilled water (DDW) at 37°C for 24 hr, after which they were dialysed against DDW to remove unbound drug. Quantitative determination of cis-Pt in the soluble complexes was performed by using the O-phenylenediamine (OPDA) reagent which exhibits a relatively high affinity towards platinum(II) and which upon interaction with cis-Pt forms a colored complex that can be determined quantitatively (8). The interaction between carrier - complexed cis-Pt and OPDA was found to provide a relative measure for the potential releasability of the drug and hence for its pharmacological activity. The quantities of total carrier-bound, namely both, non-releasable and reversibly bound drug, was determined by OPDA measurements of undialysed (unbound plus reversibly bound) and dialysed (reversibly bound) samples of the cis-Pt-carrier reaction mixtures following 24 hr incubation at 37°C.

The pharmacological activity of the complexes was assessed by their capacity to inhibit DNA synthesis in tumor cells, as compared to inhibition caused by the free drug (Figure 2). Their chemotherapeutic activity was tested in Balb/c mice that received intraperitoneal (i.p.) inoculation of F-9 embryonal carcinoma cells one or two days prior to the administration of free drug or drug complexes (9).

According to these parameters, the various cis-Pt-carrier complexes that were characterized (10), can be categorized as follows:

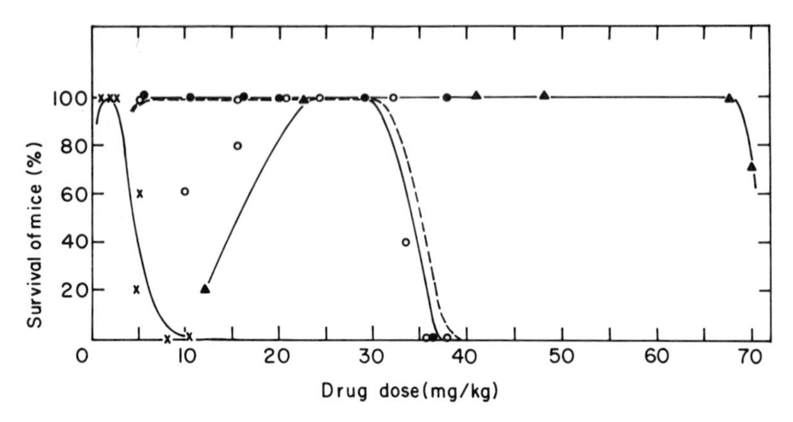

Fig. 3. Antitumor activity and toxicity of free cis-Pt (x) or cis-Pt complexed to p-L-Glu (●), p-D-Glu (o) or P-L-Asp (▲). The data represent cumulative results of several experiments using the F9 embryonal carcinoma in Balb/C mice.

(a) Highly stable complexes of DNA with a maximal drug load of 1 mole drug per 2-3 mole base. These cis-Pt-DNA complexes did not allow drug release (non-reactive with OPDA) due to their high binding affinity and were inactive against tumor cells both in-vitro and in-vivo.
(b) Low-capacity complexes such as those obtained with proteins (bovine serum albumin, immunoglobulin). These complexes were of a low drug content, only part of which being reversibly bound. Incidentally, this observation excludes the possibility of binding cis-Pt directly to antitumor antibodies for the purpose of immunotargeting and emphasizes the need of introducing an adequate intermediate carrier for drug binding.
(c) Low affinity complexes of relatively low drug content as in the case of hyaluronic acid. These complexes which can bind 1 mole drug per 3-4 mole glucoronic acid residues, were not different from free cis-Pt, both in-vitro and in-vivo.
(d) Complexes of poly-amino acids such as poly-L-glutamic acid (p-L-Glu), poly-D-glutamic acid (p-D-Glu), and poly-L-aspartic acid (p-L-asp) which can bind 1 mole cis-Pt per 5 mole amino acid residues, (higher drug loads resulted in complex precipitation) 40-50% of which is reversibly bound. These complexes were poorly active in-vitro, but when tested in-vivo they displayed a high therapeutic efficacy due to their broad range of effective dose (Figure 3). P-L-Glu complexs of Mr-13,000 and Mr-58,000 manifested a therapeutic dose range of 10-35 and 10-30 mg/kg, respectively. At this range no mortality occurred either as a result of tumor growth or due to the toxic manifestations of the drug (this is in comparison to a narrow dose range of 3-6mg/kg of the free drug). The activity of a p-D-Glu complex (Mr-34,000) was similar to that of the p-L-Glu, with a therapeutic range of 6-32 mg/kg, indicating that carrier biodegradation was not essential for drug activity. The p-L-Asp complex (Mr-11,500) was far less active, but still displayed a relatively wide therapeutic dose range (23-60 mg/kg).

(e) Moderately stable complexes, of a relatively high content of releasable drug, as those of carboxymethyl dextran (CM-dex), which bind 1 mole drug per 5 mole glucose units of dextran. The in-vitro activity of cis-Pt carboxymethyl dextran (cis-Pt-CM-dex) complexes of different molecular weights indicated that most of the complexed drug was pharmacologically active irrespective of the molecular size. However, their in-vivo therapeutic activity was dependent upon the molecular size of the CM-dex carrier. Thus, a T-10 complex was less toxic, but also less active, as compared to the free drug and its therapeutic dose range was not significantly improved, whereas complexes of T-70 and T-250 manifested increased levels of toxicity. The T-40 complex was found to be optimal for therapeutic use, since its in-vitro activity was similar to that of free cis-Pt but its therapeutic dose range was extended as a result of reduced toxicity. This complex was also found suitable for attachment to antitumor antibodies due to its appropriate molecular size and high content of reversibly bound drug.

Pharmacokinetic Studies on Cis-Pt-CM-dex Complexes

Previous pharmacokinetic studies on free cis-Pt have demonstrated a biphasic decline following intraveneous (i.v.) administration that was initiated by a rapid plasma clearance of the drug. The biphasic decline was followed by a slower process of irreversible binding to serum proteins, which resulted in drug inactivation (11,12). In order to get a better insight into the mechanism of action of cis-Pt-CM-dex complexes, a pharmacokinetic study was

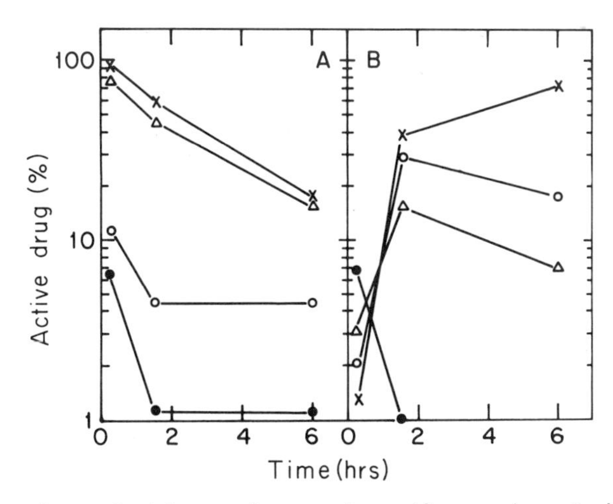

Fig. 4. Active drug in the circulation following i.v. (A) or i.p. (B) administration, into Balb/C mice, of 6.8 mg/kg of free cis-Pt (●), or cis-Pt complexed to CM-dex T-10 (o), T-40 (Δ) or T-250 (x). Samples of blood (plasma) were monitored for the presence of cis-Pt using an antitumor in-vitro assay against 38C-13 cells.

carried out comparing cis-Pt in the complexed and free forms (13). Blood levels following i.v. or i.p. administration and loss of drug activity due to protein binding following incubation with mouse serum were monitored. The monitoring was carried out using an antitumor in-vitro assay, thus taking into consideration only active drug species. The results showed that, in contrast to the extremely rapid clearance of free cis-Pt (5% after 15 min) the complexed-drug was eliminated at a slower rate. This effect was especially evident when the drug was complexed to CM-dex T-40 or T-250 (50-60% after 100 min and 15% at 6 hr). (Figure 4A).

Intraperitoneal administration of the free drug resulted in its transient appearance in the circulation shortly after injection (7%) followed by rapid elimination (Figure 4B). The complexes, on the other hand, reached the circulation at a slower rate but were maintained there at higher levels (30-50%) and for prolonged periods. These results emphasize the advantage of the complexes for intracavitary chemotherapy, since their retention in the peritoneal cavity is prolonged and their transport to the circulation delayed. Furthermore, after reaching the circulation, the complexes are potentially available for further pharmacological effects as a result of their increased retention of pharmacological activity, as described above.

Interaction of cis-Pt with serum proteins, which results in further loss of drug activity in the circulation, was also a much slower process when the drug was complexed to CM-dex, irrespective of its molecular size. After 5 hr of incubation with serum, only 12% of the free drug was active, whereas the complexes still retained 70-100% of their initial activity. This activity has, however, declined to 5-10% after 24 hr. These results suggest that the administration of cis-Pt complexed to an adequate macromolecular carrier may provide a pool for a slow and sustained release of active drug.

IMMUNOTARGETING OF CIS-PT

Immunotargeting of cis-Pt complexed to CM-dex T-40 was studied on a human oral epidermoid carcinoma (KB) grown in nude mice.

The antibody used for targeting was monoclonal antibody 108 (mAb108) raised against the extracellular domain of the epidermal growth factor receptor that is overexpressed in KB cells (14). This antibody was shown to localize on KB xenographts in nude mice and to inhibit tumor growth (15). Attempts to prepare the mAb108-cis-Pt-CM-dex conjugate by attaching CM-dex to the mAb108 and subsequently complexing cis-Pt to the conjugate, failed to produce immunoreactive preparations, due to the sensitivity of the monoclonal antibody to both, chemical conjugation and interaction with cis-Pt. Direct attachment of cis-Pt-CM-dex to mAb108, was also unsuccessful, since the reversibly bound cis-Pt in the complex was capable of interacting gradually with the antibody portion of the conjugate, thus shifting the balance towards an inactive derivative.

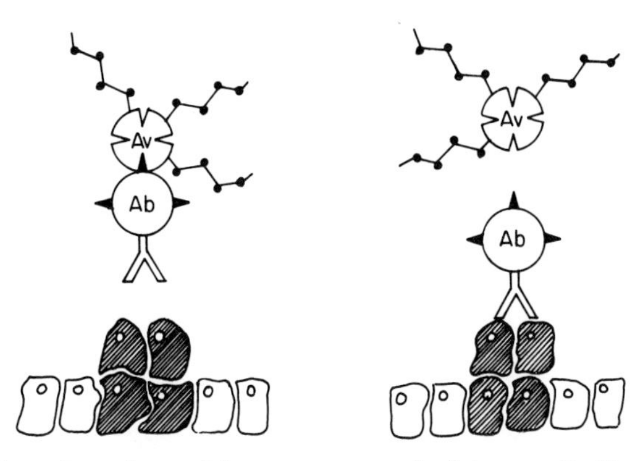

Fig. 5. A schematic representation of the one (left) or two-step (right) system for targeting pt-dex-Av via biotinylated antitumor antibody.

To overcome this problem we have introduced a novel indirect immunotargeting procedure using the biotin-avidin as an intermediary system. Both avidin and streptavidin, due to their remarkable affinity to biotin ($10^{15}M^{-1}$) have become in recent years extremely useful and versatile intermediates in a variety of biological and analytical procedures such as immunochemical and histological assays, isolation procedures, etc., (16). More recently, the use of these two systems has been extended to include in-vivo assays such as radioimmunodetection and radioimaging. These studies have shown that target/non-target radioactivity ratios and radioimmunoimaging analysis may be significantly improved by introducing a two-step targeting system, in which radiolabeled avidin (or streptavidin) is administered following the injection of biotinylated antibody. Alternatively, radiolabeled biotin can be administered following an avidin-antibody conjugate (17,18). Our studies comparing the in-vivo biodistribution of avidin and streptavidin have shown that retention of steptavidin in blood and organs was higher and more persistent, as compared to the rapid clearance of avidin (19). Hence, in order to avoid prolonged undesirable tissue retention of the drug, we have used only avidin in the targeting experiments described below. As is illustrated in the scheme (Figure 5), cis-Pt-CM-dex conjugated to avidin (Pt-dex-Av) can either be complexed to biotinylated mAb108 (b-mAb108) prior to in-vivo administration (left), or it can be injected following b-mAb108 in a two-step targeting system (right).

Preparation and Characterization of Pt-dex-Av and b-mAb108

Cis-Pt-CM-dex T-40 was conjugated to avidin at a molar ratio of 3 to 1, using water soluble carbodiimide (20). Analysis of the product indicated that close to 100% of the cis-Pt-CM-dex was conjugated to avidin and that the avidin in the conjugate retained 80% of its binding activity to biotin. The pharmacological activity of cis-Pt in the Pt-dex-Av conjugate (60-80 mole per mole avidin) has been preserved, since its activity in a 24 hr in-vitro assay against KB cells, was similar to that of free cis-pt.

Biotinylated mAb108 was prepared by coupling biotin-hydrazide to the oxidized antibody. Different preparations of b-mAb contained 3.2 to 5.2 biotin groups/mAb, which were available for interaction with Pt-dex-Av. The antibody binding capacity of b-mAb108 to KB cells was essentially unaffected by the biotinylation procedure. Furthermore, complexes between b-mAb108 and Pt-dex-Av were also capable of interacting with KB cells, preserving 60-70% of the antibody binding capacity.

Antitumor Effects

The targeting efficacy of Pt-dex-Av administered together with or following b-mAb108 was studied in-vivo using the KB tumor system. Athymic nude mice carrying established subcutaneous (s.c.) tumors were given three i.v. injections of b-mAb108, and the Pt-dex-Av was administered 24 hr after each antibody injection. A second group of mice received three injections of b-mAb108 complexed via its biotin residue to avidin in Pt-dex-Av (at a molar ratio of 1:1.6). In a third group the b-mAb108 injected 24 hr prior to Pt-dex-Av was replaced by unmodified mAb108. As shown in Figure 6A effective inhibition of tumor growth was achieved only after the two-step administration of b-mAb108 followed by Pt-dex-Av. Complexes between Pt-dex-Av and b-mAb108 were less effective apparently due to the formation of large complexes, that were either inaccessible to the tumor, or were rapidly eliminated from the circulation. Such complexes could be formed as a result of the multiple biotinylation sites on the mAb108 and the four avidin binding sites for biotin. The control treatment with unbiotinylated mAb108 followed by Pt-dex-Av, which accounts for the cumulative contribution of the specific antibody and the complexed drug to the inhibitory activity in a non-targeted system, also inhibited the KB tumors to a lesser extent. The specificity of the two-step targeting system could further be demonstrated by comparing it to additional control treatments (Figure 6B). Thus, replacing b-mAb108 by b-mAb of a different specificity, injecting mAb108 and cis-Pt alone, or mAb108 followed by cis-Pt, all resulted in non-significant inhibitory effects.

The two step treatment was further tested using a lung metastases model. Mice were given an i.v. inoculation of KB cells, and the treatments, as described above, were given twice, at 9 and 12 days following tumor inoculation. Histological analysis of the lungs removed 60 days following tumor inoculation, showed that metastatic nodules have developed in control mice but also in mice treated with the Pt-dex-Av--b-mAb108 complex, as well as in the control groups treated with mAb108 followed by cis-Pt and unbiotinylated mAb108, or b-mAb of a different specificity followed by Pt-dex-Av (Fig 7). Only the group that received Pt-dex-Av 24 hr after the administration of b-mAb108, did not develop any metasteses.

The KB tumor used in this study is effectively inhibited by both cis-Pt and mAb108 (21). Demonstration of targeting efficacy was therefore confined to low drug and antibody doses, which by themselves were non-inhibitory. It is expected that the two-step targeting procedure could be

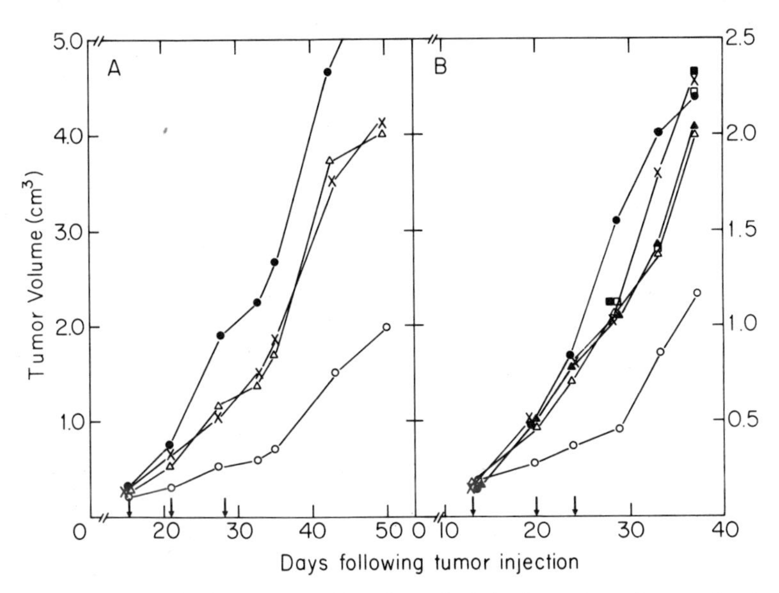

Fig. 6. The antitumor effect of Pt-dex-Av targeted to KB xenografts via b-mAb108. Nude mice (7-8/group) carrying established s.c. tumors received three i.v. injections (depicted by arrows) of (A) b-mAb108 (o) or mAb108 (▲) each followed by an i.v. injection of Pt-dex-Av 24 hr later, or mAb108 already complexed to Pt-dex-Av (x) (120μg Ab, 35μg free of complexed cis-Pt per injection). (B) b-mAb108 (o), b-mAb anti-DNP (x) or mAb108 (△) each followed by Pt-dex-Av 24 hr later, or mAb108 followed by cis-Pt (▲) (90μg mAb 30μg free or complexed cis-Pt). Three treatments by mAb108 alone (■) or cis-pt alone (□) are also depicted. Control mice were injected with phosphate buffered saline (PBS) (●).

even more pronounced in systems where the antibody can target to but not inhibit the tumor. In such a case, antibody doses may be manipulated by introducing higher levels to achieve saturation at the tumor site, thus increasing the local concentration of the drug.

CONCLUSIONS

The present study deals with the development of polymeric drug systems for controlled delivery of the chemotherapeutic drug cis-Pt. The chemical properties of this drug and its ability to form spontaneous complexes with polymeric carriers have been utilized to construct polymeric antitumor agents of improved therapeutic efficacy. Among these were complexes of the polycarboxylic-poly-amino acids, p-L-Glu, p-D-Glu and p-L-Asp, which belong to the category of "slow releasers", correlating with extended therapeutic dose ranges. Treatment by this complexes provides a wide

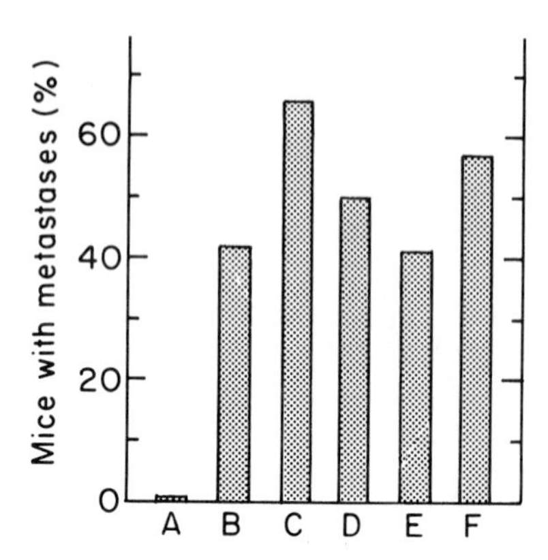

Fig. 7. Effect of treatment on the development of lung metastases. Nude mice were inoculated i.v. with 10^6 KB cells and treatment was given 9 and 12 days later. Groups of mice received i.v. administrations of the following reagents: A, b-mAb108; B, mAb108; C, b-mAb anti-DNP; each followed by Pt-dex-Av 24 hr via the same route; D, b-mAb108 complexed to Pt-dex-Av; E, mAb108 followed by cis-Pt; and F, PBS.

"safety zone", which may enable the introduction of higher, more effective, non-toxic drug doses. Complexes of CM-dex comprise a group of "fast releasers" and among these the T-40 complex seemed to fall into an optimal size range which rendered it as active but less toxic than free cis-Pt. In comparison to the narrow and inconsistent dose range of the free drug, treatment by these complexes was considerably "safer" and more efficacious.

Using this cis-Pt-CM-dex T-40 complex in combination with antitumor antibodies we have demonstrated the potential advantage of a novel two-step targeting system, in which the initial administration of a biotinylated antitumor-antibody to a tumor-bearing subject is followed by the administration of a drug-carrier-avidin conjugate. Assuming that the time interval between the two administrations allows accumulation of antibody at the tumor site and clearance of most irrelevant (unassociated with tumor) antibodies from the circulation and tissues, the two step system offers several potential advantages over the direct method: (a) easy preservation of antibody immunoreactivity following biotinylation; (b) manipulating the system by increasing antibody but not drug doses; (c) simplifying the system in which one batch of drug-carrier-(e.g. dextran)-avidin can be used for different biotinylated-antibody preparations; (d) ensuring a rapid and stable bond formation between biotinylated antibody and drug-avidin, even at low reagent concentrations, due to the high biotin-avidin binding constant (which is a million fold greater than most antigen-antibody interactions); (e) amplification of the system due to both the high drug/avidin ratios in the drug-carrier-avidin conjugates, and the possibility that 3-4 molecules of avidin may bind to each biotinylated antibody; (f) increasing specific drug localization at the tumor site which may result from the use of a fast clearing carrier such as avidin.

REFERENCES

1. A.W. Prestayko, J.C. D'Aoust, B.F. Issel and S.T. Crooke, Cisplatin (cis-diamminedichloroplatinum II). Cancer Treat. Rev., 6: 17 (1979).

2. B.J. Corden, R.L. Fine, R.F. Ozols and J.M. Collins. Chemical pharmacology of high-dose cisplatin. Cancer Chemother. Pharmacol., 14: 38 (1985).

3. R.F. Ozols and R.C. Young. High-dose cisplatin therapy in ovarian cancer. Semin. Oncol. 21:12 (1985).

4. J.J. Roberts. Mechanism of action of antitumor platinum compounds. In: Molecular mechanisms of carcinogenic and antitumor activity. C. Chagas and B. Pullman, eds. Pontificia Academia Scientiarvm. (1986).

5. M.E. Howe-Grant and S.J. Lippard. Aqueous platinum(II) chemistry: Binding to boilogical molecules. In: "Metal ions in biological systems". H. Sigel, ed. Marcel Dekker, New York. 11:63 (1980).

6. D.A. Juckett and B. Rosenberg. Actions of cis-diamminedichloroplatinum on cell surface nucleic acids in cancer cells as determined by cell electrophoresis techniques. Cancer Res, 42:3562 (1982).

7. S. Shionoya, Y. Lu and K.J. Scanlon. Properties of amino acid and transport systems in K562 cells sensitive and resistant to cis-diamminedichloroplatinum(II). Cancer Res, 46:3445 (1986).

8. B.Schechter, R. Pauzner, R. Arnon and M. Wilchek. Cis-platinum(II) complexes of carboxymethyl dextran as potential antitumor agents. I. Preparation and characterization. Cancer Biochem. Biophys. 8: 277 (1986).

9. B.Schechter, R. Pauzner, M. Wilchek, and R. Arnon. Cis-platinum(II) complexes of carboxymethyl dextran as potential antitumor agents. II. In-vitro and in-vivo activity. Cancer Biochem Biophys. 8: 289 (1986).

10. B. Schechter, A. Neumann, M. Wilchek and R. Arnon. Soluble polymers as carriers of cis-platinum. J. of Controlled Release. 10: 75 (1989).

11. J.J. Gullo, C.L. Litterst, P.J. Maguire, B.I. Sikie, D.F. Hoth and P.V. Woolley. Pharmacokinetics and

protein binding of cis-dichlorodiammine platinum(II) administration as a one hour, or as a twenty hour infusion. Cancer Chemother. Pharmacol, 5: 21 (1980).

12. C.L. Litterst, T.E. Gram, R.L. Dedrick, A.F. Leroy and A.M. Guarino. Distribution and disposition of cis-diamminedichloroplatinum(II). (NCS - 1198-75) in dogs. Cancer Res, 36: 2340 (1976).

13. B. Schechter, M.A. Rosing, M. Wilchek, and R. Arnon. Blood levels and serum-protein binding of cis-platinum(II) complexed to carboxymethyl-dextran. Cancer Chememother. Pharmacol, 24: 661 (1989).

14. J. Schlessinger. Allosteric regulation of the epidermal growth factor receptor kinase. J. Cell Biol, 103: 2067 (1986).

15. E. Aboud-Pirak, E. Hurwitz, M.E. Pirak, F. Bellot, J. Schlessinger, and M. Sela. Efficacy of antibodies to epidermal growth factor receptor against KB carcinoma in-vitro and in nude mice. J. Natl. Cancer Inst. 80: 1605, (1988).

16. M. Wilchek and E.A. Bayer. The avidin and biotin complex in biological applications. Analytical Biochemistry. 171: 1 (1988).

17. D.J. Hnatowich, F. Virzi and M. Rusckowski. Investigations of avidin and biotin for imaging applications. J. Nuclear Med, 28: 1294 (1987).

18. G. Paganelli. S. Prerez, A.G. Siccardi, G. Rowlinson, G., Deleide, F. Chiolerio, M. Malcovati, G.A. Scassellati, and Epenetos, A.A. Intraperitoneal radio-localization of tumors pre-targeted by biotinylated monoclonal antibodies. Int. J. Cancer, 45: 1184 (1990).

19. B. Schechter, R. Silberman, R. Arnon and M. Wilchek. Tissue distribution of avidin and streptavidin injected to mice. Eur. J. Biochem, 189:327 (1990).

20. B. Schechter, R. Pauzner, R. Arnon, J. Haimovich, and M. Wilchek. Selective cytotoxicity against tumor cells by cisplatin complexed to antitumor antibodies via carboxymethyl dextran. Cancer Immunol. Immunother, 25: 225 (1987).

RADIOLABELED ANTIBODY THERAPY OF HUMAN B CELL LYMPHOMAS

Oliver W. Press, Janet F. Eary, Christopher C. Badger, Paul J. Martin, Frederick R. Appelbaum, Wil B. Nelp, Ron Levy, Richard Miller, Darrell Fisher, Dana Matthews, and Irwin D. Bernstein

Departments of Medicine, Nuclear Medicine, and Pediatrics, University of Washington, Seattle, WA 98195, The Fred Hutchinson Cancer Research Center, Seattle, WA 98104, Stanford University, School of Medicine, Stanford, CA 94305, The Battelle Pacific Northwest Laboratories, Richland, WA 99352, and IDEC Pharmaceuticals Corporation, Mountain View, CA 94043

INTRODUCTION

Modern chemotherapy regimens are capable of curing the majority of patients with newly diagnosed diffuse aggressive lymphomas[1], but few patients with low-grade lymphomas or relapsed aggressive lymphomas can be cured with conventional therapy. High dose chemoradiotherapy in conjunction with marrow transplantation affords long term disease-free survival for approximately 25% of such patients[2,3], but toxicity is formidable and post-transplant relapse and treatment-related complications result in the death of most such patients. Targeted radiotherapy using radiolabeled antibodies (RAb) offers a potentially effective new approach for refractory malignancies as demonstrated by a variety of animal and human studies[4-13]. In this report we summarize our experience administering radiolabeled anti-B cell antibodies to patients with refractory non-Hodgkin's lymphomas in a phase I dose escalation trial.

EXPERIMENTAL DESIGN AND METHODS

Patient Selection: Patients with B cell lymphomas expressing one of the antigens listed in Table I were eligible for this study if they had failed previous conventional therapy, had normal renal and hepatic function, had not received chemotherapy or radiation for four weeks, had no other active medical problems, had an expected survival of at least 60 days, had <25% of their marrow infiltrated with

Immunobiology of Proteins and Peptides VI
Edited by M.Z. Atassi, Plenum Press, New York, 1991

lymphoma, and signed an informed consent approved by the Human Subjects and Radiation Safety Committees of the University of Washington and the Fred Hutchinson Cancer Research Center. Patients were excluded if they had circulating anti-mouse antibodies or extensive prior radiation therapy to any normal organ. All patients underwent autologous bone marrow harvesting, and had their marrow purged with a cocktail of anti-B cell antibodies (anti-CD20 and anti-CD10) and complement prior to cryopreservation.

Antibodies: The following antibodies were evaluated in two or more patients.

Table I. Antibodies for Radioimmunotherapy of B cell Lymphomas

Antibody	Antigen	Isotope	Immuno-reactivity	Avidity
Anti-id	Surface Ig	varies	70-90%	varies
MB-1	CD37	IgG1	90%	1×10^9
1F5	CD20	IgG2a	80%	1.5×10^7
B1	CD20	IgG2a	80%	2×10^8
LYM-1	Class II	IgG2a	90%	1×10^9

Antibodies were manufactured in sterile, pyrogen-free form by Idec Pharmaceutical Co., Mountainview CA (MB-1, anti-idiotypic antibodies), Coulter Inc., (B1), Oncogen Inc., Seattle WA (1F5), and the American Cyanamid Corporation (LYM-1). Antibodies were radiolabeled by the chloramine T method as described by Eary et. al.[14]. Labeled antibodies were characterized by electrophoresis to determine radiochemical purity. Immunoreactivity, avidity and dissociation constants were determined using solid phase radioimmunoassay and pyrogenicity was assayed using the Limulus lysate assay.

Baseline Laboratory Studies: Patients underwent standard staging including chest x-rays, computed tomography (CT) of the chest and abdomen, bilateral bone marrow aspirates and biopsies, lymph node biopsies, routine blood tests (CBC, platelets, chemistry panel, liver function tests, thyroid function tests, serum protein electrophoresis, quantitative serum immunoglobulin levels), and quantitation of B and T cells in blood and in lymph nodes by flow cytometry. Binding of antibodies (MB1, B1, and shared anti-idiotypic antibodies) to tumor cells was assessed by flow cytometry and by immunoperoxidase staining.

Biodistribution Studies: The biodistribution of I-131 labeled antibodies was determined for each patient following infusion of 0.5, 2.5, and 10 mg/kg antibody trace-labeled with 5-10 mCi I-131. A constant dose (0.2 mg/kg) of an isotype-matched, irrelevant murine antibody labeled with 2 mCi I-125 was mixed with the specific antibody and infused simultaneously as a control. All patients received SSKI (5 drops daily) to block thyroidal radioiodine uptake starting three days prior to the first labeled antibody infusion, and continuing for 21 days following the last administration. Antibodies were infused in normal saline at a rate of 50-250 mg/hr with monitoring of vital signs every 15-30 minutes. If significant toxicity was observed, the infusion rate was decreased by half. The biodistribution of antibodies was determined by serial gamma camera imaging (over 7 days) and by measurement of I-131 and I-125 content by gamma counting specimens of blood (drawn daily), urine (collected continuously), marrow (biopsied 24 hours after infusion), and tumor (biopsied 48-72 after infusion). Radiation doses were estimated for solid tumor and critical normal organs, (including liver, lung, kidney, and bone marrow) according to the Mirdose method [15].

Therapy Studies: Patients were treated with high dose radioimmunotherapy on the treatment arm of the protocol if biodistribution studies showed that all evaluable tumor sites would receive more radiation than any critical non-hematopoietic organ (liver, lung, kidney). The amount of I-131 administered was individualized based on the biodistribution and estimated radiation dosimetry in that patient. The 10 treated patients received escalating doses of radiolabeled antibodies estimated to deliver 500 cGy to 1675 cGy to the normal organ receiving the most radiation with all evaluable tumor sites receiving higher doses of radiation. All patients were hospitalized until total body activity fell <30 mCi and then were followed in the outpatient clinic. External tumor measurements were made weekly following therapy, and followup CT scans were performed one month, three months, and one year after treatment. Patients who experienced profound myelosuppression (absolute neutrophil count <200/mm^3) or thrombocytopenia (<20,000/mm^3) were rehospitalized and had bone marrow biopsies performed. If the marrow was aplastic, their cryopreserved autologous marrow was reinfused.

Evaluation of therapeutic efficacy: Tumor response was evaluated by standard criteria as follows: complete response, total disappearance of tumor; partial response, ≥50% reduction of all tumor sites; and no response, <50% reduction of tumor sites.

RESULTS AND DISCUSSION

Studies on 26 patients (using primarily anti-CD37 and anti-CD20 antibodies) have demonstrated that patients are most likely to achieve favorable RAb biodistributions if they have tumor burdens <0.5 kg and no splenomegaly, and that high doses of RAb (2.5-10 mg/kg) are required to achieve optimal RAb biodistributions. In all patients, higher antibody doses afforded better MoAb biodistributions than lower doses (10mg/kg >2.5mg/kg >0.5 mg/kg). All nine evaluable patients who received therapeutic infusions of RAb have had major tumor responses (eight complete remissions and one partial remission) with minimal toxicity except for predictable myelosuppression. Typical patterns of tumor regression are shown in Figure 1.

The only significant toxicity encountered so far has been myelosuppression, with nadir neutrophil counts (generally <500 per cu. mm) and platelet counts (generally <20,000 per cu. mm) usually occurring 3-5 weeks after radioimmunotherapy (see Figure 2). Approximately half the patients required reinfusion of autologous, purged cryopreserved marrow to shorten the duration of myelosuppression.

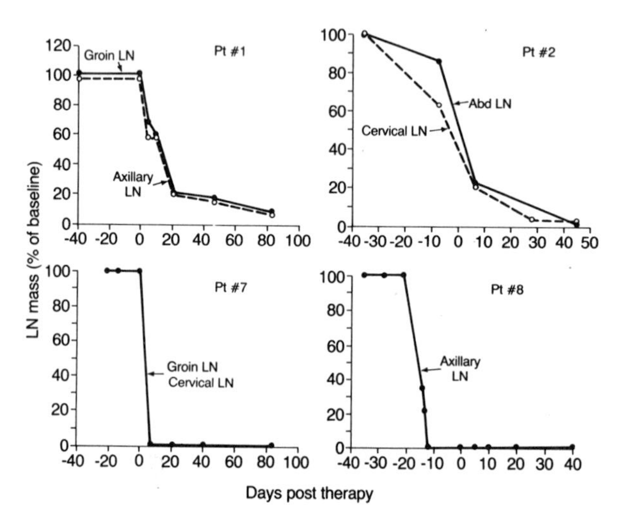

Figure 1. Rates of tumor regression in four representative patients following infusion of therapeutic doses of I-131-MB-1 on day 0. Biodistribution studies with trace radiolabeled MB-1 were performed on days -40 to 0. Two patients (No. 2 and 8) had tumor responses during the biodistribution infusions. Reproduced with permission from the Journal of Clinical Oncology 7: 1027-1038, 1989.

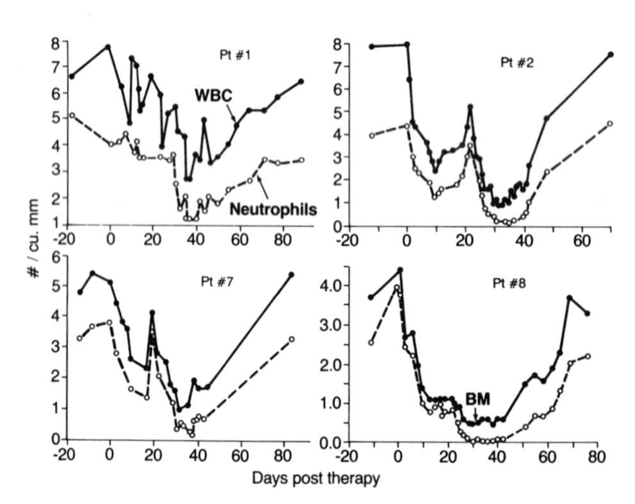

Figure 2. Leukopenia and neutropenia in four representative patients after therapeutic infusion of I-131-MB-1 on day 0. Reproduced with permission from the Journal of Clinical Oncology 7: 1027 -1038, 1989.

Our preliminary studies have shown that favorable biodistributions are achievable in patients with tumor burdens <0.5 kg who lack massive splenomegaly and receive relatively high doses of antibody (≥2.5 mg/kg). Virtually all patients demonstrating favorable RAb biodistributions will achieve complete remissions with minimal non-hematopoietic toxicity after high dose radioimmunotherapy. The observation that most of these remissions last less than a year, however, emphasizes the necessity for further RAb dose escalation to achieve long-term disease free survival. The absence of non-myeloid toxicities to date indicates that we have not yet approached the maximally tolerated dose of RAb therapy, and emphasizes the desirability of more rapid dose escalation.

In spite of these encouraging preliminary results, it must be appreciated that only 40% of patients studied so far have met our stringent criteria for therapeutic RAb infusion. A major focus of our future studies will be to identify interventions which will improve the antibody biodistributions in patients who would otherwise not be candidates for radioimmunotherapy because the amount of radiation absorbed by tumor would be less than the amount absorbed by at least one critical normal organ. Potential interventions include splenectomy for patients with massive splenomegaly, cytoreduction with chemotherapy for patients with large tumor burdens but without splenomegaly, administration of alternative antibodies, or preinfusion of "cold" (unlabeled) antibody prior to radiolabeled antibodies[16]. We hope that these approaches will allow all patients with relapsed B cell lymphomas to receive radioimmunotherapy with a reasonable expectation of achieving a durable complete remission.

References

1. P. Klimo. Chemotherapy for aggressive non-Hodgkin's lymphomas, in "Cancer: Principles and Practice of Oncology--Update 2 (no. 9)," pp. 1-12, DeVita VT, Hellman S, Rosenberg SA, eds., Lippincott, Philadelphia (1988).
2. F. Appelbaum, K. Sullivan, C. D. Buckner, and E. D. Thomas, Treatment of malignant lymphoma in 100 patients with chemotherapy, total body irradiation, and marrow transplantation, J. Clin. Oncol. 5: 1340-1347 (1987).
3. F. B. Petersen, F. R. Appelbaum, R. Hill, L. D. Fisher, C. L. Bigelow, J. E. Sanders, K. M. Sullivan, W. I. Bensinger, R. P. Witherspoon, R. Storb, R. A. Clift, A. Fefer, O. W. Press, P. L. Weiden, J. Singer, E. D. Thomas, and C. D. Buckner, Autologous marrow transplantation for malignant lymphoma: a report of 101 cases from Seattle, J. Clin. Oncol. 8: 638-647 (1990).
4. C. Badger, K. Krohn, A. Peterson, and I. D. Bernstein, Experimental radiotherapy of murine lymphoma with I-131-labeled anti-Thy 1.1 monoclonal antibody, Cancer Res. 45: 1536-1544 (1985).
5. C. Badger, K. Krohn, H. Shulman, and I. D. Bernstein, Experimental radioimmunotherapy of murine lymphoma with I-131-labeled anti-T cell antibodies, Cancer Res. 46: 6223-6228 (1986).
6. F. R. Appelbaum, P. Brown, and B. Sandmaier, Antibody-radionuclide conjugates as part of a myeloablative preparative regimen for marrow transplantation, Blood (in press).
7. S. Order, G. Stillwagon, and J. Klein, Iodine-131 antiferritin, a new treatment modality in hepatoma: A Radiation Therapy Oncology Group study, J. Clin. Oncol. 3: 1573-1582 (1985).

8. R. Lenhard, S. Order, and J. Spunberg, Isotopic immunoglobulin: A new systemic therapy for advanced Hodgkin's Disease, J. Clin. Oncol. 3: 1296-1300 (1985).
9. S. Rosen, A. Zimmer, and R. Goldman-Leikin, Radioimmunodetection and radioimmunotherapy of cutaneous T cell lymphomas using an I-131-labeled monoclonal antibody: An Illinois Cancer Council study, J. Clin. Oncol. 5: 562-573 (1987).
10. S. DeNardo, G. DeNardo, and L. O'Grady, Pilot studies of radioimmunotherapy of B cell lymphoma and leukemia using I-131 Lym-1 monoclonal antibody, Antibod. Immunoconjugates Radiopharmaceut. 1: 17-33 (1988).
11. A. Epenetos, A. Munro, and S. Stewart, Antibody-guided irradiation of advanced ovarian cancer with intraperitoneally administered radiolabeled monoclonal antibodies, J. Clin. Oncol. 5: 1890-1899 (1987).
12. L. Lashford, D. Jones , and J. Pritchard, Therapeutic application of radiolabeled monoclonal antibody UJ13A in children with disseminated neuroblastoma, NCI Monogr. 3: 53-57 (1987).
13. O. W. Press, J. Eary, C. C. Badger, F. A. Appelbaum, P. J. Martin, R. Levy, R. Miller, S. Brown, W. B. Nelp, K. A. Krohn, D. Fisher, K. De Santes, B. Porter, P. Kidd, E. D. Thomas, and I. D. Bernstein, Treatment of patients with refractory non-Hodgkin's lymphoma with radiolabeled MB-1 (anti-CD37) antibody, J. Clin. Onc. 7: 1027-1038 (1989).
14. J. Eary, O. Press, C. Badger, P. Martin, F. Appelbaum, K. Krohn, R. Levy, S. Brown, R. Miller, D. Fisher, W. Nelp, and I. Bernstein, Imaging and treatment of B-cell lymphoma, J. Nucl. Med. 31:1257-1268 (1990).
15. Society of Nuclear Medicine: MIRD Primer for absorbed dose calculation, Washington D.C., Society of Nuclear Medicine (1988).
16. J. A. Bianco, B. Sandmaier, P. A. Brown, C. Badger, I. Bernstein, J. Eary, L. Durack, F. Schuening, R. Storb, and F. Appelbaum, Specific marrow localization of an I-131-labeled anti-myeloid antibody in normal dogs: effects of a "cold" antibody pretreatment dose on marrow localization, Exp. Hematol. 17: 929-934 (1989).

ACTIVATION OF PRODRUGS BY ANTIBODY-ENZYME CONJUGATES

Peter D. Senter, Philip M. Wallace, Hakan P. Svensson, David E. Kerr, Ingegerd Hellstrom and Karl Erik Hellstrom

Oncogen
3005 First Avenue, Seattle, WA 98121

Monoclonal antibodies (MAbs) against human tumor antigens have been the subject of extensive investigation as carriers of cytotoxic agents to tumor cells. Promising results, both *in vitro* and *in vivo*, have been obtained in studies involving immunoconjugates such as MAb-toxins (Vitetta et al., 1987; Blakey et al., 1988), MAb-drugs (Pietersz, 1990; Koppel, 1990), and radiolabeled MAbs (Goldenberg, 1990). Based on these findings, many such conjugates are currently being evaluated in the clinic.

The use of immunoconjugates for the treatment of solid tumors is complicated by the fact that solid tumors are often poorly vascularized (Sands et al., 1988) and have barriers to penetration by macromolecules (Clauss and Jain, 1990). In addition, the heterogeneity of target antigen expression (Vallera et al., 1983) and limitations in the potency of the targeted agent (Thorpe, 1985) can result in sub-optimal antitumor effects of the immunoconjugate.

With this in mind, we have directed our attention to the development of an approach in which MAbs are used to deliver enzymes to tumor cell surfaces (Senter, 1990). The enzymes are selected for their abilities to convert relatively non-toxic drug precursors (prodrugs) into active anticancer drugs. The drugs formed in this manner can then penetrate into nearby tumor cells, many of which may be inaccessible to the conjugate (Fig. 1).

ENDOGENOUS ENZYMES

The enzyme, alkaline phosphatase (AP, bovine intestinal), was chosen for our initial investigations to test the concept depicted in Fig. 1 (Senter et al., 1988). AP is a highly active enzyme that acts on a broad range of substrates. We began by

Prodrug (pd)
Enzyme
Drug (d)
pd
d

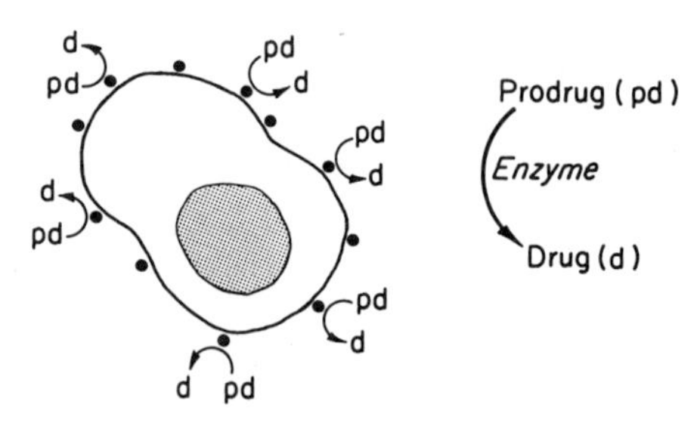

Figure 1. The activation of prodrugs by targeted enzymes

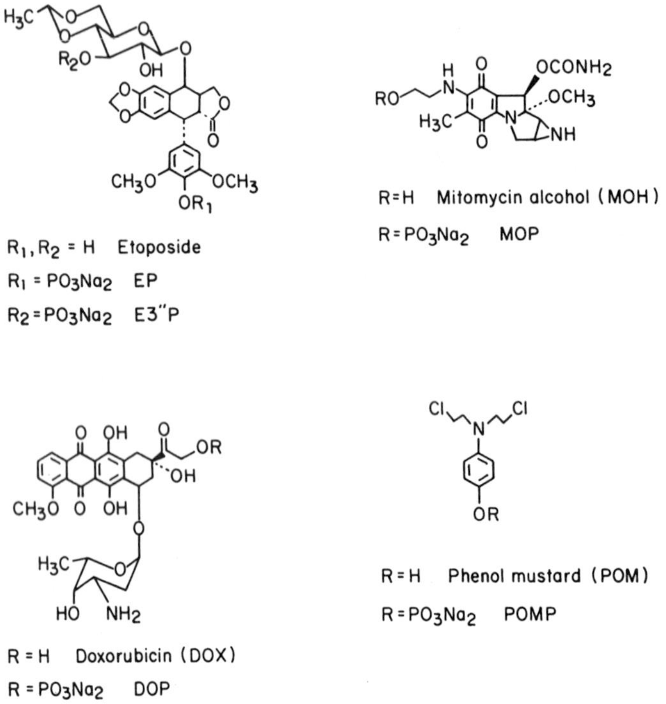

R1, R2 = H Etoposide
R1 = PO3Na2 EP
R2 = PO3Na2 E3"P
R=H Mitomycin alcohol (MOH)
R=PO3Na2 MOP
R = H Doxorubicin (DOX)
R = PO3Na2 DOP
R=H Phenol mustard (POM)
R=PO3Na2 POMP

Figure 2. Structures of drugs and their corresponding phosphorylated derivatives

phosphorylating a variety of antitumor agents which differed in their modes of action and spectra of activities. These agents are shown in Fig. 2.

Using *in vitro* cytotoxicity assays, it was shown that the phosphorylated drugs were significantly less toxic to the H2981 human lung adenocarcinoma cells than the drugs which were formed after phosphate hydrolysis (Table 1). This reduced toxicity may be due to the inability of the charged prodrugs to penetrate through cell membranes.

The MAbs, L6 (anticarcinoma, Hellstrom, et al., 1986), IF5 (anti CD-20, Clark et al., 1985) and P1.17 (mouse myeloma immunoglobulin) were covalently attached to enzymes through stable thioether bonds. After purification, the MAb-AP conjugates were largely free of aggregates and unconjugated proteins. Fluorescence-activated cell sorter analysis indicated that L6 and L6-AP bound to the H2981 adenocarcinoma line and that the control antibodies bound much more weakly. IF5 conjugates bound well to Daudi, a B-cell lymphoma line.

In vivo experiments with MAb-AP/prodrug combinations were performed on mice that had subcutaneous transplants of either H3347 human colon carcinoma or H2981 human lung adenocarcinoma tumors (Senter et al., 1988; Senter et al., 1989). Therapy was initiated 1-2 weeks after tumor implantation by first injecting the conjugates and then waiting 24 hours before treatment with the prodrugs at their maximally tolerated doses. The antitumor effects were compared to those of the drugs or prodrugs given without prior conjugate treatment.

TABLE 1. IC_{50} Values of Drugs and Drug Phosphates on H2981 Cells

Agent	IC_{50} (μM)
Etoposide	1
EP	>100
E3"P	>100
MOH	6
MOP	>100
Doxorubicin	0.1
DOP	1.5
Phenol Mustard	1
POMP	>100

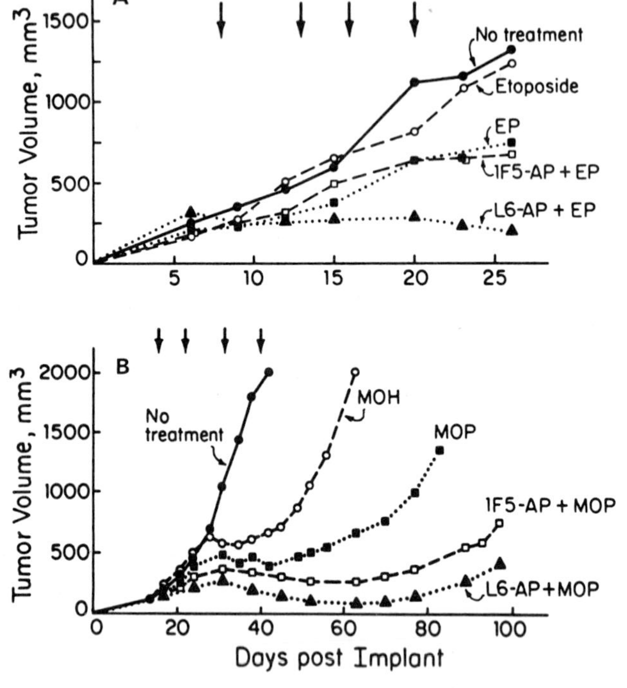

Figure 3. Effects of drugs, prodrugs, and MAb-AP conjugates with prodrugs on tumors in nude mice. Arrows indicate drug treatment schedule. Conjugates were administered 24 h earlier; A) H3347 colon carcinoma; B) H2981 lung adenocarcinoma.

Significant antitumor activities were observed for the L6-AP/prodrug combinations (Fig. 3). In the H3347 colon carcinoma model (Fig. 3A) the response to L6-AP plus EP exceeded that observed for the control conjugate IF5-AP plus EP, suggesting immunological specificity. A similar trend on the H2981 lung adenocarcinoma was observed in the mitomycin series (Fig. 3B). Treatment of mice with L6-AP followed 24 hr later with MOP resulted in an antitumor effect that was far greater than that observed for the prodrug alone. It was estimated that the L6-AP/MOP combination eliminated more than 99.99% of the H2981 tumor cells.

The control conjugate, IF5-AP, also enhanced the activity of MOP, but to a slightly lesser degree than L6-AP. This was most likely due to the rather low ratio of specific tumor uptake of L6-AP to IF5-AP in this experiment. Higher degrees of immunological specificity may be obtained by prolongation of the time interval between conjugate and prodrug administration, use of lower molecular weight $F_{(ab')}$ or $F_{(ab')2}$ conjugates, or by clearance of the unbound conjugate with agents such as anti-idiotypic (Hellstrom et al., 1990) or anti-enzyme antibodies (Bagshawe, 1989). These experiments are currently underway.

It was encouraging to learn that AP conjugates could significantly enhance the activity of phosphorylated drugs *in vivo*. This was unexpected, since AP is known to be present in many biological tissues which could lead to prodrug hydrolysis before the tumor site is reached.

One approach to overcome the breakdown of phosphorylated drug derivatives has involved the use of more stable drug phosphates. E3"P (Fig. 2) was found to be more stable than EP, yet was still a substrate for AP. This increased stability may be due to a combination of steric hindrance and the nature of the phosphate ester that is cleaved. *In vivo* studies (manuscript in preparation) have shown that L6-AP/E3"P combinations are active.

It was previously reported that phenol mustard phosphate (POMP) (Fig. 2) was more than 10 times less toxic to mice than phenol mustard (POM, Bukhari et al., 1972). We have prepared these nitrogen mustards, and have shown that L6-AP/POMP combinations are more active than POMP treatment alone (Table 2). The antitumor effect of MAb-AP conjugates in combination with POMP was immunologically specific, since L6-AP/POMP was superior to IF5-AP/POMP. The use of more stable phosphates, such as E3"P and POMP may help circumvent the problem of non-specific breakdown by endogenous phosphatases.

Arylsulfatase is another enzyme which is currently under investigation. Although arylsulfatases are present in the body, the activity is low and the enzymes

TABLE 2. Effects of MAb-enzyme/prodrug combinations on the growth of subcutaneous H2981 tumors in nude mice (L6 positive; IF5, P1.17 negative).

Conjugate	Drug (dose)	Days to reach 500 mm^3
-	-	31
-	Phenol mustard (9 mg/kg)	26[1]
-	POMP (65 mg/kg)	30
L6-AP	POMP (45 mg/kg)	45
IF5-AP	POMP (45 mg/kg)	30
-	-	36
-	5FU (32 mg/kg)	36
-	5FC (80 mg/kg)	36
L6-CDase	5FC (60 mg/kg)	47
P1.17-CDase	5FC (60 mg/kg)	35

[1]The control in this experiment took 22 days to reach 500 mm^3

have considerable substrate specificity. Several sulfated anticancer drugs have been prepared which are hydrolyzed by arylsulfatases from microbial sources. For example, phenol mustard sulfate (Fig. 4) is hydrolyzed by commercially available arylsulfatases and arylsulfatases from Streptomyces. The product upon hydrolysis, phenol mustard (IC_{50} 1μM), is much more toxic to H2981 cells than phenol mustard sulfate (IC_{50}> 50μM). *In vivo* studies have shown that phenol mustard sulfate is approximately 25 times less toxic to mice than phenol mustard, thus indicating that this prodrug has considerable *in vivo* stability (Bukhari, et al., 1972). Because of the availability of a wide variety of arylsulfatases and sulfated prodrugs, this system may have many of the advantages of AP without several of the disadvantages.

NON-ENDOGENOUS ENZYMES

The use of enzymes that carry out reactions not ordinarily occurring in the body may be advantageous when compared to enzymes like AP, since prodrugs based on such reactions would be expected to be more stable and less toxic. Some of the enzymes under active investigation along with their corresponding substrates are shown in Fig. 4.

Penicillin-V amidase (PVA) is an industrially used enzyme that cleaves a variety of phenoxyamide-containing molecules. The phenoxyamide derivative of

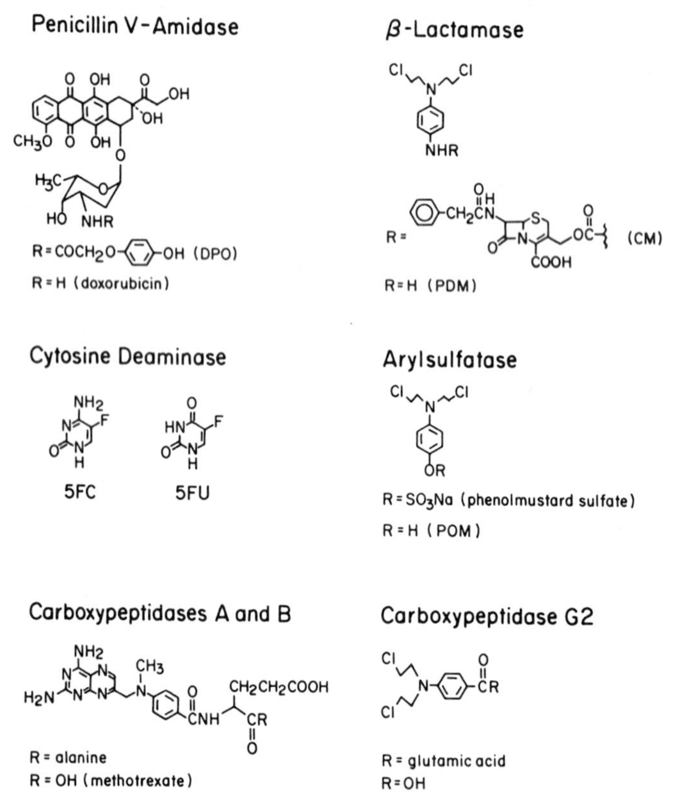

Figure 4. Enzymes and their corresponding substrates and prodrugs.

doxorubicin (DPO) was shown to be 80-fold less toxic to H2981 cells than doxorubicin (Kerr et al., 1990). The cytotoxic effect of DPO was substantially increased when the H2981 cells were pretreated with L6-PVA. The cytotoxic effect was antigen specific, since only the binding MAb-PVA conjugate increased the cytotoxicity of the prodrug.

The enzyme, cytosine deaminase (CDase), catalyzes the hydrolysis of cytosine to uracil. 5-Fluorocytosine (5FC), a clinically approved antifungal agent, is also a substrate for the enzyme (Nishiyama et al., 1985) and is converted to the anticancer drug 5-fluorouracil (5FU). CDase has been isolated from a variety of microorganisms, but in contrast to AP, has not yet been found in mammalian cells. In light of the fact that both 5FC and 5FU are clinically used, L6 antibody binds well to colon carcinomas (Hellstrom et al., 1986) and is currently in clinical trials, L6-CDase/5FC may be well suited to test the targeted enzyme/prodrug strategy in the clinic.

In vitro studies indicated that 5FC was non-toxic to H2981 cells up to a concentration of 200 μM. IC_{50} values of approximately 15 μM were obtained when cells were treated with 5FU or L6-CDase plus 5FC. The non-binding conjugate, IF5-CDase, did not increase the cytotoxic effect of 5FC.

5FU, 5FC and the non-binding control conjugate P1.17-CDase plus 5FC had little effect on the growth of H2981 tumor xenografts in nude mice (Table 2). Significant tumor growth delays were observed in animals receiving L6-CDase plus 5FC. This is most likely due to increased levels of intratumoral 5FU in mice receiving L6-CDase/5FC compared to when 5FU is given systemically. Studies are now underway to optimize the therapeutic effects of L6-CDase/5FC and to explore the scope of activity on colon carcinoma tumor models.

Several other enzyme/prodrug combinations are currently being explored for antitumor activities. Among them are ß-lactamase enzymes of microbial origin. ß-Lactamases may be ideally suited for the targeting strategy described here, since there are many different forms of the enzyme that are able to hydrolyze substituted cephalosporins. Cephalosporin mustard (CM, Fig. 4) has been found to be a substrate for 3 different ß-lactamases. The product upon hydrolysis, phenylenediamine mustard (PDM) is about 50-fold more cytotoxic than CM on H2981 cells. *In vivo* experiments are now in progress. Other groups have reported promising *in vitro* and *in vivo* results using conjugates of carboxypeptidases A and B (Kuefner et al., 1989), carboxypeptidase G2 (Bagshawe et al., 1988; Bagshawe, 1989) (Fig. 4) and glucose oxidase (Stanislawski et al., 1989).

CONCLUSIONS

On the basis of many of the experiments described here, there are reasons to believe that the site-specific activation of prodrugs by MAb-enzyme conjugates will have considerable potential for the treatment of cancer. Eventual clinical trials of this two-stage strategy will require the evaluation of methods to obtain optimal target to non-target ratios of the immunoconjugates and to gain a greater understanding of how drug potency and enzyme turnover rates influence therapeutic efficacy. Studies along these lines are currently underway.

REFERENCES

Bagshawe, K.D., 1989, Towards generating cytotoxic agents at cancer sites, Br. J. Cancer, 60:275.

Bagshawe, K.D., Springer, C.J., Searle, F., Antoniw P., Sharma, S.K., Melton, R.G., Sherwood, R.F.A., 1988, Cytotoxic agent can be generated selectively at cancer sites, Br. J. Cancer, 58:700.

Blakey, D.C., Wawrzynczak, E.J., Wallace, P.M., Thorpe, P.E., 1988, Antibody toxin conjugates: a perspective, Prog. Allergy, 45:50.

Bukhari, M.A., Everett, J.L., and Ross, W.C.J. Glucuronic, Sulphuric and Phosphoric Esters of p-Di-2-chloroethylaminophenol, Biochem. Pharm., 21:963.

Clark, E.A., Shu, G., Ledbetter, J.A., 1985, Role of the bp35 cell surface polypeptide in human B-cell activation, Proc. Natl. Acad. Sci. USA, 82:1766.

Goldenberg, D.M., 1990, Cancer Imaging with Radiolabeled Antibodies, Kluwer Academic Publishers, Boston.

Hellstrom, K.E., Yelton, D.E., Fell, H.P., Beaton, D., Gayle, M., MacLean, M., Kahn, M., Hellstrom, I., 1990. Epitope mapping and use of anti-idiotypic antibodies to the L6 monoclonal antibody, Cancer Res., 50: 2449.

Hellstrom, I., Horn, D., Linsley, P., Brown, J.P., Brankovan, V., Hellstrom, K.E., 1986, Monoclonal mouse antibodies raised against human lung carcinoma, Cancer Res., 46:3917.

Kerr, D.E., Senter, P.D., Burnett, W.V., Hirschberg, D.L., Hellstrom, I., and Hellstrom, K.E., 1990, Antibody-penicillin-V-amidase conjugates kill antigen-positive tumor cells when combined with doxorubicin phenoxyacetamide, Cancer Immunol. Immunother., 31:202.

Koppel, G.A., 1990, Recent advances with monoclonal antibody drug targeting for the treatment of human cancer, Bioconj. Chem., 1:13.

Pietersz, G.A., 1990, The linkage of cytotoxic drugs to monoclonal antibodies for the treatment of cancer, Bioconj. Chem., 1:89.

Sands, H., Jones, P.L., Shah, S.A., Plame, D., Vessella, R.C., and Gallagher, B.M., 1988, Correlation of vascular permeability and blood flow with monoclonal antibody uptake by human clouser and renal cell xenografts, Cancer Res., 48:188.

Senter, P.D., Activation of prodrugs by antibody-enzyme conjugates: A new approach to cancer therapy, FASEB, 4:188.

Senter, P.D., Saulnier, M.G., Schreiber, G.J., Hirschberg, D.L., Brown, J.P., Hellstrom, I., Hellstrom, K.E., 1988, Anti-tumor effect of antibody-alkaline phosphatase conjugates in combination with etoposide phosphate, Proc. Natl. Acad. Sci. USA, 85:4842.

Senter, P.D., Schreiber, G.J., Hirschberg, D.L., Ashe, S.A., Hellstrom, K.E., Hellstrom, I., 1990, Enhancement of the *in vitro* and *in vivo* antitumor activities of phosphorylated mitomycin C and etoposide derivatives by monoclonal antibody-alkaline phosphatase conjugates, Cancer Res., 49:5789.

Stanislawski, M., Rousseau, V., Goavec, M., Ito, H., 1989, Immunotoxins containing glucose oxidase and lactoperoxidase with tumoricidal properties: *In vitro* killing effectiveness in a mouse plasmacytoma cell model, Cancer Res., 49:5497.

Thorpe, P.E., 1985, Antibody carriers of cytotoxic agents in cancer therapy: a review, in: "Monoclonal antibodies '84: biological and clinical applications," Editrice Kurtis, Milan.

Vitetta, E.S., Fulton, R.J., May, R.D., Till, M., Uhr, J.W., 1987, Redesigning nature's poisons to create anti-tumor reagents, Science, 238:1098.

CANCER IMAGING AND THERAPY WITH RADIOLABELED ANTIBODIES

David M. Goldenberg

Garden State Cancer Center and
Center for Molecular Medicine and Immunology
Newark, New Jersey 07103

Introduction

Functional and constituent markers of neoplasms are being sought to elucidate differences between normal and malignant cells for diagnostic and therapeutic measures, and to provide leads to understanding mechanisms of cancer development, progression, and, ultimately, prevention. Enzymes, hormones, receptors, oncogene products, and other cancer-related molecules, including the extensive family of tumor-associated and oncofetal antigens, have been the focus of study for differentiating tumor from normal tissues and for monitoring the progression of disease (1). Although truly tumor-specific markers have remained elusive, those which are quantitatively increased with neoplasia have been used successfully as diagnostic or monitoring tools (2). In 1973 and 1974, we demonstrated, in a human colonic carcinoma xenograft model, that radioactive goat antibodies against carcinoembryonic antigen (CEA) can target and image CEA-producing tumors (3,4). Thereafter, in 1978, we extended this approach to humans, demonstrating that CEA-producing tumors can be targeted and imaged with ^{131}I-labeled, affinity-purified, goat antibodies against CEA (5). This result indicated that truly cancer-specific substances are not required as targets for antibody localization, and that radiolabeled antibodies can be used to detect and visualize tumors having the appropriate antigens by external scintigraphic methods. We termed this method cancer radioimmunodetection (RAID), and since then several thousand patient studies have been performed with different antibodies, labels, and imaging methods on a diverse group of tumors (6,7). The purpose of this paper is to summarize the status of RAID and to present a perspective on the prospects for cancer therapy with radioimmunoconjugates, or radioimmunotherapy (RAIT).

Radioimmunodetection (RAID)

The initial studies of RAID with CEA antibodies resulted in the following conclusions (5,8): (a) Despite elevated levels of blood CEA, doses as low as 0.2 mg anti-CEA IgG labeled with ^{131}I could disclose CEA-producing tumors of 2 cm and larger; (b) the formation of complexes between circulating CEA and the anti-CEA IgG did not hinder RAID; (c) imaging could be successful within 48 h if blood pool

Immunobiology of Proteins and Peptides VI
Edited by M.Z. Atassi, Plenum Press, New York, 1991

and interstitial background radioactivity is reduced by dual-isotope subtraction methods; (d) non-CEA-containing tumors failed to show localization; (e) in tumor-positive patients with normal blood CEA titers, RAID could disclose the tumors; (f) true-positive (sensitivity) and true-negative (specificity) detection and accuracy rates of at least 85 percent, on a tumor-site basis, could be achieved; (g) RAID revealed occult tumor sites missed by conventional radiological methods, which were confirmed subsequently by clinical/radiological followup and/or surgery/biopsy; and (h) radioactivity count ratios of tumor/adjacent normal tissue of about 2.5 were measured in surgical specimens. These results were obtained in digestive tract cancers, as well as in other cancer types, including lung, breast, ovarian, cervical, and endometrial carcinomas with the same CEA antibody radioconjugates (8), and were generally confirmed by studies with CEA (9) and other antibodies (10). The introduction of murine monoclonal antibodies (MAbs), following these initial successful trials with purified (monospecific) polyclonal antibodies, spurred intensive interest in further improvements in RAID, and involved different antibody forms, different labels, and different imaging procedures.

Most MAb used for targeting cancer in humans have proved to be fairly wide in their distribution, of a pan-carcinoma nature, as is evidenced by the representative list given in Table 1. Regardless of the cell type or antigen used to develop the MAb, diverse tumor types appear to be targeted by such MAbs, indicating that tumor cell-type-specificity is an unusual event. Although many of the target antigens are also present, albeit in reduced quantities, in many normal tissues, a preferential localization to tumor is achieved because the injected antibodies do not usually have access to normal structures, such as to mucosal epithelium when the basement membrane is intact. When tumors invade and gain access to neighboring structures and blood vessels, a new and different vasculature and permeability appear to allow greater access to injected macromolecules. This increased accretion of macromolecules in tumors, including immunoglobulins, has been recognized since at least 1939 (11). The vascular network of a growing neoplasm is both important with regard to the tumor gaining access to materials needed for sustenance and growth and for antineoplastic agents, as well as a potential means to control tumor proliferation, such as with agents that show a selectivity for the vascular endothelium associated with cancer. Also of importance in gaining access to the tumor is the nature of the immunoglobulin, particularly for tumor penetration. The nature of the antigen target, whether a cytoplasmic, membrane, extra-membrane, or and/or receptor molecule also influences tumor accretion and internalization. An example is epidermal growth factor receptor (EGFR), which is a specific cell-membrane receptor that mediates the action of EGF, a polypeptide which affects the proliferation and differentiation of many different cell types (12), such as gliomas, breast carcinomas, and squamous cell carcinomas of the lung (12-15). EGFR antibodies have shown targeting of tumors in animal models and in humans (16-18), and the internalization of such antibodies has stimulated the use of electron capture radionuclides, such as ^{125}I, conjugated to EGFR antibodies for internalization and nuclear irradiation (19). This presupposes, however, that all tumor cells will be bound by antibody, since the crossfire effect of radionuclides with more penetrating energies is not the same for such electron capture agents.

F(ab')$_2$ Fab and Fab', and smaller units of fragments penetrate better than whole immunoglobulin (20), and show, therefore, more rapid targeting and higher tumor/non-tumor localization ratios at earlier times than for whole immunoglobulin

Table 1

CANCERS TARGETED CLINICALLY WITH RADIOLABELED ANTIBODIES

Cancer Type	Abs	Cancer Type	Abs
Colorectal	CEA 17-1A 19-9 791T/36 CSAp (MU-9) B72.3	Ovarian	CEA HMFG-2 791T/36 SM3 OC-125 OV-TL3 B72.3
Prostate	PAP PSA	Lymphomas	Ferritin Lym-1 Immu-LL2 T-101 MB-1
Melanoma	p97 96.5 9.2.27 225.28S ZME 018	Breast	CEA B72.3 HMFG MA-5
Lung	CEA NR-LU-10 EGFR	Liver	AFP Ferritin
Trophoblast & Germ Cell	AFP HCG	Neuroblastoma	3F8 BW575/9

(20). However, whole immunoglobulin appears to bind better and for longer to tumor than its fragments. For imaging purposes, therefore, where high tumor/background ratios achieved soon after antibody injection are desireable, antibody fragments appear to be the preferred delivery vehicles (21). Although all forms of foreign immunoglobulins will result in patients making antibodies to isotype and idiotype regions of the antibodies (22), particularly after repeated doses, Fab and Fab' reagents appear to be the least immunogenic (23), thus again supporting their suitability as imaging agents.

The subject of MAb dose is complex and sometimes confusing, primarily because the effects are interpreted in terms of radionuclide uptake in tumor and other organs, which may in fact not always reflect the antibody's biodistribution in contrast to that of the radionuclide.

The issue for RAID is less what the absolute accretion of MAb in tumor is than the ratio of tumor to background counts. Animal and clinical studies indicate that, with radioiodine labels, no advantage is achieved with higher doses of MAb (24). There is some clinical evidence that with the use of an ^{111}In label with melanoma and CEA MAbs, a high dose of unlabeled antibody mixed with 1-2 mg of labeled MAb yields improved targeting of tumor, particularly in the liver (25). However, other studies have claimed good results at the 1-2 mg dose without using unlabeled antibody (26), but these have relied on "cold" images of tumors in the liver, which is not truly antibody targeting.

Most studies agree that single-photon emission computed tomography (SPECT) improves image contrast and resolution, as compared to planar scanning (27). However, this can also increase the rate of false-positive results, since reconstruction artifacts can occur with SPECT. Therefore, the recent use of combined SPECT and CT image comparisons (28) promises to aid in the level of confidence in the interpretation of results by both imaging modalities.

Based upon many of these considerations, as well as a double-blind, multicenter trial of different doses of two anti-CEA MAb fragments labeled with ^{123}I, we concluded that early and accurate imaging of small lesions, including liver metastases, could be achieved with the Fab' MAb form at a 1 mg dose (29). Since ^{123}I is expensive to procure, being shipped from the cyclotron source for each use, ^{99m}Tc was preferred because of its ready availability from on-site generators, its excellent photon energy for imaging, and its low cost. A direct, simple (1-step, 5-min) labeling method was developed (30), and we evaluated ^{99m}Tc-labeled anti-CEA Fab' at a 1 mg dose i.v. in patients with CEA-producing tumors by planar and SPECT imaging. It was found that tumors as small as 0.5 cm could be disclosed by SPECT, that imaging could be successful as early as 2-5 h after injection of the radioimmunoconjugate, that over 90% of known tumor sites, as well as additional occult tumors, could be revealed, and that detecting tumors in the liver at early imaging times, which has been a problem for ^{111}In-labeled antibodies (25,26), could be achieved (31). An expanded, multicenter, clinical trial of colorectal cancer imaging with this ^{99m}Tc-labeled anti-CEA Fab' agent is now being completed, and it is already apparent that the basic findings are being confirmed (32). At the present time, the major clinical indications for RAID are confirmation of viable tumor detected by other anatomically-based radiological methods and disclosure of recurrence or metastasis in patients with a high suspicion of disease progression, particularly when there is a rising tumor marker titer in the blood. Other indications under study are presurgical staging of extent of disease, postsurgical evaluation of residual disease, and confirmation of tumor targeting of an antibody to be used for therapy. The overall current status of cancer imaging with radiolabeled antibodies is summarized in Table 2, which indicates that RAID is already shown to be a safe and useful procedure to complement other radiological approaches, such as computed tomography and magnetic resonance imaging.

Radioimmunotherapy (RAIT)

The use of radiolabeled MAbs for the treatment of cancer is a logical extension of the targeting and imaging achieved with these reagents. A major advantage that RAIT has over antibody-targeted drugs or toxins is that a tumor cell without the appropriate target antigen can be killed by the radiation crossfire from adjacent cells having the determinant and binding the radioactive antibody. This crossfire, or

Table 2

CURRENT STATUS OF ANTIBODY IMAGING

SUBJECT	FINDINGS
Safety	Radioactive antibodies have been found to be safe in over 7,000 patients studied worldwide.
Sensitivity	On a tumor-site basis, results between 60 and over 90 percent sensitivity and specificity have been reported, with the highest accuracy rates found for MAbs labeled with ^{131}I, ^{123}I, or ^{99m}Tc.
Small tumors	Tumor as small as 0.4 to 0.5 cm have been disclosed with ^{99m}Tc MAbs, especially with emission tomography, but resolution usually is in the range of 1.0 to 2.0 cm.
Occult tumors	Tumors missed by other methods, including CT, have been revealed by antibody imaging.
Serum antigen	Antibody imaging can be positive even before the antigen titer in the blood is elevated. Complexation with circulating antigen does not compromise antibody imaging.
HAMA	Repeated injections of animal antibodies result in human antibodies against these foreign proteins, that can compromise antibody targeting.

bystander effect, will depend on a number of factors, such as the nature and energy of the isotope, the spatial arrangement of the actively proliferating malignant cells with respect to the labeled antibody, and the antibody's penetration. RAIT is certainly beset with many more difficulties than RAID, including a low uptake of antibody and radiolabel by the tumor, thus delivering relatively low radiation doses, bone marrow toxicity, heterogeneity of antigen-expressing tumor cells, and human antibody response against mouse antibodies (HAMA). While amount and heterogeneity of antibody uptake in tumors is of less concern for RAID (only the tumor-to-background ratios are important for imaging), these can have considerable implications for RAIT. In this regard, the choices of antibody, isotope, and dose schedule are important considerations.

In terms of the target, a large variety of cell constituents, secretory products, or growth factors have been used for RAID or RAIT, as depicted in Table 2 for various tumor types. It is apparent from the overlap that none of the antibodies is truly tumor type-specific, but they appear to react with a number of different cancers. Possible exceptions are lymphoma antibodies, which do not appear to bind to carcinoma cells. This also may be true for certain ectodermal tumor antibodies, but more clinical data are needed to support this contention. Heterogeneity of expression of the antigen targets, their density and location on or in the tumor cell, and the accessibility of the binding antibody to these antigen determinants (especially when the antigen target may be present in nontumor tissues or circulating) are all important concerns. The observed antigenic heterogeneity of tumors may not be absolute, but instead a range of concentration of antigen. Many tumor cells may have the tumor-associated antigen, but at low levels of expression. If the antibody binds to a small number of antigen sites, but is heavily conjugated with a radiation emitter that traverses a large distance, then a high cell kill might be achieved. Similarly, methods to enhance antigen expression by tumor cells are being developed (33). The use of a target antigen that is internalized, such as the epidermal growth factor receptor (EGFR), is another means to achieve a high level of tumor cell destruction, especially if EGFR is increased on certain tumor cells (19).

The nature and form of the targeting antibody that is to deliver toxic doses of radiation to the tumor cells is also of paramount importance. High levels of radioactivity should be delivered as quickly as possible and should be retained in the tumor relative to dose-limiting normal tissues as long as possible. DeNardo et al. (34) have suggested that the physical half-life of the radionuclide should be similar to the biological half-life of the antibody in the tumor in order to maximize the ratio of the tumor-to-nontumor dose. Experimental studies have indicated that antibody fragments, particularly $F(ab')_2$ have distinct advantage in RAIT, but there is as yet no clinical evidence for the increased value of fragments over whole IgG. Larson et al. (35) did report some tumor responses in melanoma patients receiving ^{131}I-labeled Fab fragments of an anti-melanoma MAb. However, antibody fragments, particularly the univalent form, are less well retained by tumor than whole IgG (20), thereby reducing the radiation dose to tumor unless a suitable repeated dose schedule is achieved. Since antibody fragments are more rapidly cleared from the body than is whole IgG (35), higher radiation doses may be given with such fragments. Moreover, the fragments, particularly the univalent form, are less immunogenic, thus mitigating or delaying HAMA responses (20).

The injected antibody can form complexes with the patient's antibody (HAMA), as described, or with the circulating tumor antigen, which can accumulate in reticuloendothelial organs, such as liver and spleen, and compromise the tumor dose. Nontargeted, circulating radioactive antibody can deliver high radiation doses to the bone marrow, which explains the dose-limiting myelotoxicity experienced with RAIT, particularly thrombocytopenia and leukopenia (36). Thus, there may be an advantage in clearing the blood of residual circulating antibody in order to reduce bone marrow irradiation. We and others have shown that this can be achieved by injecting a second antibody directed against the first, radiolabeled, antibody, leading to a deposition of the labeled antibody in the spleen and liver (37,38). This could also result in an increased clearance of antibody from the tumor, so timing of the administration of the second, anti-antibody will be important (39). Other measures to control or mitigate bone marrow toxicity include autologous bone marrow transplantation, and the use of cytokines and growth factors (40,41).

Iodine-131 has proved its cytotoxic capacity in the treatment of thyroid cancer, and because it is readily available, inexpensive, and the chemistry of its conjugation to antibodies is well developed, it has become the most extensively studied radionuclide in clinical RAIT (42). Other interesting therapeutic radionuclides in clinical trials include ^{90}Y, ^{186}Re, ^{188}Re, and ^{67}Cu as beta-emitters, and ^{125}I as an Auger electron emitter (43). Alpha-emitters, such as astatine-211 and bismuth-212, have also been recommended for RAIT (44). Whereas beta-emission fractionation of therapy permits time for normal tissue repair to occur, this is not the case for alpha-particle irradiation. Also, both Auger electrons and alpha-particles need to be bound to, or targeted within the cell, near to the nuclear DNA, since their ranges span from nanometers to micrometers, whereas beta-emitters can span millimeters. Thus, beta energy is more appropriate for large tumors, and those isotopes emitting energy traversing shorter distances are better suited for the treatment of very small tumors or clusters of tumor cells.

Order and coworkers have shown that when combined with other therapies, ^{131}I-anti-ferritin polyclonal antibodies made in different animals can show a 40 percent response rate in patients with hepatocellular carcinoma (45). Response rates in colorectal and other gastrointestinal cancer patients treated with ^{131}I-labeled CEA and other MAbs have been generally poor. Tumor doses achieved usually are below 2000 cGy, whereas the minimum required for response is probably more than 5000 cGy (42).

^{131}I-labeled antibodies to EGFR or to placental alkaline phosphatase have been used to treat 10 patients with recurrent grade III or IV glioma intravenously or by internal carotid artery infusions, and 6 showed clinical improvement lasting from 6 months to more than 3 years (17). Neuroblastoma- and melanoma-targeting MAbs have also shown antitumor effects when labeled with ^{131}I (46). As mentioned already, ^{125}I-labeled EGFR MAbs have been used with some success in the treatment of patients with glioma (19).

Perhaps the most consistent successes in RAIT have been achieved in patients with lymphomas and leukemias, because of the radiosensitivity and good vascularization of these tumors (47). These patients also have a poorer or infrequent HAMA response, thus permitting repeated doses (48). A number of B-cell lymphoma antibodies have been developed and used clinically, with good response rates (47). Our own studies with a new B-cell specific antibody, EPB-2 (or LL2), have demonstrated very good targeting in non-Hodgkin's B-cell lymphoma, as well as tumor responses to low does of ^{131}I-labeled MAb (47). A ^{99m}Tc-LL2 Fab' imaging kit has also shown good localization of lymphoma sites as a potential staging agent (49).

It is apparent that RAIT is still limited by a number of problems involving the choice of antibodies, choice of isotopes, and the relatively low tumor uptake of the radioimmunoconjugate. Since repeated doses are likely, human or humanized MAbs to reduce host antibody responses to the immunoglobulin vehicle are preferred. In order to reduce toxicity to adjacent, normal tissues, use of short-range irradiation for small tumors, even micrometastases being treated in an adjuvant setting, is of interest. Antibody accretion in tumors may be enhanced by affecting tumor vascularization and vascular permeability, by combining with other biological response modifiers, and by locoregional application. All of these approaches are under investigation, and it therefore appears reasonable to predict that RAIT will advance to becoming an additional important modality for the treatment of cancer.

REFERENCES

1. S. Sell, Cancer Markers of the 1990s, Clinics Lab. Med. 10:1 (1990).

2. R. B. Herberman and D. W. Mercer, eds., "Immunodiagnosis of Cancer," Marcel Dekker, Inc., New York (1990).

3. F. J. Primus, R. H. Wang, D. M. Goldenberg, et al., Localization of human GW-39 tumors in hamsters by radiolabeled heterospecific antibody to carcinoembryonic antigen, Cancer Res. 33:2977 (1973).

4. D. M. Goldenberg, D. F. Preston, F. J. Primus, et al., Photoscan localization of GW-39 tumors in hamsters using radiolabeled anticarcinoembryonic antigen immunoglobulin G, Cancer Res. 34:1 (1974).

5. D. M. Goldenberg, F. DeLand, E. Kim, et al., Use of radiolabeled antibodies to carcinoembryonic antigen for the detection and localization of diverse cancers by photoscanning, N. Engl. J. Med. 298:1384 (1978).

6. D. M. Goldenberg, Current status of cancer imaging with radiolabeled antibodies, J. Cancer Res. Clin. Oncol. 113:203 (1987).

7. S. M. Larson, Clinical radioimmunodetection, 1978-1988: overview and suggestions for standardization of clinical trials, Cancer Res. 50:892 (1990).

8. D. M. Goldenberg, E. E. Kim, F. H. DeLand, et al., Radioimmunodetection of cancer with radioactive antibodies to carcinoembryonic antigen, Cancer Res. 40:2984 (1980).

9. K. R. Hine, A. R. Bradwell, T. A. Reeder, et al., Radioimmunodetection of gastrointestinal neoplasms with antibodies to carcinoembryonic antigen, Cancer Res. 40:2993 (1980).

10. P. A. Farrands, M. V. Pimm, M. J. Embleton, et al., Radioimmunodetection of human colorectal cancers by an anti-tumour monoclonal antibody, Lancet 2:397 (1982).

11. F. Duran-Reynolds, Studies on the localization of dye and foreign proteins in normal and malignant tissues, Am. J. Cancer 35:98 (1939).

12. G. Carpenter and J. G. Zendegui, Epidermal growth factor, its receptor, and related proteins, Exp. Cell Res. 164:1 (1986).

13. T. A. Libermann, N. Razon, A. D. Bartel, et al., Expression of epidermal growth factor receptors in human brain tumours, Cancer Res. 44:753 (1984).

14. R. Perez, M. Pascual, A. Macias, et al., Epidermal growth factor receptors in human breast cancer, Breast Cancer Res. Treat. 4:189 (1984).

15. B. Ozanne, C. S. Richards, F. Hendler, et al., Overexpression of the EGF receptor is a hallmark of squamous cell carcinomas, J. Pathol. 149:9 (1986).

16. A. Goldenberg, H. Masui, C. Divgi, et al., Imaging of human tumor xenografts with an indium-111-labeled anti-epidermal growth factor receptor monoclonal antibody, J. Natl. Cancer Inst. 81:1616 (1989).

17. H. P. Kalofonos, T. R. Pawlikowska, A. Hemingway, et al., Antibody guided diagnosis and therapy of brain gliomas using radiolableled monoclonal antibodies

against epidermal growth factor receptor and placental alkaline phosphatase, J. Nucl. Med. 30:1636 (1989).

18. C. R. Divgi, S. Welt, M. Kris, et al., Phase I and imaging trial of indium 111-labeled anti-epidermal growth factor receptor monoclonal antibody 225 in patients with squamous cell lung carcinoma, J. Natl. Cancer Inst. 83:97 (1991)

19. L. W. Brady, A. M. Markoe, and D. V. Woo, Iodine-125-labeled anti-epidermal frowth factor receptor-425 in the treatment of glioblastoma multiforme, in: "The Present and Future Role of Monoclonal Antibodies in the Management of Cancer. Frontiers of Radiatrion Therapy and Oncology," pp. 151-160, J. M. Vaeth, J. L. Meyer, eds., Karger, Basel (1990).

20. R. D. Blumenthal, R. M. Sharkey, and D. M. Goldenberg, Current perspectives and challenges in the use of monoclonal antibodies for imaging and therapy, in: "Advanced Drug Delivery Reviews," pp. 279-318, Elsevier Science Publishers B.V., Amsterdam (1990).

21. G. Buraggi, A. Turrin, E. Bombardieri, et al., Immunoscintigraphy of colorectal carcinoma with $F(ab')_2$ fragments of anti-CEA monoclonal antibody, Cancer Detect. Prevent. 10:335 (1987).

22. R. Schroff, K. Foon, S. Beatty, et al., Human anti-murine immunoglobulin responses in patients receiving monoclonal antibody therapy, Cancer Res. 45:879 (1985).

23. J. A. Carrasquillo, K. A. Krohn, P. Beaumier, et al., Diagnosis of and therapy for solid tumours with radiolabelled antibodies and immune fragments, Cancer Treatment Reports 68:317 (1984).

24. R. M. Sharkey, F. J. Primus, and D. M. Goldenberg, Antibody protein dose and radioimmunodetection of GW-39 human colon tumor xenografts, Int. J. Cancer 39:611 (1987).

25. Y. Z. Patt, L. M. Lamki, T. M. Haynie, et al., Improved tumor localization with increasing dose of indium-111 labeled anticarcinoembryonic antigen monoclonal antibody ZCE 025 in metastatic colorectal cancer, J. Clin. Oncol. 6:1220 (1988).

26. J. D. Beatty, D. M. Hyams, B. A. Morton, et al., Impact of radiolabeled antibody imaging on management of colon cancer, Am. J. Surg. 157:13 (1989).

27. E. L. Kramer, J. J. Sanger, C. Walsh, et al., Contribution of SPECT to imaging of gastrointestinal adenocarcinoma with ^{111}In-labeled anti-CEA monoclonal antibody, AJR 151:697 (1988).

28. E. L. Kramer, M. E. Noz, J. J. Sanger, et al., CT-SPECT fusion to correlate radiolabeled monoclonal antibody uptake with abdominal CT findings, Radiol. 172:861 (1989).

29. D. M. Goldenberg, H. Goldenberg, R. M. Sharkey, et al., Imaging of colorectal carcinoma with radiolabeled antibodies, Sem. Nucl. Med. 19:262 (1989).

30. H. J. Hansen, A. L. Jones, R. M. Sharkey, et al., Preclinical evaluation of an "instant" ^{99m}Tc-labeling kit for antibody imaging, Cancer Res. 50:794 (1990).

31. D. M. Goldenberg, H. Goldenberg, R. M. Sharkey, et al., Clinical studies of cancer radioimmunodetection with carcinoembryonic antigen monoclonal antibody fragments labeled with ^{123}I or ^{99m}Tc, Cancer Res. 50:909 (1990).

32. C. M. Pinsky, D. M. Goldenberg, T. J. Wlodkowski, et al., Detection of occult metastases of colorectal cancer by the use of anti-CEA Fab' fragments labeled with Tc-99m (abstr.), J. Nucl. Med. (in press).

33. J. W. Greiner, F. Guadagni, P. Horan Hand, et al., Augmentation of tumor antigen expression by recombinant human interferons: enhanced targeting of monoclonal antibodies to carcinomas, in: "Cancer Imaging with Radiolabeled Antibodies," pp. 413-432, D. M. Goldenberg, ed., Kluwer Publishers, Norwell (1990).

34. L. M. Cobb and J. L. Humm, Radioimmunotherapy of malignancy using antibody targeted radionuclides, Br. J. Cancer 54:863 (1986).

35. S. M. Larson, J. A. Carrasaquillo, K. A. Krohn, et al., Localisation of ^{131}I-labelled p97-specific Fab fragments in human melanoma as a basis for radiotherapy, J. Clin. Invest. 72:2102 (1983).

36. S. E. Order, J. L. Klein, and P. K. Leichner, Anti-ferritin IgG antibody for isotopic cancer therapy, Oncology 38:154 (1981).

37. R. H. J. Begent, K. D. Bagshawe, and R. B. Pedley, Use of second antibody in radioimmunotherapy, NCI Monogr. 3:59 (1987).

38. D. M. Goldenberg, R. M. Sharkey, and E. Ford, Anti-antibody enhancement of iodine-131 anti-CEA radioimmunodetection in experimental and clinical studies, J. Nucl. Med. 28:1604 (1987).

39. R. M. Sharkey, J. Mabus, and D. M. Goldenberg, Factors influencing anti-antibody enhancement of tumor targeting with radioantibodies, Cancer Res. 48:2005 (1988).

40. I. D. Bernstein, J. F. Eary, C. C. Badger, et al., High dose radiolabeled antibody therapy of lymphoma, Cancer Res. 50:1017 (1990).

41. R. D. Blumenthal, R. M. Sharkey, L. M. Quinn, et al., Use of hematopoietic growth factors to control myelosuppression caused by radioimmunotherapy, Cancer Res. 50:1003 (1990).

42. S. E. Order, Presidential address: systemic radiotherapy - the new frontier, Int. J. Radiat. Oncol. Biol. Phys. 18:981 (1990).

43. B. W. Wessels and R. D. Rogus, Radionuclide selection and model absorbed dose calculations for radiolabeled tumour associated antibodies, Medical Physics 11:638 (1984).

44. R. M. Macklis, B. M. Kinsey, A. I. Kassis, et al., Radioimmunotherapy with alpha-particle-emitting immunoconjugates, Science 240:1024 (1988).

45. S. E. Order, G. B. Stillwagon, J. L. Klein, et al., I-131 antiferritin: a new treatment modality in hepatoma: a Radiation Therapy Oncology Group study, J. Clin. Oncol. 3:1573 (1985).

46. N-K. V. Cheung, H. Lazarus, F. D. Miraldi, et al., Ganglioside GD2 specific monoclonal antibody 3F8: a Phase I study in patients with neuroblastoma and malignant melanoma, J. Clin. Oncol. 5:1430 (1987).

47. D. M. Goldenberg, J. A. Horowitz, R. M. Sharkey, et al., Targeting, dosimetry, and radioimmunotherapy of B-cell lymphomas with Iodine-131-labeled LL2 monoclonal antibody, J. Clin. Oncol. 9:548 (1991).

48. C. C. Badger and I. D. Bernstein, Prospects for monoclonal antibody therapy of leukemia and lymphoma, Cancer 58:584 (1986).

49. S. Murthy, D. M. Goldenberg, R. E. Lee, et al., Initial targeting studies in lymphoma patients with a new Tc-99m-labeled monoclonal antibody kit (abstr.), J. Nucl. Med. (in press).

GENETIC APPROACHES TO A VACCINE FOR PERTUSSIS

Luciano Nencioni, Mariagrazia Pizza,
Gianfranco Volpini, Audino Podda,
Rino Rappuoli

Sclavo Research Center
Via Fiorentina 1, 53100 Siena, Italy

INTRODUCTION

Whooping cough is a severe disease which each year affects over 60 million children and is responsible for over one million deaths. The vaccine presently available, composed of killed bacteria, is very effective in preventing the disease. However, it is not yet widely used in developing countries while in Western countries the fear of adverse reactions has decreased the vaccine acceptance with a resulting increase of morbidity and mortality due to the disease (Moxon and Rappuoli, 1990).

The cellular vaccine which was introduced approximately 50 years ago, gives over 80% of protection against whooping cough but, at the same time, can induce fever, redness, persistent crying and, although rarely, permanent neurological damages.

These local and systemic side effects are mainly caused by the active pertussis toxin (PT) which is contained in the vaccine together with many other different proteins.

Pertussis toxin is, in fact, the major virulence factor produced by Bordetella pertussis (B. pertussis) but it is also the most immunogenic vaccine component capable of inducing protective immunity.

The pressure to have a new and safer vaccine led to the development of a number of acellular vaccines composed of purified antigens. Indeed, a vaccine composed of chemically detoxified PT has been shown to be effective in preventing whooping cough during a large-scale clinical trial carried out in Sweden (Ad hoc group, 1988). This showed that PT-based vaccines could be effectively used, provided that we find better ways to detoxify pertussis toxin. In fact, the chemical methods used so far to detoxify pertussis toxin, either do not inactivate completely the toxin which can then revert to toxicity (a feature observed in the vaccines used in Sweden (Storsaeter et al., 1988) or modify the toxin so heavily that it loses immunogenicity (Nencioni et al., 1990, manuscript submitted).

The complete and stable detoxification of PT is crucial for the development of safer vaccines because some of the severe side effects associated with pertussis vaccination might be due to trace amounts of active toxin (Steinman et al., 1985).

While a variety of chemical methods to detoxify PT were undertaken by vaccine manufacturers and research laboratories, we decided to tackle the problem from a different point of view and we cloned and sequenced

Immunobiology of Proteins and Peptides VI
Edited by M.Z. Atassi, Plenum Press, New York, 1991

the pertussis toxin operon with the aim of developing a new vaccine by recombinant DNA technology.
In fact, with the advent of biotechnology, molecular genetics has provided new tools to inactivate toxins by modifying one or more codons in their genes. These molecules do not need chemical treatment and have no risk of reversion to toxicity. We have recently constructed genetically detoxified PT mutants and tested them in preclinical as well as in phase I and II clinical studies as potential vaccines against whooping cough.
Pertussis toxin is a complex bacterial toxin composed of five subunits named S1 through S5 (Tamura et al., 1982). Subunit S1 is an enzyme which intoxicates cells by ADP-ribosylating eukaryotic GTP-binding proteins involved in signal transduction thus modifying the cellular response to exogenous stimuli. Subunits S2, S3, S4 and S5 present in a 1:1:2:1 ratio are involved in binding the receptors on the surface of eukaryotic cells and delivering the toxic S1 across the cell membrane so that it can reach its target.
The genes coding for the five subunits of pertussis toxin were found to be clustered in a fragment of DNA of 3,200 base pairs and to be organized as an operon (Nicosia et al., 1986; Locht and Keith, 1986). Since E. coli turned out to be unable to express the entire pertussis toxin operon and assemble the holotoxin, we decided to express separately each individual subunit. Eventually, large amounts of each subunit were expressed by us as fusion proteins in E. coli (Nicosia and Rappuoli, 1987) and as non fused proteins in E. coli (Burnette et al, 1988) and B. subtilis (Runeberg-Nyman et al., 1990) by other laboratories. In each case the recombinant S1 subunits were found to maintain their enzymatic activity suggesting that the folding of the recombinant proteins was identical to that of the natural molecule. This suggested that the recombinant molecules could be effective vaccines since monoclonal antibodies raised against the holotoxin but recognizing the S1 subunit, were able to neutralize the toxin in vitro and to protect mice against infection with virulent B. pertussis (Sato et al., 1984; Sato et al., 1987).
However, to our surprise, none of the recombinant molecules, including S1, induced toxin neutralizing antibodies or protected mice against infection with virulent bacteria (Nicosia and Rappuoli, 1987). Similar results were obtained with recombinant subunits made by other laboratories.
In marked contrast, a rabbit serum raised against the holotoxin was able to neutralize the toxin even when diluted 1/1000. These suggested that the natural toxin and the recombinant S1 subunits were inducing two different populations of antibodies, and that only those against the natural toxin were protective. Indeed, further studies showed that the recombinant subunits have a conformation different from the natural molecule because the protective epitope is formed by three regions which are contiguous in the tridimensional structure of the natural S1 subunit but not in the recombinant S1 subunit (Bartoloni et al., 1988; Pizza et al., 1989).

CONSTRUCTION OF GENETICALLY

DETOXIFIED PT MUTANTS

The pioneering work of Uchida et al. (1971) had shown that the best way to detoxify a toxin is the modification of its gene in order to obtain non toxic molecules which are devoid of enzymatic activity. Therefore, we mutagenized the S1 gene in order to identify aminoacids essential for the enzymatic activity. To engineer the B. pertussis strain W28, the 0.9 Kb KpnI-XbaI DNA fragment encoding for the S1 subunit of the PT operon was subcloned in the Bluescript KS plasmid vector (Stratagene, San Diego, CA, USA) and the single-stranded DNA was mutaganized using oligonucleotide primers. After in vitro mutagenesis, the recombinant fragment was substituted in the chromosome of B. pertussis W28 by allelic exchange, using the pRTP1 plasmid vector. Three aminoacids whose substitution reduced the enzymatic activity of the S1 subunit, were identified by our group (Pizza et al., 1988), and several others were described by other groups (Barbieri et al., 1988, Burnette et al., 1988). These include Arg9, Asp11, Arg13, Trp26, and Glu129. When the genes containing the combination of some mutations (9/129, 13/129 or 26/129) were introduced in the chromosome of B. pertussis, we obtained strains producing pertussis toxin molecules with a toxicity lower than 0.0001% as compared with the wild type toxin (Pizza et al., 1989).

PRECLINICAL STUDIES

The results of preclinical studies have been described in detail in the paper by Nencioni et al. (1990) and summarized in table 1. Briefly, the genetically detoxified molecule PT-9K/129G containing the aminoacid substitutions Arg9-->Lys, Glu129-->Gly in the sequence of the S1 subunit and representative of all the above-mentioned double mutants, did not show any of the toxic properties typical of PT such as lymphocytosis, histamine sensitivity, potentiation of anaphylaxis, hyperinsulinaemia and acute toxicity.

Conversely, the double mutant maintained all the physicochemical and antigenic properties of the native toxin confirming that the mutations introduced had not altered the immunological properties of the molecule. In fact, PT-9K/129G was able to be recognized by human T cell clones against immunodominant epitopes of the S1 subunit, to induce toxin-neutralizing antibodies and to protect mice from the intracerebral infection with virulent B. pertussis.

Genetically inactivated molecules are not only safer theoretically in that there is no possibility of subsequent reversion to toxicity but, results so far obtained have demonstrated that they are also superior immunogenes compared to chemically detoxified molecules. In fact, as shown in table 2, the treatment of PT-9K/129G with increasing doses of formaldehyde (0.07 and 0.42%), decreased its affinity for anti-PT γ-globulins and masked the epitope recognized by the protective monoclonal antibody 1B7. Furthermore the formalin-treated molecules induced anti-PT antibodies which were unable to neutralize the native toxin in vitro and their protective activity in the intracerebral challenge assay, was also dramatically decreased. These results suggested that formaldehyde treatment greatly affected the protective B-cell epitopes naturally present in the molecules and therefore very high doses of antigens are required to produce significant levels of neutralizing antibodies.

In light of these results, we conclude that the classical formalin detoxification should be replaced, when possible, by genetic manipulations which do not change the physicochemical and immunological properties of the native molecules.

Such recombinant DNA techniques are of extreme interest for the in vitro modification of the PT gene which can then be reintroduced under appropriate conditions in the natural host for high level expression of the protein to be included in a safer acellular pertussis vaccine.

However, if on the one hand it seems quite clear that these new vaccines should contain detoxified PT as major pertussis component, on the other the final vaccine formulation is still undergoing international debate.

In fact, to achieve a higher protection against pertussis, additional virulence antigens from B. pertussis can be eventually proposed for

Table 1. *In vivo* and *in vitro* properties of PT-9K/129G compared with purified native PT

Property	Dose	PT	PT-9K/129G
CHO cell clustered growth	(ng/ml)	0.005	> 5,000
ADP ribosylation	(μg)	0.001	> 20
Histamine sensitization	(μg/mouse)	0.1-0.5	> 50
Leukocytosis stimulation	(μg/mouse)	0.02	> 50
Anaphylaxis potentiation	(μg/mouse)	0.04	> 7.5
Enhanced insulin secretion	(μg/mouse)	< 1	> 25
In vivo acute toxicity	(ug/kg)	N.D.	> 1,500
Mitogenicity	(μg/ml)	0.1-0.3	0.1-0.3
Hemagglutination	(μg/well)	0.1	0.1
Affinity constant (anti-S1)	[Ka(L/mol)]	2.4×10^8	6.1×10^8
Affinity constant (anti-PT)	[Ka(L/mol)]	2.0×10^{10}	9.8×10^9

N.D. = not determined

Table 2. Effect of formaldehyde treatment on some *in vitro* and *in vivo* properties of PT-9K/129G

Formaldehyde	Affinity constant		Immunogenicity		IC Challenge
(%)	Goat gamma globulins anti-PT	MoAb 1B7	Guinea pigs immunized with 3 μg		Mice immunized with 5 μg (Survivors)
	[Ka(L/Mol)]		ELISA Titer[a]	CHO Titer[b]	
0.00	5.2×10^9	5.5×10^7	3.5	1/1,280	13/16
0.07	5.2×10^8	-	3.1	1/160	8/16
0.42	6.7×10^7	-	3.3	1/80	1/16

[a] ELISA levels of antibodies raised in guinea pigs are expressed as absorbance values at 405 nm of undiluted sera.

[b] Neutralizing titers of antibodies raised in guinea pigs are expressed as reciprocals of the highest serum dilutions resulting in 100% inhibition of the clustering effect on CHO cells induced by 120 pg of wild-type PT tested in triplicate.

vaccine purpose. These antigens which include the filamentous hemagglutinin (FHA), the agglutinogens, the pili, and an outer membrane protein of 69 kilodaltons (69 Kd), are normally expressed by the engineered strain of B. pertussis producing the genetically detoxified pertussis toxin mutants.

CLINICAL STUDIES

After the successful preclinical studies which have shown that the double mutant PT-9K/129G alone is safe and immunogenic (Pizza et al., 1989; Nencioni et al., 1990), we have tested in adult volunteers two different formulations of acellular pertussis vaccine composed respectively of the double mutant alone (Lot D5/FA) or combined with FHA and 69Kd proteins (Lot PFK/2) as summarized in table 3.

Table 3. Vaccine composition per dose (0.5 ml)

Lot N°	PT-9K/129G (μg)	FHA (μg)	69Kd (μg)	$Al(OH)_3$ (mg)	Thimerosal (mg)
D5/FA	15	-	-	0.5	0.05
PFK/2	7.5	10	10	0.5	0.05

All healthy adult volunteers recruited for their low anti-PT antibody titers were ramdonly attributed to receive, in double blind, one dose of vaccine or placebo. After 6 weeks, each group of volunteers received a second dose of vaccine or placebo with the same procedure used for the first injection.
The results of this study are summarized in tables 4 and 5 and described in detail in the paper by Podda et al. (1990). Total and neutralizing antibody titers are expressed respectively as geometric mean of ELISA units (EU)/ml and of reciprocals of serum dilution inhibiting the clustering effect on CHO cells induced by 120 pg of native PT. 95% confidence intervals are reported in parenthesis.
Briefly, both vaccines did not induce either local or systemic adverse reactions and did not cause any changes in the parameters tested, including the leukocyte, insulin and IgE levels that are usually altered by wild type PT (Wardlaw and Parton, 1983).
These results indicate that PT-9K/129G can be safely used for human immunization and confirm the results previously obtained in animal models (Pizza et al., 1989; Nencioni et al., 1990) and summarized in table 1.
One injection of both vaccines was capable of inducing high titers of antibodies specific for each vaccine components with strong neutralizing activity. Since the U.S. Reference Pertussis Antiserum (a kind gift of the Laboratory of Pertussis, FDA, U.S.A.) used in our ELISA and CHO assays allows reasonable comparision with previous studies, it can be concluded that either 15 or 7.5 μg of PT-9K/129G which are contained respectively in D5/FA and PFK/2 vaccines, produce a humoral response higher than that obtained using 50 or 25 μg of chemically detoxified PT (Sekura et al., 1988; Winberry et al., 1988).
This confirms in vivo with humans that genetic detoxification maintains the natural epitopes of the proteins which, conversely, may be lost or altered with chemical detoxification.

Table 4. Serum antibody responses in adult volunteers receiving placebo or D5/FA vaccine

Time	ANTI-PT ELISA TITER (EU/ml)			
	Placebo		Vaccine	
	IgG	IgA	IgG	IgA
Day 0	6.5 (3.5-12.2)	2.8 (1.1-7.0)	6.7 (4.6-9.8)	2.6 (1.3-5.1)
Day 30	8.1 (4.2-15.5)	2.4 (0.8-7.4)	496.5 (199-1,233.3)	293.9 (130-665)
Day 70	7.9 (4.5-13.9)	2.4 (0.5-3.2)	522.4 (216.8-1,258)	143.5 (72.1-286)

	NEUTRALIZING TITER (CHO CELLS)	
	Placebo	Vaccine
Day 0	12.8 (7.5-21.6)	13.6 (8.6-21.5)
Day 30	12.8 (6.7-24.9)	1,810.2 (737-4,446)
Day 70	11.7 (5.9-23.0)	1,974.0 (820-4,751)

Alternatively, the ratio between toxin neutralizing titers and total anti-PT ELISA titers, may suggest that a not chemically modified molecule is able to induce antibodies with higher affinity for the native PT as demonstrated in animal models (Table 2).

The second dose of vaccine did not cause a significant further increase of humoral responses, indicating that all adult volunteers had been previously exposed to *B. pertussis* and consequently gave a secondary response to the first immunization.

Interestingly, as reported in table 4, all adult volunteers receiving 15 µg of PT-9K/129G showed high levels of anti-PT IgA antibodies, as assessed in the ELISA assay.

Other investigators have observed a significant serum IgA response to PT in patients with pertussis (Thomas et al., 1989) but a negligible response after vaccination of adults with whole cell (Thomas et al., 1989) and acellular pertussis vaccines (Granstrom et al., 1987).

It is therefore conceivable that PT-9K/129G which maintains intact the property of the native molecule to bind to cellular receptors, can mimic

the natural infection and is able to stimulate the immunity induced by the native PT at the mucosal level during the disease.
In summary, if on the one hand the frequency and the magnitude of vaccine-induced humoral IgA response may be also dependent on the prior exposure to the antigen on the other, no doubt that a major role is played by the nature of the immunizing antigen.
Although this study in human volunteers was only the first clinical step toward a new acellular vaccine against whooping cough, the results obtained so far are very encouraging in terms of safety and immunogenicity for PT-9K/129G. Since the efficacy of PT in preventing pertussis disease has already been established in the Swedish clinical trial (Ad Hoc Group for the Study of Pertussis Vaccines, 1988; Olin and Storsaeter, 1989), the presence of a molecule which eliminates all problems associated with chemical detoxification of PT, suggests that PT-9K/129G should be the antigen of choice for future vaccines against whooping cough. These vaccines may contain either PT-9K/129G alone or combined with other antigens, such as FHA and 69Kd proteins. More generally, it can be concluded that genetic manipulation can be efficiently used to modify the properties of natural molecules in order to make them suitable for human use, and that safe acellular pertussis vaccines are now feasible.
After the successful phase I studies in adult volunteers, both vaccines

Table 5. Serum antibody responses in adult volunteers receiving placebo or PFK/2 vaccine

Group	Antigen	ELISA IgG TITERS (EU/ml)		
		Day 0	Day 42	Day 72
Placebo	PT	10.9 (4.3-27.6)	11.8 (4.8-28.9)	14.6 (8.6-24.9)
	FHA	18.2 (5.6-58.9)	18.1 (5.4-60.2)	19.9 (6.3-62.9)
	69Kd	2.6 (0.4-19.2)	2.4 (0.3-18.8)	2.6 (0.4-16.9)
Vaccine	PT	7.7 (1.9-31.9)	1,465.4 (415.8-5,163.5)	1,931.3 (661.0-5,642.0)
	FHA	5.5 (1.9-16.7)	835.5 (195.2-3,575.8)	436.9 (112.9-1,690.5)
	69Kd	1.2 (0.2-7.8)	162.5 (13.1-2,009.7)	115.7 (8.1-1,649.4)
		NEUTRALIZING TITERS (CHO CELLS)		
Placebo		20.0 (9.4-42.7)	28.3 (15.8-50.5)	22.4 (11.4-44.3)
Vaccine		15.8 (7.8-32.4)	3,620.4 (1,393.9-9,403)	3.600.5 (1,392-9,404)

are now been testing in 3- and 15-month old children. The preliminary results of phase II clinical trials confirm that these acellular pertussis vaccines are safe and are able to induce high levels of total and toxin-neutralizing antibodies.

REFERENCES

1. Ad Hoc Group for the study of Pertussis Vaccines, 1988, Placebo-controlled trial of two acellular pertussis vaccines in Sweden: Protective efficacy and adverse events, Lancet, i:955.
2. Barbieri, J.T., and Cortina, G., 1988, ADP-ribosyltransferase mutations in the catalytic S1 subunit of pertussis toxin, Infect. Immun., 56:1934.
3. Bartoloni, A., Pizza, M., Bigio, M., Nucci, D., Ashworth, L.A., Irons, L.I., Robinson, A., Burns, D., Manclark, C., Sato, H., and Rappuoli, R., 1988, Mapping of a protective epitope of pertussis toxin by in vitro refolding of recombinant fragments, Bio/Technology, 6:709.
4. Burnette, W.N., Cieplak, W., Mar, V.L., Kaljot, K.T., Sato, H., and Keith, J.M., 1988, Pertussis toxin S1 mutant with reduced enzyme activity and a conserved protective epitope, Science, 242:72.
5. Granstrom, M., Thoren, M., Blennow, M., Tiru, M., and Sato, Y., 1987, Acellular pertussis vaccine in adults: adverse reactions and immune response, Eur. J. Clin. Microbiol., 6:18.
6. Locht, C., and Keith, J.M., 1986, Pertussis toxin gene: nucleotide sequence and genetic organisation, Science, 232:1258.
7. Moxon, R., and Rappuoli, R., 1990, Modern vaccines: Haemophilus influenzae infections and whooping cough, Lancet, i:1324.
8. Nencioni, L., Pizza, M.G., Bugnoli, M., De Magistris, M.T., Di Tommaso, A., Giovannoni, F., Manetti, R., Marsili, I., Matteucci, G., Nucci, D., Olivieri, R., Pileri, P., Presentini, R., Villa, L., Kreeftenberg, J.G., Silvestri, S., Tagliabue, A., and Rappuoli, R., 1990, Characterization of genetically inactivated pertussis toxin mutants: candidates for a new vaccine against whooping cough, Infect. Immun., 58:1308.
9. Nencioni, L., Volpini, G., Peppoloni, S., Bugnoli, M., De Magistris, M.T., Marsili, I., and Rappuoli, R., 1990, Properties of the pertussis toxin mutant PT-9K/129G after formaldehyde treatment, Infect. Immun. (Manuscript submitted).
10. Nicosia, A., Perugini, M., Franzini, C., Casagli, M.C., Borri, M.G., Antoni, G., Almoni, M., Neri, P., Ratti, R.,and Rappuoli, R., 1986, Cloning and sequencing of the pertussis toxin genes: operon structure and gene duplication, Proc. Natl. Acad. Sci., USA, 83:4631.
11. Olin, P., and Storsaeter, J., 1989, The efficacy of a cellular pertussis vaccine, JAMA, 261:560.
12. Pizza, M., Bartoloni, A., Prugnola, A., Silvestri, S., and Rappuoli, R., 1988, Subunit S1 of pertussis toxin: mapping of the regions essential for ADP-ribosyl transferase activity. Proc. Natl. Acad. Sci., USA, 85:7521.
13. Pizza, M., Covacci, A., Bartoloni, A., Perugini, M., Bugnoli, M., Manetti, R., Nencioni, L., and Rappuoli, R., 1989, Pertussis toxin CRMs and new vaccines against whooping cough, in: "Proceedings of 4th European workshop on bacterial protein toxins, Urbino, Italy," Gustav Fischer Verlag, Stuttgart, p. 507.
14. Pizza, M., Covacci, A., Bartoloni, A., Perugini, M., Nencioni, L., De Magistris, M.T., Villa, L., Nucci, D., Manetti, R., Bugnoli, M., Giovannoni, F., Olivieri, R., Barbieri, J.T., Sato, H., and Rappuoli, R., 1989, Mutants of pertussis toxin suitable for vaccine development, Science, 246:497.

15. Podda, A., Nencioni, L., De Magistris, M.T., Di Tommaso, A., Bossu', P., Nuti, S., Pileri, P., Peppoloni, S., Bugnoli, M., Ruggiero, P., Marsili, I., D'Errico, A., Tagliabue, A., and Rappuoli, R., 1990, Metabolic, humoral, and cellular responses in adult volunteers immunized with the genetically inactivated pertussis toxin mutant PT-9K/129G, J. Exp. Med., 172:861.
16. Runeberg-Nyman, K., Olander, R., Karvonen, M., Muotiala, A., Hyvarinen, T., Nurminen, M., Wahlstrom, E., Himanen, J.P., and Sarvas, M., 1990, Immunogenicity and protective capacity of pertussis toxin subunits produced in B. subtilis, in: "Vaccines 90. Modern approaches to new vaccines, including prevention of AIDS," F. Brown, R.M. Chanock, H.S. Ginsberg, and R.A. Lerner, eds, Cold Spring Harbor: Cold Spring Harbor Laboratory, p. 425.
17. Sato, H., and Sato, Y., 1984, Bordetella pertussis infection in mice; correlation of specific antibodies against two antigen, pertussis toxin and filamentous hemagglutinin with mouse protectivity in intercerebral or aerosol challenge system, Infect. Immun., 46:415.
18. Sato, H., Sato, Y., and Ito, A., 1987, Effect of the monoclonal antibody to pertussis toxin on toxin activity, Infect. Immun., 55:909.
19. Sekura, D.R., Zhang, Y., Roberson, R., Acton, B., Trollfors, B., Tolson, N., Shiloach, J., Bryla, D., Muir-Nash, J., Koeller, D., Scheerson, R., and Robbins, J.B., 1988, Clinical, metabolic, and antibody responses of adult volunteers to an investigational vaccine composed of pertussis toxin inactivated by hydrogen peroxide, J. Pediatr., 113:806.
20. Steinman, L., Weiss, A., Adelman, N., Lim, M., Zuniga, R., Ochlert, J., Hewlett, E., and Falkow, S., 1985, Pertussis toxin is required for pertussis vaccine encephalopathy, Proc. Natl. Acad. Sci., USA, 82:8733.
21. Storsaeter, J., Olin, P., Renemar, B., Lagergard, T., Norberg, R., Romanus, V., and Tiru, T., 1988, Mortality and morbidity from invasive bacterial infections during a clinical trial of acellular pertussis vaccines in Sweden, Pediatr. Infect. Dis. J., 7:637.
22. Tamura, M., Nogimori, K., Murai S., Yajima, M., Ito, K., Katada, T., UI, M., and Ishii, S., 1982, Subunit structure of the islet activating protein, pertussis toxin, in conformity with the A-B model, Biochemistry, 21:5516.
23. Thomas, M.G., Ashworth, L.A.E., Miller, E., and Lambert, H.P., 1989, Serum IgG, IgA, and IgM responses to pertussis toxin, filamentous hemagglutinin, and agglutinogens 2 and 3 after infection with Bordetella pertussis, and immunization with whole-cell pertussis vaccine, J.Infect. Dis., 160:838.
24. Uchida, T., Gill, D.M., and Pappenheimer, A.M.Jr., 1971, Mutation in the structural gene for diphtheria toxin, carried by the temperate phage beta, Nature New Biol., 233:8.
25. Wardlaw, A.C., and Parton, R., 1983, Bordetella pertussis toxins, Pharmacol. & Ther., 19:1.
26. Winberry, L.R., Walker, R., Cohen, N., Todd, C., Sentissi, A., and Siber, G., 1988, Evaluation of a new method for inactivating pertussis toxin with tetranitromethane, in: "Abstract of the International Workshop of B. pertussis" Keith, J., ed., Rocky Mountains Laboratory, Hamilton, Montana.

STRATEGIES FOR TYPE-SPECIFIC GLYCOCONJUGATE VACCINES OF STREPTOCOCCUS PNEUMONIAE

Beth Arndt [+] and Massimo Porro [*]

[+] Praxis Biologics Inc. Rochester N.Y.- USA 14623
[*] Biosynth Srl, Rapolano Terme, Siena-ITALY 53040

INTRODUCTION

Prevention of bacterial infections in infants by Gram-positive and Gram-negative encapsulated microrganisms is still a world-wide concern. According to reliable estimates (US National Institute of Allergy and Infectious Diseases, 1986) the infant mortality by Streptococcus pneumoniae approaches to seven million cases annually and the mortality by encapsulated bacteria (Streptococcus pneumoniae, Haemophilus influenzae type B and Neisseria meningitidis) accounts for about 38% of the total burden by infectious diseases.

The highly purified polysaccharide vaccines introduced in the seventies (Gotschlich et al., 1969; Austrian, 1979) have been demonstrated to be efficacious in adults and older children but not significantly immunogenic in young infants (Teele et al., 1981; Makela et al., 1981; Douglas and Miles, 1984). The main reason for the poor immunogenicity of purified capsular polysaccharides in infants seems related to the immunological properties of the polysaccharides (helper T-cell independent antigens or TI-2 antigens) as well as to the immaturity of the host's immune system.

In this respect, data collected from immunological experiments in animals and humans lead to some insight in the activation mechanism of the host's immune system by polysaccharides. Specifically, capsular polysaccharides of Streptococcus pneumoniae have been reported to bind and activate different lymphocyte populations. Activation of B cells mainly results in the secretion of antibodies of IgM and IgG2 isotype in man and IgM and IgG3 in mice (Rijkers and Mosier, 1985); activation of regulatory T-cells (suppressor and contrasuppressor) results in the suppression of amplifier and helper T-cells leading, as ultimate result, to the inhibition or very low induction of IgG isotype antibodies (Paul et al., 1971; Braley Mullen, 1986a; 1986b; Baker, 1990); activation of Natural Killer (NK) cells, in infant mice, results in age-related down-regulation of IgM isotype (Khater et al., 1986).

Taken together, the data suggest that a polysaccharide antigen induces modest levels of IgM antibodies due to the concomitant activation of B and NK cells and low levels of IgG antibodies, which do not show a secondary immune response following a booster dose of the same antigen, because the activity of suppressor T-cells results in the inhibition of amplifier and helper T-cells.

A strategy to overcome the problems associated with the polysaccharide vaccines involves the use of glycoconjugates, that is covalently linked carbohydrate antigens and haptens to protein carriers working as T-helper dependent antigens. This strategy goes back to the pioneering and elegant studies of immunochemistry performed by Avery and Goebel in the thirties (Avery and Goebel,1929; Goebel, 1938) and it is now taking advantage of the newer technologies developed in recent years.

An oligosaccharide-based conjugate vaccine (Anderson et al, 1986) and a polysaccharide conjugate vaccine (Lepow et al, 1987) for prophylaxis of Haemophilus influenzae type b infections in infants have been recently licensed by the U.S. Food and Drug Administration. A comparison of immunological activity for the two basically different strategies used in the synthesis of a glycoconjugate vaccine (the use of short-chain oligosaccharides versus long-chain polysaccharides) will be determined on the basis of efficacy in field trials and this will require some time.

It is the purpose of this work to investigate the immunogenic activity in animal models, of glycoconjugates synthesized according to the same protocol, with the same carrier protein but using carbohydrate haptens of two different chain lenght: short-chain versus long chain oligosaccharides. In an attempt to get significant informations from similar work, we have considered four polysaccharide structures of Streptococcus pneumoniae with monosaccharide sequences completely unrelated to each other (type 6A,14,19F and 23F). The choice of these four serotypes has been done on the basis of the observation that they were found to be poor immunogens in respect to other polysaccharides of Streptococcus pneumoniae in the pediatric population (Douglas et al, 1983) and that at least one of them (type 6A) has been reported as the poorest antigen among the serotypes present in the combined polysaccharide-based pneumococcal vaccine (Makela et al,1981). The carrier protein CRM197 has been chosen because of its high immunogenicity as a carrier which is also able to induce protective levels of neutralizing antibodies to diphtheria toxin (Porro et al, 1985; Porro et al, 1986; Porro,1987).

MATERIALS AND METHODS

Preparation of oligosaccharides

Purified capsular polysaccharides of Streptococcus pneumoniae type 6A, 14, 19F and 23F had immunochemical characteristics comparable to those previously reported (Porro et al, 1983). Oligosaccharides of different sizes were generated by controlled acid hydrolysis in the following conditions: polysaccharides were solubilized at 2-5 mg/ml in 10^{-2} M acetic acid (type 6A), $5x10^{-1}$ M trifluoroacetic acid (type 14), $1.5x10^{-2}$ M phosphoric acid (type 19F) and 10^{-1} M acetic acid (type 23F). Hydrolysis proceeded in sealed vials at 100°C (type 6A and 23F), T=70°C (type 14) and T=50°C (type 19F) for various times in order to get oligosaccharides of desired size according to Fig.1. The rate of hydrolysis of the capsular polysaccharides were investigated by the molar ratios: end-reducing group/total phosphorus (type 6A and 23F), end-reducing group/N-Acetyl glucosamine (type 14) and ester-bound phosphate group/total phosphorus (type 19F), according to the structures of the capsular polysaccharides type 6A (Rebers and Heidelberger, 1961), type 14 (Lindberg et al, 1977), type 19F (Jennings et al, 1980; Ohno et al, 1980) and type 23F (Richards and Perry, 1988). Chemical methods for assays of phosphorus (Chen et al, 1956), end-reducing groups (Porro et al, 1981), hexosamine (Ashwell, 1957) and ester-bound phosphate groups by alkaline phosphatase hydrolysis (Eby R.,personal communication), were used. The sites of hydrolysis in each polymer have been investigated by ^{13}C and ^{1}H-Nuclear Magnetic Resonance (NMR) as previously reported (Porro et al, 1985) and by Gas-chromatography (Dick W.E., personal communication). The homogeneity of the oligosaccharide preparations were tested by gel chromatography on Sephadex G-15 and G-50 (Pharmacia, Uppsala, Sweden) by chemical analysis of the eluates, using the above reported methods.

Immunochemical characterization of the oligosaccharides

Negatively-charged oligosaccharides (type 6A, 19F and 23F) were tested for immunochemical reactivity against their specific pneumococcal reference rabbit antisera (Statens Seruminstitut, Copenhagen, Denmark) by inhibition of immunoprecipitation in differential immunoelectrophoresis (Porro et al, 1985) using the homologous capsular polysaccharides as reference antigens. The neutral type 14 oligosaccharide and the homologous reference polysaccharide were analyzed by inhibition of immunoprecipitation in radial immunodiffusion. The immunochemical specificity for each purified oligosaccharide, with respect to that detected for the reference homologous capsular polysaccharide, was calculated on the basis of the ratio between the minimal inhibitory concentration found for the polysaccharide antigen (MIC Ag) and that found for the selected oligosaccharide hapten (MIC Hp), as formerly described (Porro et al, 1985).

Synthesis of glycoconjugates

The purified oligosaccharides (5 mg/ml) of the selected sizes for the four serotypes, were reacted at pH=9.0 (0.2M KH_2PO_4) T=100°C for 15 min, with diaminoethane in a molar excess 10:1 with the estimated amount of end-reducing groups. The temperature was then reduced at T=50°C and pyridine borane at a molar excess 25:1 with respect to the amount of end-reducing groups in the oligosaccharide was used in reaction. Reduction followed for 48hs. Purification of the amino-activated oligosaccharides followed on Sephadex G-15 superfine equilibrated in $1.5x10^{-2}$ M sodium chloride. The purified oligosaccharides were freeze-dried and then transformed to monosuccinimidyl ester of adipic acid as previously reported (Porro et al, 1985). Conjugation of the activated oligosaccharides to the protein CRM197 (purified and characterized as described by Porro et al, 1985) occurred overnight a T=4°C in a 50% v/v solution dimethylsulfoxide (DMSO)/10^{-1} M sodium bicarbonate buffer (pH=8.0), using a molar ratio activated oligosaccharide/amino groups of CRM197 (titered according to Habeeb, 1966) 1:2.

Finally, each glycoconjugate was purified on Sephadex G-100 superfine or Sephacryl S-200 according to the size of the glycoconjugate synthesized and estimated by SDS-PAGE analysis in the conditions published (Porro et al, 1985).

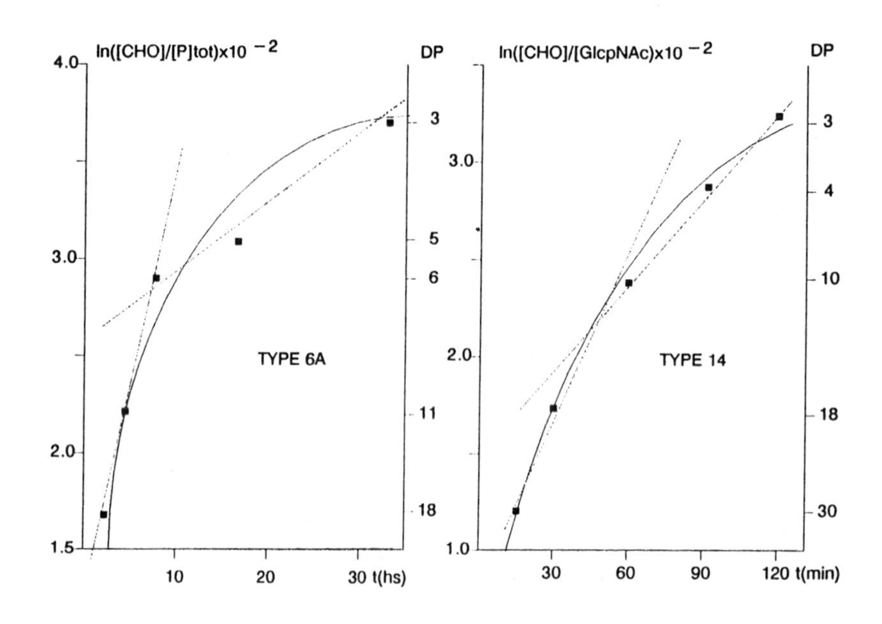

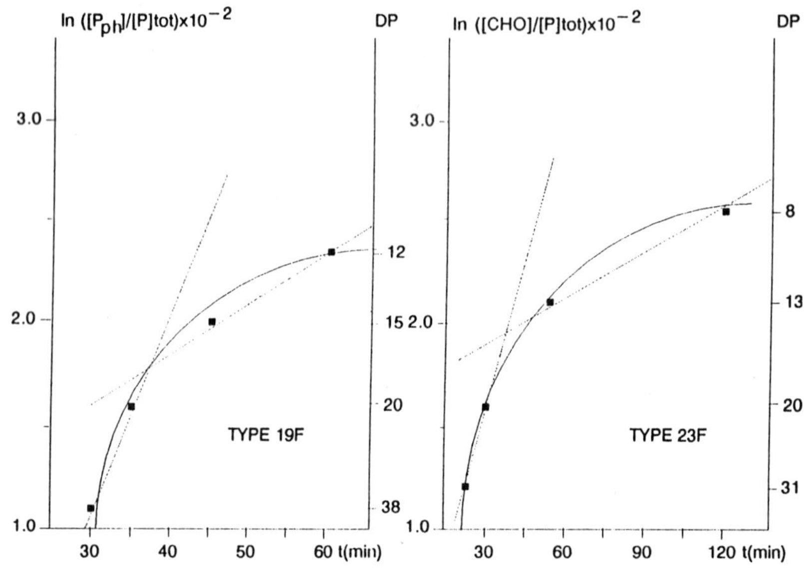

Figure 1. Rate of hydrolysis of S.pneumoniae type-specific polysaccharides. (For explanation see text, Methods section)

Physical-Chemical and Immunochemical characterization of the glycoconjugates

The analytical procedures used for characterization of the four type-specific glycoconjugates, were basically the same previously reported (Porro et al, 1985).

Immunization procedures and immunological analysis

For each of the four type-specific capsular antigens of *Streptococcus pneumoniae* , two kinds of glycoconjugate vaccines were prepared, different only in the length of the oligosaccharide chain (DP=3-6 and DP=10-14, see the Results section). Each kind of glycoconjugate vaccine was formulated as mixture of the four glycoprotein antigens in either soluble or aluminium hydroxide-adsorbed form. Each immunizing dose contained 2.5 μg of conjugated type-specific oligosaccharide in 0.5 ml volume of PBS pH=7.2. The adsorbed formulation contained 1 mg/dose of aluminium hydroxide and all the conjugate material was adsorbed under these conditions.

Groups of five New Zealand rabbits (Av.wt.= 2 kg) and five 8 week-old Swiss-Webster mice were immunized s.c. with each vaccine formulation respectively at week 0,4 and 8 (bleeding at week 0,4,6 and 10) and at week 0 and 4 (bleeding at week 0,4 and 6).Three groups of five guinea pigs each, were injected with the purified protein CRM197 according to the schedule previously reported (Porro et al, 1980) with the following amount: priming dose of 30,60 and 120 μg followed by a booster dose, 6 weeks later, respectively of 60,120 and 240 μg.The "range" of the CRM197 doses injected included 0.5,1 and 2 times the maximum dose (50 μg) of conjugated CRM197 present in the mixture of the four type-specific glycoconjugates.

IgG antibodies to type specific capsular polysaccharides of *Streptococcus pneumoniae* have been titered for each rabbit and for the pool of mice sera by ELISA assay, basically in the conditions published (Porro et al, 1985) coating the PVC plates with type-specific polysaccharides at 5 μg/ml. Titers were detected as the highest dilution yielding an optical density (OD) value at 405 nm twice of the background level and were expressed as geometric mean of the titers determined in each rabbit. Inhibition-ELISA was performed by pre-incubating (2hs at r.t.) rabbit sera (at constant dilution 1:5,000 v/v) with serial concentrations of competitors (homologous or heterologous type-specific polysaccharides and oligosaccharides). The molar concentration of each competitor was calculated on the basis of the molecular weight of the basic repeating unit present in its structure, in order to avoid inaccurate evaluation of the molecular weight of pneumococcal polysaccharide structures which appear as size-distributed systems by gel-chromatography and by antibody recognition of the eluates (Porro et al, 1983). Neutralizing antibodies to diphtheria toxin raised in rabbits by the conjugates, have been titered on sera pool using Vero cells (Strain P 142) culture in COSTAR plates, basically according to Zucher and Murphy 1984, and using a diphtheria toxin (LIST BIOLOGICS lot #06 1/28/82) concentration of $5x10^{-13}$ M with a reference anti DT antiserum (U.S.Standard #47 09/04/88,Office of Biologics, Food and Drug Administration). Neutralizing antibodies to diphtheria toxin in guinea pigs have been titered as previously published (Porro et al,1980). Statistical analysis was performed by the Student's t test .

RESULTS

Physical-chemical and immunochemical characteristics of oligosaccharide haptens type 6A,14,19F and 23F

The capsular polysaccharides of Streptococcus pneumoniae type 6A,14,19F and 23F are complex polymers of basic repeating units whose structures have been already investigated and clarified.

Depolymerization of the four type-specific polysaccharides by chemical hydrolysis in mild conditions,as reported in the METHODS section, was investigated by the rate of hydrolysis.

Type 6A is a polymer with the linear structure:

$$\left[\xrightarrow{2} \text{D-Galp} \xrightarrow[\alpha]{1,3} \text{D-Glcp} \xrightarrow[\alpha]{1,3} \text{L-Rhap} \xrightarrow[\alpha]{1,3} \text{D-Ribitol} \xrightarrow{5} \text{O-}\overset{\overset{\displaystyle O}{\|}}{\underset{\underset{\displaystyle O^-}{|}}{P}}\text{-O-} \right]_n$$

(Rebers and Heidelberger,1961)

Type 14 is a polymer with branched structure:

$$\left[\xrightarrow{4} \text{D-Glcp} \xrightarrow[\beta]{1,6} \underset{\underset{\displaystyle \text{D-Galp}}{\overset{4}{\underset{1}{|}}\ \beta}}{\text{D-GlcpNAc}} \xrightarrow[\beta]{1,3} \text{D-Galp} \xrightarrow[\beta]{1} \right]_n$$

(Lindberg et al, 1977)

Type 19F is a polymer with linear structure:

$$\left[\xrightarrow{4} \text{D-ManpNAc} \xrightarrow[\beta]{1,4} \text{D-Glcp} \xrightarrow[\alpha]{1,2} \text{L-Rhap} \xrightarrow[\alpha]{1} \text{O-}\overset{\overset{\displaystyle O}{\|}}{\underset{\underset{\displaystyle O^-}{|}}{P}}\text{-O-} \right]_n$$

(Jennings et al, 1980; Ohno et al, 1980)

Type 23F is a polymer with branched structure:

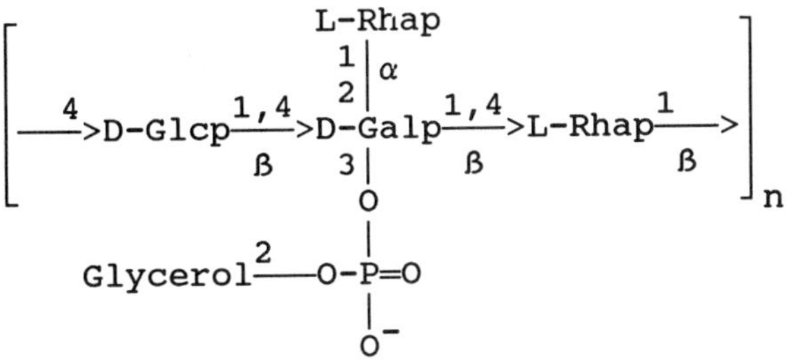

```
                    L-Rhap
                     1|α
                     2|
 [ —4—>D-Glcp—1,4—>D-Galp—1,4—>L-Rhap—1—> ]
             β      3|      β         β    n
                     O
                     |
       Glycerol—2—O-P=O
                     |
                     O-
```

(Richards and Perry, 1988).

The following molar ratios were calculated for each purified oligosaccharide in order to estimate the degree of polymerization (DP) attained: end-reducing groups (CHO)/total phosphorus $(P)_{tot}$ (type 6A); end-reducing groups (CHO) / N-Acetyl Glucosamine (Glcp-NAc) (type 14); alkaline phosphatase-hydrolyzed phosphates (Pph)/total phosphorus $(P)_{tot}$ (type 19F); end-reducing groups (CHO)/total phosphorus $(P)_{tot}$ (type 23F). In Fig.1 the relationship between time of hydrolysis and DP is reported. For type 19F the ratio of alkaline phosphatase-hydrolyzed phosphate/total phosphorus was used in the place of end-reducing groups/total phosphorus since the rhamnose residue (Rhap) (where hydrolysis occurs) is α-linked via the C-2 to the D-Glcp residue and it shows a lower reducing activity than a Rhap residue not substituted at the near carbon (C-2) of the end-reducing anomeric carbon (C-1). In Fig.2, the same relationship is referred to the amount of end-reducing groups estimated per mg of carbohydrate material: this allowed a reliable and fast method for selection of type-specific oligosaccharide of a desired DP.

The sites of hydrolysis for the type-specific polymers have been detected in the following sequences:

L - Rhap $\xrightarrow[\alpha]{1,3}$ D - Ribitol (type 6A)

D - Glcp $\xrightarrow[\beta]{1,6}$ D - GlcpNAc
D - Galp $\xrightarrow[\beta]{1,4}$ D - Glcp (type 14)

L-Rhap $\xrightarrow[\alpha]{1}$ O-P(=O)(O⁻)-O- (type 19F)

L-Rhap $\xrightarrow[\beta]{1,4}$ D-Glcp (type 23F)

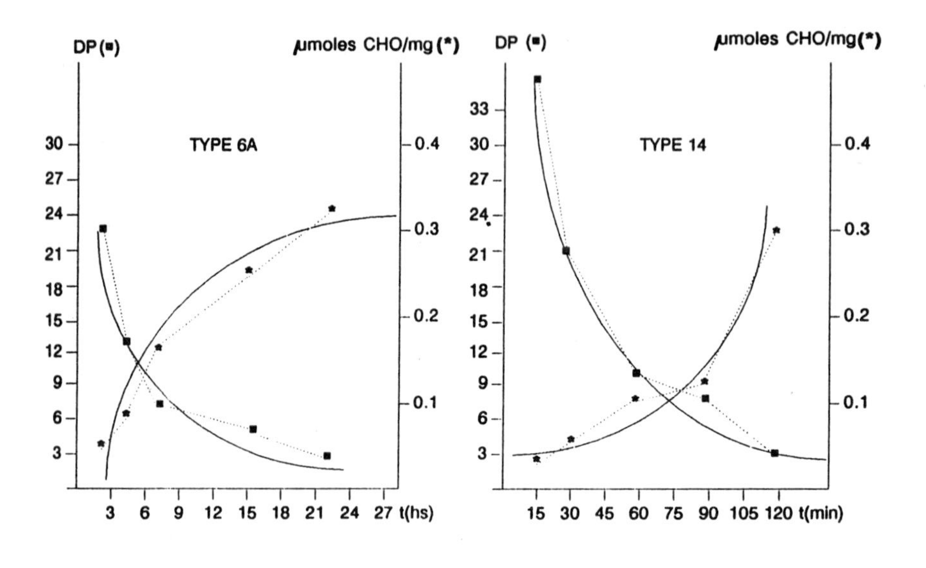

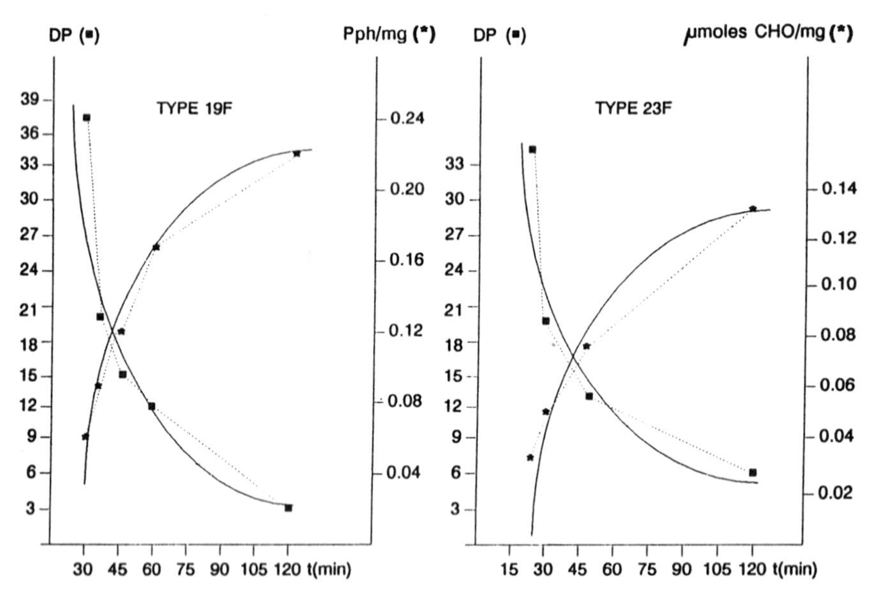

Figure 2. Rate of hydrolysis of S.pneumoniae type – specific polysaccharides. (For explanation see text, Methods section)

With the exception of type 14, which shows two equally sensitive linkages to mild acid conditions, the other three type-specific polymers have shown a preferred site of hydrolysis. This homogeneity is mirrored by the comparable behaviour of the type-specific polysaccharides in the rate of hydrolysis reported in Fig.1. The size-distribution of the oligosaccharides generated by acid hydrolysis tends toward the exponential form as the random degradation theory (random hydrolysis) for a polysaccharide would predict (Mazur, 1984). Thus, the theory and the data reported suggest that the oligosaccharides generated by mild acid hydrolysis in controlled conditions are the result of random degradation of size-distributed carbohydrate systems (type specific polysaccharides) through cleavage of preferred sites of hydrolysis. Accordingly, the obtained oligosaccharide structures behaved on gel-chromatography with a size-distribution similar in shape to the one characteristic for the homologous polysaccharide from which they were derived (data not shown). Thus, our definition of degree of polymerization (DP) for a given purified oligosaccharide is the expression of its average DP, that is the mean of the size-value within the oligosaccharide size-distribution (mean of a bell-shaped normal distribution).

When the purified type-specific oligosaccharides of different size (DP) have been analyzed by antibody recognition using the reference polyclonal rabbit antiserum, the exponential function shown in Fig.3 was obtained. The immunochemical specificity detected by the ratio MIC Ag/MIC Hp of oligosaccharides compared to their homologous type - specific polysaccharide gave evidence for a logarithmic increase (10^{-6} - 10^{-1}) of specificity beginning from some simple monosaccharide present in the basic repeating unit of the polymers (immunodominant sugar) up to a size for the oligosaccharides of DP=5. In the size-range equivalent to DP=10 and DP=25, the calculated specificity was approaching the one estimated for a type-specific polysaccharide with a DP value several orders of magnitude higher (DP≥1000). From this analysis, two significantly different sizes of type-specific oligosaccharides have been selected for conjugation to the protein carrier: oligosaccharides in the size-range DP=3-6 and DP=10-14, the former having shown a specificity value between 10 and 100 times lower than the homologous polysaccharides, while the latter were mimicking the complete antigenic repertoire of the native polymers.

Physical-chemical and immunochemical characteristics of glycoconjugates type 6A,14,19F and 23F

In Table I, the properties of the two kinds of glycoconjugates are compared. In Fig.4, the qualitative aspect of the four type-specific DP=3-6 glycoconjugates in SDS-PAGE analysis is shown. The DP=10-14 glycoconjugates did not significantly penetrate the 9% (w/w) acrylamide running gel (data not shown). In all cases tested, conjugation of the oligosaccharides resulted in their increased immunochemical specificity, an observation previously reported and explained (Porro et al, 1985;1987).

Immunology of the glycoconjugates

Comparison of the immunological properties found in rabbits for the two kinds of glycoconjugates, short-chain vs long-chain conjugated oligosaccharides, is shown in Fig.5. The IgG titers induced by the short-chain conjugated oligosaccharides were found at higher levels than those estimated for the long-chain conjugated oligosaccharides. For each serotype, in either the priming or booster activity, the IgG titers of the former aluminium adsorbed conjugate model exceeded the titers obtained with the latter-one at a statistically significant level $P<0.005$ (Fig.5A vs Fig.5C). For the same model of conjugate, the booster activity of each serotype was significantly higher than the priming-one at a statistically significant level $P<0.01$ (Fig.5A and Fig.5C). Titers obtained with each of the two models, either against the type-specific capsular polysaccharides or against the carrier protein, were significantly higher in the formulation employing aluminium hydroxide, at a level $P<0.01$ (Fig.5A vs Fig.5B and Fig.5C vs Fig.5D). This result confirms the one previously reported for type 6A conjugate (Porro et al,1985).

TABLE I. Characterization of the Glycoconjugates

TYPE	HAPTENS DP	HAPTENS MW	MW OF GLYCO-CONJUGATES (SDS-PAGE)	RATIO OLIGOS/PROT (moles/mol)	SPECIFICITY of RABBIT Ab for HAPTENS free	conj.
6A	3	$2.1x10^3$	$77.6x10^3$	7	10^{-3}	10^{-2}
14	5	$3.5x10^3$	$85.1x10^3$	6	$5x10^{-2}$	10^{-1}
19F	3	$1.9x10^3$	$69.2x10^3$	4	10^{-3}	10^{-2}
23F	6	$4.5x10^3$	$85.0x10^3$	5	10^{-1}	10^0
6A	10	$7.1x10^3$	$>10^5$	6	10^{-1}	10^0
14	12	$8.3x10^3$	$>10^5$	10	10^{-1}	10^0
19F	10	$6.1x10^3$	$>10^5$	5	10^{-1}	10^0
23F	14	$11.1x10^3$	$>10^5$	10	10^{-1}	10^0

Type 19F and 23F (DP=10-14) oligosaccharide conjugates were immunologically silent in either soluble or adsorbed for mulation (Fig.5C and Fig. 5D). In all experiments reported, higher IgG titers to type-specific polysaccharides were detected when the anti-carrier protein immune response was also higher as expected for a true T-helper dependent antigen. Significantly higher anti-carrier protein titers were observed with the short-chain conjugated oligosaccharide model of the vaccine (Fig.5A vs Fig.5C and Fig.5B vs Fig.5D). It is important to note, that the protein carrier CRM197, although "in vitro" it is antigenically cross-reactive with diphtheria toxin, is not significantly immunogenic in animals unless a specific treatment increasing its stability to proteolytic enzymes is performed (Pappenheimer et al, 1972; Porro et al, 1980). As Table II shows, even injecting into guinea pigs up to four times the amount of CRM197 present in the formulations of the combined glycoconjugate vaccines, no significant or very low anti-diphtheria toxin titers have been elicited in comparison to those seen with CRM197 in conjugated form. These data, combined with those mentioned above for CRM197, strongly suggest that chemical glycosylation of the carrier protein CRM197 results in improved stability to "in vivo" proteolysis.

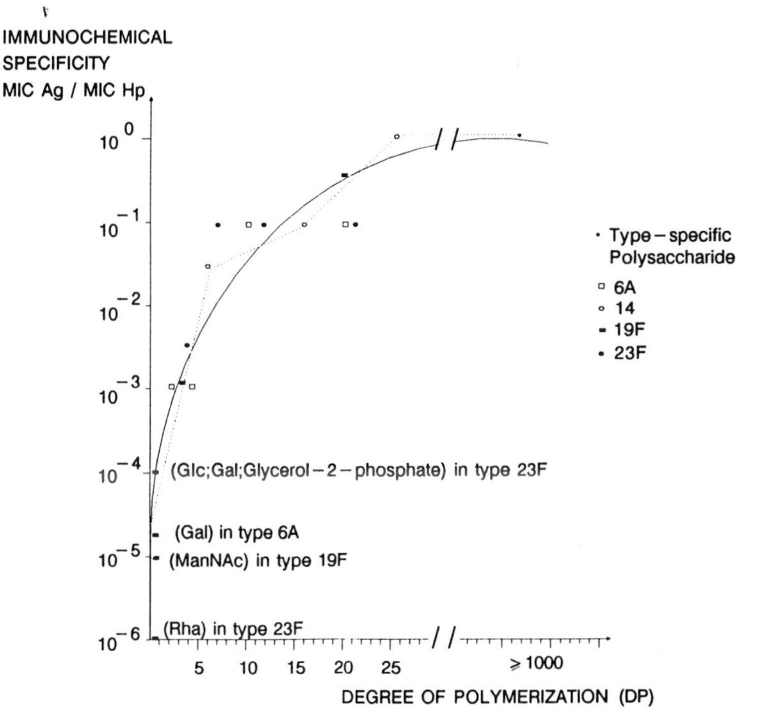

Figure 3. Oligosaccharides generated from controlled hydrolysis of homologous polysaccharides of S.pneumoniae type 6A, 14, 19F and 23F : immunochemical specificity of reference rabbit antisera as estimated by inhibition of immunoprecipitation in differential Immuno-electrophoresis. (For explanation see text, Methods Section)

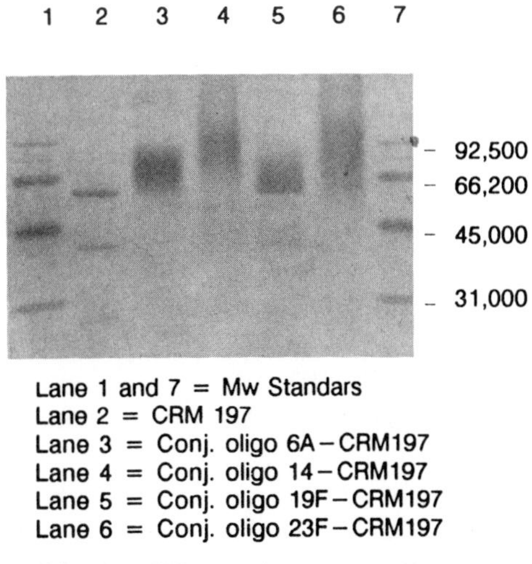

Lane 1 and 7 = Mw Standars
Lane 2 = CRM 197
Lane 3 = Conj. oligo 6A-CRM197
Lane 4 = Conj. oligo 14-CRM197
Lane 5 = Conj. oligo 19F-CRM197
Lane 6 = Conj. oligo 23F-CRM197

Figure 4. Qualitative aspect of the four (DP=3-6) type-specific glycoconjugates in SDS-PAGE. All glycoconjugates were loaded in equal amount of protein (2µg). Gel was silver stained as reported previously (Porro et al, 1985).

Immunological analysis in mice for two out of the four type specific short-chain (DP=3-6) oligosaccharide conjugates (6A and 14) has been performed in order to investigate the optimal dose-response curve in this animal model (Fig.6).

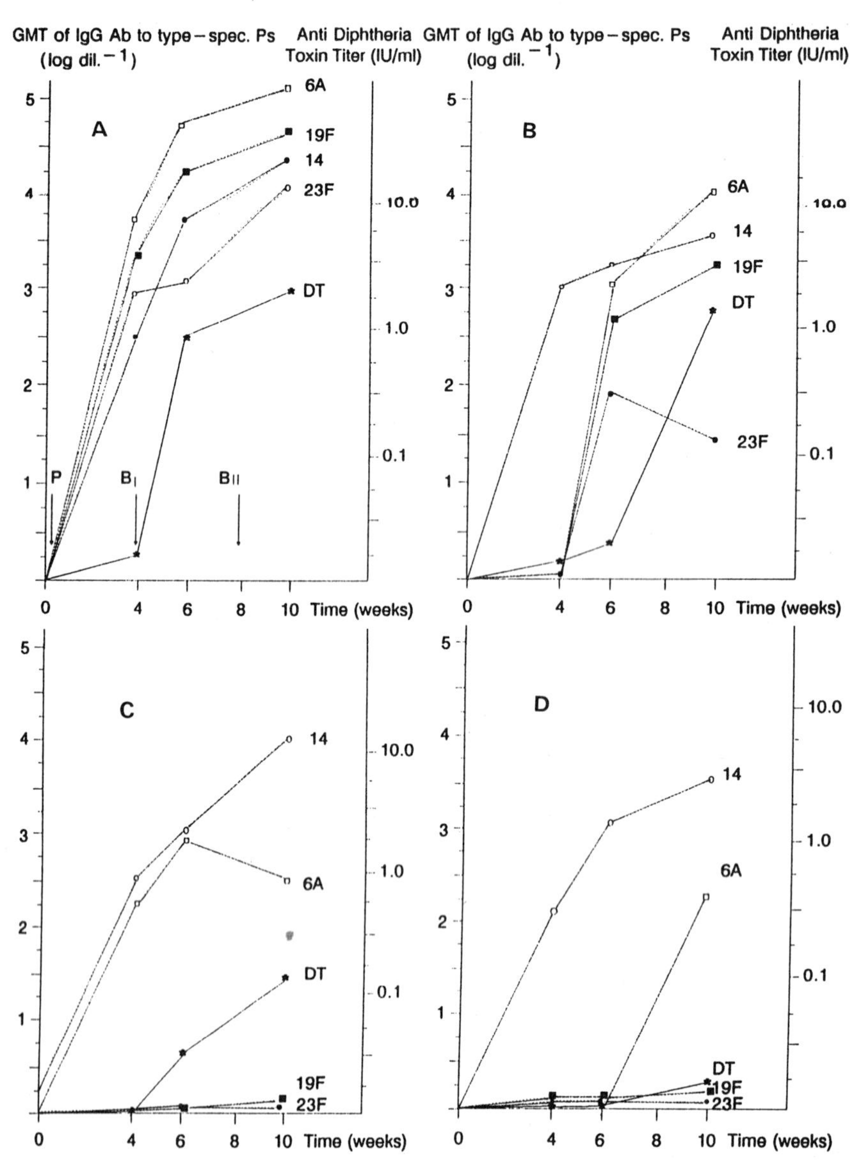

Figure 5. Comparison of the immunological properties in rabbit of the two models of glycoconjugates specific for S. pneumoniae type 6A, 14, 19F and 23F.

A : DP 3 – 6 oligo–CRM197 conjugates $Al(OH)_3$ –adsorbed
B : DP 3 – 6 oligo–CRM197 conjugates soluble
C : DP 10 – 14 oligo–CRM197 conjugates $Al(OH)_3$ –adsorbed
D : DP 10 – 14 oligo–CRM197 conjugates soluble

TABLE II. Neutralizing antibodies to Diphtheria Toxin (DT) in sera pool of guinea pigs immunized with CRM197

DOSE OF ANTIGEN (μg)		NEUTRALIZING Ab TO DT (IU/ml SERUM)
PRIMING	BOOSTER	AFTER BOOSTER DOSE
30	60	< 0.01
60	120	< 0.01
120	240	0.01

NOTE: pre-immunization sera contained titers of less than 0.01 IU/ml. The estimated protective level in man is 0.01 IU/ml (Wilson and Miles, 1975).

In the case of type 14 oligosaccharide conjugate, priming and boosting activity for IgG antibodies has been detected in the range 2.5-30 μg of conjugated oligosaccharide. In the case of type 6A oligosaccharide conjugate, the priming activity was seen only at the highest dose of oligosaccharide (30 μg). In both cases, increasing the dose of conjugate resulted in the increase of the IgG titers, either after the first or the second dose injected.No plateauing of the IgG titers was seen in the dose-range adopted for the two type-specific conjugates.

Estimation of affinity constant value of the rabbit IgG induced by one (type 6A) out of the four (DP=3-6) oligosaccharide conjugates injected, is reported in Fig.7. Inhibition of the ELISA reaction specifically occurred by type 6A purified capsular polysaccharide and by type 6A (DP=10) oligosaccharide but no inhibition occured by either a heterologous type 14 (DP=12) oligosaccharide activated by the molecular spacer adipic acid or conjugated to the carrier protein CRM197.These data confirmed the specificity of the immune response induced and the absence of neodeterminants introduced in the oligosaccharide structures via the chemical activation procedure. Such a comparable result has been previously reported for other conjugates using oligosaccharides activated by the succinimidyl ester of adipic acid (Porro, 1987).

Inhibition by 50% of the ELISA-detected IgG titers to type 6A capsular polysaccharide occurred at the same molar value for either the homologous polymer or the type 6A (DP=10) oligosaccharide: $Ka=4.0x10^9\ M^{-1}$, when the carbohydrate concentration was referred to the basic repeating unit of the structure. The estimated value falls in the range of affinity expected for a highly specific IgG antibody population.

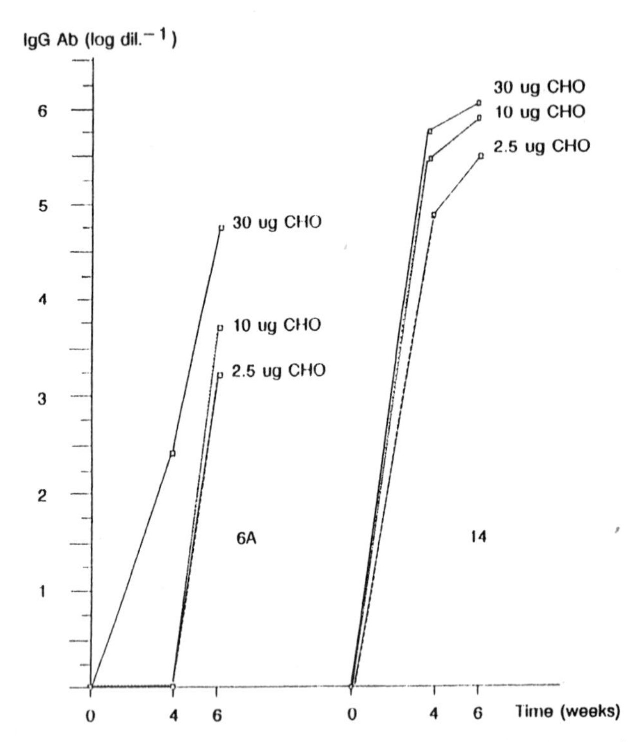

Figure 6. IgG antibody response in sera pool of mice to (DP=3–6) oligosaccharide–CRM197 conjugates for serotype 6A and 14: dose–response analysis.

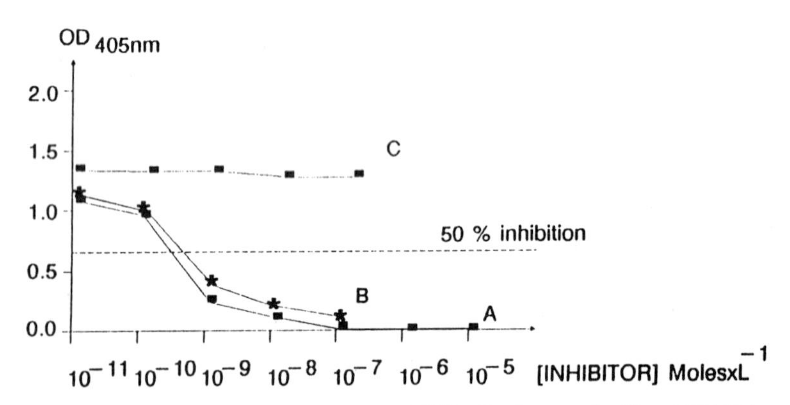

Figure 7. Rabbit IgG immune response to type 6A (DP=3) oligosaccharide–CRM197 conjugate: affinity value of IgG isotype induced to the capsular polysaccharide as estimated by inhibition –ELISA.

A– Type 6A capsular polysaccharide

B– Type 6A oligosaccharide (DP=10) in free form or conjugated to the carrier CRM197

C– Type 14 oligosaccharide (DP=12) activated by the molecular spacer or conjugated to CRM197

DISCUSSION

Although the idea to improve the immunogenicity of saccharide haptens through covalent coupling to a carrier protein was first proposed and realized about half-century ago, only recently glycoconjugate vaccines have been introduced in clinical medicine as new prophylactic agents to overcome the problems associated with the use of polysaccharide vaccines. Neverthless, basic questions about the immunological mechanism by which a carbohydrate structure does acquire the immunological characteristics peculiar of protein antigens, still remain to be answered.

A significant immunological feature of protein antigens is their capability in activating helper and amplifier T-cells showing priming and amplification of the immune response serologically detectable in mammalians as anamnestic induction of IgG isotype antibodies. Carbohydrates do not share such a property neither in animals nor in humans (Kabat, 1961;Paul et al, 1971; Gotschlich, 1975).

From the chemical point of view, a given glycoconjugate can be synthesized using different methods of coupling and carbohydrates of significantly different molecular size, so that an obvious question to be raised is whether any model of glycoconjugate synthesized can exhibit the desired immunological properties. In particular, it would be important to understand the immunological role of the chain length of a saccharide hapten in a given glycoconjugate as related to the acquired T-cell dependency and consequently to the anamnestic induction of specific IgG antibodies.

With this specific desire in mind, we have undertaken the present work and the results obtained with two models of glycoconjugates comparable in terms of chemistry and carrier protein adopted but different in the length of the carbohydrate chain coupled, have given some significant answers at least in animal models. The model synthesized by short-chain (DP=3-6) type specific oligosaccharide haptens has shown a significantly higher activity either in the primary or in the secondary immune response when compared to the model synthesized by long-chain (DP=10-14) oligosaccharides. It is noteworthy to observe that the immunochemical specificity estimated "in vitro" for the latter model of glycoconjugate by type-specific reference antisera, was significantly higher than the one estimated for the former model. Thus, since the "in vivo" results were obtained with each of the four different saccharide structures, it seems evident that the size of the oligosaccharides is responsible for the different immunological activity of the glycoconjugates found in animal immunizations.

The oligosaccharide size seems to directly affect the expression of immunogenicity of the carrier protein, since any enhancement of the anti-carrier immune response was paralleled by the enhancement of the type-specific IgG response to the carried oligosaccharide. In the case of the vaccine model synthesized by short-chain oligosaccharides the anti-carrier immune response was significantly higher than the one detected using the long-chain conjugated oligosaccharides. This modulating activity related to the carbohydrate moiety of the glycoconjugates, may be interpreted by the following hypothesis. Since pneumococcal polysaccharides do not significantly activate helper T-cells and do not amplify the immune system response because the activation of suppressor T-cells, a longer carbohydrate chain in a given glycoconjugate could limit the extent of the immune response induced by the protein carrying the carbohydrate. In other words, several degrees of T-cell dependency might exist for a glycoconjugate antigen, resulting from the balance of the specific immunological features of the two structurally different components. Although other hypothesis may also be considered, like a "covering up" phenomenon occurring by a long-chain oligosaccharide on specific epitopes of the carrier protein responsible for activation of helper and amplifier T-cells, on the basis of the antibody specificity detected for the anti-carrier immune response we believe that the former hypotesis better fits with the experimental results.

About the specificity and affinity of the anti-carbohydrate IgG induced in rabbit by the type 6A (DP=3) conjugate, the comparable (sigmoidal) shape of the binding curves obtained when either the homologous capsular polysaccharide or oligosaccharide were competing for the same specific antigen-antibody complex, is demonstrating the identical chemical nature of the antigenic site present in the two saccharide structures. Thus, the antibody response to a conjugated short-chain oligosaccharide hapten generated from the homologous native polymer, results in IgG antibodies having complete immunochemical specificity for the native antigen. Furthermore, the same affinity constant value of the IgG antibody population found for the oligosaccharide and the homologous polysaccharide, is demonstrating identity between the antigenic sites recognized and not cross-reactivity. Since both specificity and affinity of the induced IgG antibody were identical for the carbohydrate structures, one can assume the IgG antibody population to be homogeneous (Berzofsky and Schechter,1981), that is all the IgG antibodies present in the rabbit sera are directed against the same determinant.

The size of the recognized determinant was not estimated because an unambiguous measure of it would require structurally defined saccharides, like synthetic oligomers of the basic repeating unit for each type-specific carbohydrate. However, since in two out of the four type-specific glycoconjugates the IgG antibodies have been induced by DP=3 oligosaccharides (ten to twelve monosaccharide residues),the minimal structure for such oligosaccharide haptens able to induce a consistent and specific secondary immune response in mammalians after conjugation to a carrier protein would most likely encompass two basic repeating unit (DP=2) or six to eight monosaccharide residues, which is roughly the size of the maximal determinant reported in literature for repeating polypeptide and polysaccharide homopolymers (Schlossman et al, 1965; Kabat,1960;1966; Van Vunakis et al, 1966) and proteins (Atassi,1975).

In conclusion, the data obtained by comparison of "in vitro" versus "in vivo" immunochemical properties of two models of glycoconjugate synthesized by making use of oligosaccharides with different chain length, have indicated that short-chain oligosaccharides are significantly better immunogens than long-chain homologous oligosaccharides, after coupling to a carrier protein.In addition, they have shown the property to induce a primary and secondary IgG immune response which is homogeneous and specifically directed to the same determinants of the native polysaccharides. On the basis of the size of the smallest oligosaccharides used in this work, the data also strongly suggest that the IgG immune response specific for the carbohydrate chain of a glycoconjugate is restricted to the linking area of the glycoconjugate and most likely confined to few monosaccharide residues of the carbohydrate structure.

ACKNOWLEDGEMENTS

The authors wish to thank Dr R.Eby, Dr W.E. Dick and the Serology group of Praxis Biologics Inc., respectively for their helpful analytical support in the analysis of oligosaccharide structures and for running the immunological assays. Also, the authors are grateful to Dr M. Velucchi and Mrs M.P. Porro of Biosynth S.r.l., for computer assistance and secretarial skill.

REFERENCES

Anderson, P.W., Pichichero, M.E., Insel, R.A.,Betts, R., Eby, R., and Smith, D.H., 1986, Vaccines consisting of periodate-cleaved oligosaccharides from the capsule of Haemophilus influenzae type b coupled to a protein: structural and temporal requirements for priming in the human infants, J. Immunol., 137:1181.

Ashwell, G.,1957, Determination of Hexosamines, in "Methods in Enzymology", 3:95, S.P. Colowick and N.O. Kaplan, Eds., Academic Press, New York.

Atassi, M.Z., 1975, Antigen structure of myoglobin, the complete immunochemical anatomy of a protein and conclusion relating to antigenic structures of proteins, Immunochemistry, 12:423.

Austrian,R., 1979, Pneumococcal vaccine: development and prospects, Am.J.Med., 67:547.

Avery,O.T., and Goebel, W.F., 1929, Chemoimmunological studies on conjugated carbohydrate proteins. II. Immunological specificity of synthetic sugar-protein antigens, J.Exp. Med., 50:533.

Baker P.J., 1990, Regulation of magnitude of antibody response to bacterial Polysaccharide antigens by thymus-derived limphocytes, Infect. Immun., 58:3465.

Berzofsky, J.A., and Schechter, A.N.,1981, The concepts of cross reactivity and specificity in immunology, Molec.Immunol.,18:751.

Braley-Mullen, H.,1986a, Requirements for activation of contrasuppressor T-cells by type III pneumococcal polysaccharide, J.Immunol., 136:396.

Braley-Mullen, H., 1986b, Characterization and activity of contrasuppressor T-cells induced by type III pneumococcal polysaccharide, J.Immunol., 137:2761.

Chen P.S., Toribara T.Y. and Warner H., 1956, Microdetermination of phosphorus, Analyt.Chem., 28:1756.

Douglas, R.M., Paton, J.C., Duncan, S.J., and Hausman,D.J.,1983,Antibody response to pneumococcal vaccination in children younger than five years of age, J.Infec.Dis.,148:131.

Douglas, R.M., and Miles, H.B.,1984, Vaccination against *Streptococcus pneumoniae* in childhood: lack of demonstrable benefit in young Australian children, J.Infect.Dis., 149:861.

Goebel, W.F., 1938, Chemo-immunological studies on conjugated carbohydrate proteins. IV. The immunological properties of an artificial antigen containing cellobiuronic acid, J. Exp. Med., 68:469.

Gotschlich, E.C., Goldschneider, I., and Artenstein, M.S.,1969, Human immunity to the meningococcus.IV. Immunogenicity of Group A and Group C meningococcal polysaccharides in human volunteers, J. Exp. Med., 129:1367.

Gotschlich, E.C.,1975, Development of polysaccharide vaccine for prevention of meningococcal disease, Allergy, 9:245.

Habeeb,A.F.S.A., 1966, Determination of free amino groups in proteins by trinitrobenzensulfonic acid, Anal. Biochem. 14:328.

Kabat, E.A., 1960, The upper limit for the size of the human antidextran combining site, J. Immunol., 84:82.

Kabat, E.A., and Mayer, M.M., 1961, Chemical nature of antigens, in " Experimental Immunochemistry" p.7-9, E.A. Kabat and M.M. Mayer, Eds., Thomas, Springfield Illinois, USA.

Kabat, E.A., 1966, The nature of an antigenic determinant J. Immunol., 97:1.

Khater, M., Macai, J., Genyea, C., and Kaplan, J., 1986, Natural killer cell regulation of age-related and type-specific variations in antibody responses to pnemococcal polysaccharides, J. Exp. Med., 164:1505.

Jennings, H.J., Rosell, K.G., and Carlo, D.J., 1980, Structural determination of the capsular polysaccharide of *Streptococcus pneumoniae* type-19(19F), Can.J. Chem., 58:1069.

Lepow, M.L., Barkin, R.M., Berkowitz, C.D., et al., 1987, Safety and immunogenicity of Haemophilus influenzae type b polysaccharide-diphtheria toxoid conjugate vaccine in adults, J. Infect. Dis., 150:402.

Lindberg, B., Lönngren, J., and Powell, D.A., 1977, Structural studies on the specific type 14 pneumococcal polysaccharide., Carbohydr.Res., 58:177.

Makela, P.H., Leinonen, M., Pukander, J., and Karma, P., 1981, A study of the pneumococcal vaccine in prevention of clinical acute attacks of recurrent otitis media, Rev.Infect. Dis., 3:S124.

Mazur, A.K., 1984,Most-probable distribution at enzyme depolymerization of polysaccharides, Biopolymers, 23:859.

Ohno, N., Yadomae, T., and Miyazaki, T., 1980, The structure of the type-specific polysaccharide of Pneumococcus type XIX. Carbohydr. Res., 80:297.

Pappenheimer, A.M.,Jr, Uchida, T., and Harper, A.A.,1972 An immunological study of the diphtheria toxin molecule,Immunochemistry, 9:891.

Paul, W.E., Katz, D.H., and Benacerraff, B., 1971, Augmented anti-SIII antibody responses to an SIII-protein conjugate, J. Immunol., 107:685.

Porro, M., Saletti, M., Nencioni, L., Tagliaferri, L., and Marsili, I., 1980, Immunogenic correlation between cross-reacting material (CRM197) produced by a mutant of C. diphtheriae and diphtheria toxoid, J.Infect.Dis, 143:716.

Porro, M., Viti, S., Antoni, G.,and Neri, P.,1981, Modifications of the Park-Johnson ferricyanide submicromethod for the assay of reducing groups in carbohydrates, Analyt. Biochem., 118:301.

Porro, M., Fabiani, S., Marsili, I., Viti, S.,and Saletti, M., 1983, Immunoelectrophoretic characterization of the molecular weight polydispersion of polysaccharides in multivalent bacterial capsular polysaccharide vaccines, J. Biol.Stand., 11:65.

Porro, M., Costantino, P., Viti, S., Vannozzi, F., Naggi, A., and Torri, G., 1985, Specific antibodies to diphtheria toxin and type 6A pneumococcal capsular polysaccharide induced by a model of semi-synthetic glycoconjugate antigen, Molec. Immunol., 22:907.

Porro, M., Costantino, P., Giovannoni, F., Pellegrini, V., Tagliaferri, L., Vannozzi, F., and Viti S., 1986, A molecular model of artificial glycoprotein with predetermined multiple immunodeterminants for Gram-positive and Gram-negative encapsulated bacteria, Molec. Immunol., 23:385.

Porro, M.,1987, Artificial glycoproteins of predetermined multivalent antigenicity as a new generation of candidate vaccines to prevent infections from encapsulated bacteria: analysis of antigenicity versus immunogenicity, in "Towards better carbohydrate vaccines" pp 279-306, R.Bell and G.Torrigiani, Eds., John Wiley & Sons, New York, USA.

Rebers, P.A., and Heidelberger, M., 1961, The specific polysaccharide of type VI pneumococcus. II. The repeating unit, J. Am. Chem., 83:3056.

Richards, J.C., and Perry, M.B., 1988, Structure of the specific capsular polysaccharide of Streptococcus pneumoniae type 23F, Can. J. Biochem., Cell Biol., 66:758.

Rijkers, G.T., and Mosier, D.E., 1985, Pneumococcal polysaccharides induce antibody formation by human B lymphocytes in vitro, J.Immunol., 135:1.

Schlossman, S.F., Yaron, A., Ben-Efraim, S & Sober, H. A., 1965, Immunogenicity of a series of α,N-DNP-L-lysines, Biochemistry, 4:1638.

Teele, D.W., Klein, J.O., and the Greater Boston Collaborative Otitis Media Study Group, 1981, Use of Pneumococcal vaccine for prevention of recurrent acute otitis media in infants in Boston, Rev. Infect. Dis., 3:S113.

United States National Institute of Allergy and Infectious Diseases, 1986, "New Vaccine Development, Establishing Priorities", vol. II pg.15, National Academy Press, Washington D.C., USA.

Van Vunakis, H., Kaplan, H., and Levine, L., 1966, Immunogenicity of polylysine and polyornithine when complexed to phosphorylated bovine serum albumin, Immunochemistry, 3:393.

Wilson, G.S., and Miles, A., 1975, Diphtheria and other diseases due to corynebacteriae, in "Topley and Wilson's Principles of Bacteriology, Virology and Immunity", p.1800-1842, E. Arnold, Ed., Arnold, London.

Zucker, D., and Murphy, J.R., 1984, Monoclonal antibody analysis of diphtheria toxin, I. Localization of epitopes and neutralization of cytotoxicity, Molec.Immunol., 21,785.

IMMUNOPROPHYLAXIS OF OTITIS MEDIA

G. Scott Giebink

Department of Pediatrics
University of Minnesota School of Medicine
Minneapolis, Minnesota

INTRODUCTION

In considering the development of a vaccine to prevent otitis media, it is important to explore the rationale for such a vaccine, the epidemiology and spectrum of the disease, as well as its impact on child health, the etiology and pathogenesis of otitis media, especially as related to middle ear microbial pathogens, and the present status of otitis media vaccine research.

Otitis media occurs early in life, and it affects almost every child at least once. Despite the widespread availability of potent antibiotics active against the bacterial pathogens that cause otitis media, many children suffer recurrent episodes, and some develop lifelong complications and sequelae. At least three of every four children have at least one episode by their third birthday, and one child has three or more episodes.[1] One in 10 children with otitis media suffers prolonged middle ear inflammation with persistent middle ear fluid and a mild to moderate hearing loss.[2,3] Children who have prolonged middle ear fluid during the first three years of life score lower on tests of speech, language and cognitive abilities than their peers who have less time with middle ear fluid.[4]

Otitis media affects the child, his or her family, and the entire health care system. In 1986-87 there were 31 million physician visits for otitis media; 16.8 million of these were for children age birth to 6 years.[5] At $91 per patient treatment, the annual cost of otitis media has been estimated to exceed $1 billion. Surgical intervention costs of $0.6 to 1.2 billion bring the total to at least $2.5 billion, or about 3% of the United States health care budget.[6] The indirect cost of parent lost work time has not been estimated.

An otitis media vaccine, therefore, has the potential of dramatically reducing childhood morbidity and health care costs. Moreover, since sinusitis and pneumonia are caused by the same bacteria that infect the middle ear, an otitis media vaccine would probably also be effective prevention of these diseases.

Immunobiology of Proteins and Peptides VI
Edited by M.Z. Atassi, Plenum Press, New York, 1991

The spectrum of otitis media is reflected by the clinical course of the disease and by the histopathology of middle ear mucoperiosteum. The clinical spectrum extends from asymptomatic to symptomatic disease and from self-limited to complicated disease with sequelae. About 50 percent of otitis media episodes occurring during the first year of life appear to be clinically asymptomatic, yet these asymptomatic episodes seem to be as important as symptomatic episodes in predicting the likelihood of future middle ear disease.[7] Middle ear inflammation affects the mucoperiosteal epithelium, subepithelial space and periosteum, and the characteristics of this inflammation vary with the stage of the disease.[8,9]

OTITIS MEDIA ETIOLOGY and PATHOGENESIS

Otitis media is a multifactorial disease with elements of susceptibility that seem to be genetically, developmentally, and environmentally determined. Genetic factors include eustachian tube anatomy and physiology, and immune response. Abnormal eustachian tube function is one of the most important factors in otitis media pathogenesis.[10] Environmental factors include exposure to specific bacterial and viral pathogens and pollutants. The intensity of these exposures, as might be determined by group size and structure in child day care, is also important. Developmental factors include immune response, and anatomic and physiologic characteristics that mature with age. The shorter eustachian tube in infants compared with adults may also be an important developmental factor. The effectiveness of an otitis media vaccine is significantly challenged by the multiple factors that predispose to the disease.

BACTERIAL PATHOGENS IN OTITIS MEDIA

Bacteria are cultured from about 70% of middle ear fluid samples obtained from children with acute, symptomatic otitis media; Streptococcus pneumoniae and Haemophilus influenzae are the predominant bacterial species cultured.[11] Other bacteria isolated less often include Moraxella (Branhamella) catarrhalis, Streptococcus pyogenes, and Staphylococcus aureus. There has been surprisingly little change in middle ear bacteriology during the last 16 years with the exception of three recent reports of increased prevalence of M. catarrhalis. However, recent studies have revealed that bacterial culture underestimates the contribution of bacteria to otitis media pathogenesis since pneumococcal capsular polysaccharide antigens and endotoxin are frequently detected in sterile acute and chronic middle ear effusions.[12-14] Recent evidence suggests that the subcellular components of pneumococcus and H. influenzae play an important role in maintaining chronic middle ear inflammation.

Fortunately, only a few of the more than 80 pneumococcal serotypes account for the majority of middle ear infections with this organism. The declining role of types 1, 3 and 5 in otitis media and the emergence of groups 6, 14, 19 and 23 has been documented in Scandinavia and North America.[15] This is noteworthy since antibody responsiveness to groups 6, 14, 19 and 23, which account for nearly 60%

of pneumococcal otitis media episodes, appears to develop later in childhood.[16] Moreover, an analysis of 704 first attacks of pneumococcal otitis media showed 61% to have occurred during the first year of life, 18% in the second and 7% in the third year.[17] Because there has been a shift during the last four decades in pneumococcal types causing invasive disease, ongoing surveillance of pneumococcal types recovered from the middle ear will be an important aspect of an otitis media vaccine program.

H. influenzae isolated from infected middle ears are almost always nontypable and are infrequently (10-15%) type b. Otitis media caused by types a, e and f have been reported but are very rare. The prevalence of H. influenzae as an otitis media pathogen is similar in younger and older patients.[18] H. influenzae strains have been characterized by three methodologies: biotyping based on biochemical differences among strains,[19] electrophoresis based on metabolic enzymes,[20] outer membrane protein analysis by patterns of protein migration on sodium dodecyl sulfate-polyacrylamide gel electrophoresis,[21] and lipooligosaccharide analysis by Western blot assay.[22] Results of these assays have revealed considerable diversity among the nontypable H. influenzae (NTHi) strains causing otitis media.

The bacteriology of recurrent otitis media is also important in developing vaccine strategies. Pneumococcal types in recurrent disease were rarely the same serotype in one large study,[17] and recurrent otitis media due to NTHi was usually associated with new strains in another study.[23] These observations indicate the importance of developing a polyvalent otitis media vaccine or a vaccine containing one or more protective epitopes common to multiple strains.

VIRAL PATHOGENS IN OTITIS MEDIA

Impaired eustachian tube opening and resulting negative middle ear pressure is a consequence of upper respiratory viral infection and often leads to otitis media.[24] Increased nasopharyngeal carriage of pneumococci has been documented after the onset of common colds,[25] and it has been estimated that three common colds occur for each acquisition of a pneumococcus.[26]

Only certain viruses, however, are strongly associated with symptomatic otitis media in children. Henderson, et al.[27] demonstrated that one-fourth to one-third of respiratory syncytial virus, adenovirus and influenza virus infections were associated with acute otitis media, whereas otitis media was no more common after rhinovirus infection than in children without a viral infection. The failure of many investigators to detect viral pathogens in middle ear fluid from children with acute otitis media is probably due to limitations of the isolation techniques used since in one recent study, antigen detection methods in addition to viral isolation yielded RSV, adenovirus, and parainfluenza virus in 34%, 7%, and 2%, respectively, of 137 middle ear fluid samples.[28] Bacterial pathogens were cultured in 45% of the cases with virus present, and virus was present in 51% of 63 bacterial culture negative effusions.

We have investigated the pathophysiology of pneumococcal otitis media associated with influenza A virus infection. In the chinchilla animal model, combined upper respiratory inoculation with wild-type influenza A virus and pneumococcus led to a high incidence of pneumococcal otitis media, whereas either pathogen alone inoculated nasally infrequently resulted in otitis media.[29] Influenza A virus strains, however, differed in chinchillas as in humans in their capacity to enhance the development of pneumococcal otitis media.[30] These experiments suggested that immunoprophylaxis of bacterial otitis media may be addressed by identifying respiratory virus species with otitis media pathogenicity.

We have begun to explore the effectiveness of cold-tolerant, attenuated and inactivated influenza A virus vaccines in this model. We observed that both vaccines elicited a significant serum antibody response in about 60% of animals. However, the cold-tolerant, attenuated vaccine was more effective than the inactivated vaccine in blocking colonization with the wild-type influenza virus and pneumococcus, and in reducing the attack rate of otitis media.[31]

POTENTIAL VACCINES FOR NONTYPABLE HAEMOPHILUS INFLUENZAE

The surface of NTHi is composed of outer membrane proteins (OMPs), lipooligosaccharide, and fimbriae or pili. Several of these surface components have been identified as potential vaccine antigens. Their success as vaccine antigens will be based on their conservation among nontypable and type b H. influenzae strains, their accessibility to the host's immune system, their immunogenicity in infants, and their requirement as a virulence factors for the organism. The observation that chinchilla antiserum raised to formaldehyde-fixed NTHi conferred passive protection against otitis media in other chinchillas suggests that one or more surface antigens hold promise as an otitis media vaccine.[32]

Several candidate OMP antigens have been identified. Protein-2 (P2) is a 36-41 kd molecule that comprises about 50% of outer membrane protein content, is the target of human bactericidal antibody, is protective in the infant rat Hib model, but shows antigen heterogeneity among strains.[33] P6 has a molecular weight of 16.6 kd, comprises 1 to 5% of the outer membrane protein, elicits bactericidal antibody, and is highly conserved among strains.[33] Two other OMPs, protein a (P1) and protein e (P4), have elicited homologous and heterologous Hib protection in the infant rat model, but results were confounded by several interesting observations, which may apply to other OMPs.[34] These investigators noted that (1) seemingly pure OMP fractions induced antibodies in rabbits that were not monospecific, (2) lipooligosaccharide absorption of the rabbit antiserum was not complete, (3) antibodies to lipooligosaccharide seemed to block the protective effect of OMP antibodies, and (4) bovine serum albumin in the Hib inoculum had some protective effect in the absence of antiserum.[34]

The field success of pilus vaccines for E. coli neonatal diarrhea in pigs[35] and for M. bovis infectious keratoconjunctivitis in cattle[36] has led to identification of four H. influenzae pili types.[37] One of these, the LKP (long, thick, hemagglutination positive) pilus, shows a different type distribution for Hib and NTHi strains, although both Hib and NTHi can express the same types. Brinton, et al.[37] determined that a 12-valent LKP vaccine would cover 69% of the NTHi otitis media strains, and a 20-valent vaccine would cover 80 to 90% of strains. These investigators found that many middle ear NTHi isolates did not express LKP pili but contained LKP pilus genes. Since NTHi isolated from the nasopharynx more frequently expressed LKP pili, they reasoned that an LKP vaccine might still prevent otitis media by preventing nasal colonization with the pilus-expressing strain. Strain-specific protection with an LKP pilus vaccine has been demonstrated in the chinchilla model, although results were somewhat confounded by a small amount of lipooligosaccharide in the vaccine that induced some agglutinating antibody.[38]

Lipooligosaccharide probably does not induce a protective response since infants with Hib meningitis have been found to have antibodies against lipooligosaccharide at the time of infection.[33] Moreover, this molecule shows enormous heterogeneity among strains.

PNEUMOCOCCAL VACCINE STUDIES IN OTITIS MEDIA

There is limited information available on the antibody responses of infants and children to natural exposure with pneumococcus. Gray, et al.[39] studied the epidemiology of pneumococcal disease and antibody responses in a cohort of children followed from birth to 24 months. Between 12 and 24 months there were no overall differences in serum antibody concentrations between pneumococcal carriers and non-carriers, but antibody concentrations varied considerably among individuals and over time. The serum antibody response to infection was quite variable among infants, and naturally occurring serum antibody did not appear to be protective. However, antibody may have contributed to recovery since in another study higher antibody concentrations in middle ear fluid were associated with more rapid clearing of infection.[40] Similar observations were made in the pneumococcal otitis media animal model.[41]

Our research has focused on immunoprophylaxis of pneumococcal otitis media. In the chinchilla model, we demonstrated that systemic administration of a human bacterial polysaccharide immune globulin, which contained levels of anticapsular immunoglobulin G similar to those found in human adults, protected chinchillas from pneumococcal otitis media.[42] Similarly, serum IgG antibody against Hib enhanced bacterial clearance in the murine lower respiratory tract.[43] These studies indicate that specific immunoglobulin G antibody may provide an important antibacterial defense of respiratory mucosal surfaces.

Active immunoprophylaxis using systemically administered pneumococcal capsular polysaccharides (PCPs) has also been studied in chinchillas. Purified PCP was shown to be immunogenic in chinchillas, and subcutaneous vaccination elicited at least a ten-fold increase in serum antibody in approximately two-thirds of vaccinated chinchillas.[44] Chinchillas with a serum antibody increase after vaccination showed an 87% reduction in vaccine-type pneumococcal otitis media, compared to a 28% reduction in vaccinated nonresponding animals.[45]

Pneumococcal vaccine trials in Boston, Alabama and Finland completed a decade ago showed that vaccination with polyvalent PCPs gave serotype-specific protection in children who produced sufficient anti-capsular serum antibody after vaccination.[46] As a result, enhancing the immunogenicity of these capsular polysaccharides has been a goal of the otitis media research effort.

Antibody responses and protection have been greatly accelerated in human infants by coupling the H. influenzae type b capsular polysaccharide antigen, both polymeric polysaccharide and derived oligosaccharides, to carrier proteins -- the so called "conjugate" vaccines. The enhancement has been attributed to eliciting T lymphocyte helper effects by the proteins[47] Similar approaches have been tried and are in progress for the pneumococcal types causing systemic infections and otitis media. The optimal composition and structural properties of these conjugates for raising antibodies in infants remain to be determined.

Schneerson, et al.[48] found that a tetanus toxoid conjugate of type 6A PCP raised both 6A and tetanus antibodies in adults. Anderson and Betts[49] studied in adult volunteers the immunogenicity of types 6A, 14, 18C, 19F and 23F saccharides of varying length and composition coupled by reductive amination to diphtheria toxoid. Immunogenicity of the type 6A conjugates was associated with higher overall saccharide content rather than chain length. Parallel changes in total anti-capsular and IgG-specific antibodies were noted. Eleven of 19 subjects who demonstrated significant antibody responses to one of the several conjugate preparations also had a significant rise in type-specific opsonic activity, and eight subjects did not.

We are currently evaluating in the chinchilla model the immunogenicity and otitis media protective efficacy of type 6B and 23F capsular polysaccharides coupled to a Neisseria meningitidis group B outer membrane protein complex (OMPC) and adsorbed with aluminum hydroxide. Mean increase in serum antibody, measured by radioimmunoassay, was 7.8 times the prevaccination level (range, 2-28) in 17 animals given a 2 μg/kg 6B PCP-OMPC intramuscular dose (geometric mean: 0.16 μg/ml prevaccination and 0.93 after 4 weeks). The response was not significantly different after 0.5 and 8.0 μg doses, and a similar four week titer but somewhat slower response was seen after subcutaneous vaccination. The mean serum antibody increase in 11 animals given 2 μg/kg 23F PCP-OMPC intramuscularly was 9.4 (geometric mean: 0.11 prevaccination and 0.78 after four weeks). Antibody responses to monovalent vaccines and bivalent 6B/23F-OMPC vaccine were similar. Antibody persisted at least eight weeks after vaccination.

Vaccinated and unvaccinated control animals were challenged by direct middle ear inoculation of type 6B pneumococci. Pneumococcal otitis media developed in all 11 control animals and all six type 23F PCP-OMPC vaccinated animals, but in only eight (27%) of 30 type 6B-OMPC vaccinated animals ($p<0.0001$). However, otitis media with sterile middle ear fluid developed in 16 (53%) of these 30 animals suggesting brief replication of pneumococci in the middle ear and attenuated middle ear inflammation before antibody-mediated killing. The titer of serum antibody before challenge was strongly associated with protection: pneumococcal otitis media developed in 82% of animals with antibody concentrations less than 1.0 μg/ml, but in only 11% with greater antibody concentrations ($p<0.0001$).

Therefore, type 6B PCP-OMPC conjugate vaccine is both immunogenic and protective for vaccine-type otitis media in the chinchilla model. Given the close correlation of this model with human otitis media and the parallel results of previous PCP vaccine trials in chinchillas and human infants, clinical trials should be initiated to determine the efficacy of this conjugate vaccine in preventing pneumococcal disease in infants and children.

SUMMARY

Prospects for an effective otitis media vaccine are bolstered by a number of encouraging observations. Results of pneumococcal polysaccharide vaccine trials beginning in 1975, the enormously enhanced immunogenicity of protein-Hib polysaccharide coupled vaccines in infants, and the apparent effectiveness of a protein-PCP coupled vaccine in experimental otitis media suggest that a pneumococcal vaccine targeted to prevent invasive and middle ear infections is not too distant. The identification of several conserved surface antigens on NTHi and demonstration of otitis media protection elicited by these antigens in an animal model give promise for the development of H. influenzae vaccines for otitis media. Evidence that attenuated influenza A virus vaccination may also be an effective strategy for otitis media prevention, at least in an animal model, suggests that priority should be given to testing the efficacy of influenza, parainfluenza and respiratory syncytial virus vaccines with respect to otitis media prevention.

It seems quite likely that not one but several immunoprophylaxis approaches will be necessary to reduce the overall incidence of otitis media given the multifactorial nature of the disease. Increasing parent and physician concern with the high incidence of otitis media and its morbidity suggests high participation rates in vaccine trials and high utilization of vaccines shown to be protective. Even if a vaccine could reduce the incidence of otitis media by 30%, an annual health care savings of $300-750 million would be achieved.

REFERENCES

1. Teele DW, Klein JO, Rosner BA, et al. Epidemiology of otitis media during the first seven years of life in children in Greater Boston: a prospective, cohort study. J Infect Dis 160:83 (1989).
2. Kempthorne J, Giebink GS. Pediatric approach to otitis media diagnosis and management. In Goycoolea MV (ed), Otitis Media: The Pathogenesis Approach. Otolaryngol Clin NA, in press (1991).
3. Fria TJ, Cantekin EI, Eichler, JA. Hearing acuity of children with otitis media with effusion. Arch Otolaryngol 111:10 (1985).
4. Teele DW, Klein JO, Chase C, et al. Otitis media in infancy and intellectual ability, school achievement, speech, and language at age 7 years. J Infect Dis 162:685 (1990).
5. Bluestone CD, Klein JO, Paradise JL, et al. Workshop on effects of otitis media on the child. Pediatrics 71:639 (1983).
6. Stool SE, Field MJ. The impact of otitis media. Pediatr Infect Dis J 8 (Suppl):S11 (1989).
7. Marchant CD, Shurin PA, Turczyk VA, et al. Course and outcome of otitis media in early infancy: A prospective study. J Pediatr 104:826 (1984).
8. Yoon TH, Paparella MM, Schachern PA, Lindgren BR. Morphometric study of the continuum of otitis media. Ann Otol Rhinol Laryngol 99 (Suppl 148):23 (1990).

9. Yoon TH, Paparella MM, Schachern PA, Lindgren BR. Histopathological study of human middle ear mucosa in otitis media. Topographical considerations of the middle ear cleft. Ann Otol Rhinol Laryngol, in press (1991).
10. Bluestone CD. Pathogenesis of otitis media related to vaccine efficacy. Pediatr Infect Dis J 8 (Suppl):S14 (1989).
11. Giebink GS. The microbiology of otitis media. Pediatr Infect Dis J 8 (Suppl):S18 (1989).
12. Luotonen J, Herva E, Karma P, et al. The bacteriology of acute otitis media in children with special reference to Streptococcus pneumoniae as studied by bacteriological and antigen detection methods. Scand J Infect Dis 113:177 (1981).
13. Leinonen MK. Detection of pneumococcal capsular polysaccharide antigens by latex agglutination, counterimmunoelectrophoresis, and radioimmunoassay in middle ear exudates in acute otitis media. J Clin Micro 11:135 (1980).
14. DeMaria TF. Endotoxin and otitis media. Ann Otol Rhinol Laryngol 97:31 (1988).
15. Austrian R. Epidemiology of pneumococcal capsular types causing pediatric infections. Pediatr Infect Dis J 8 (Suppl):S21 (1989).
16. Paton JC, Toogood IR, Cockington RA, Hansman D. Antibody response to pneumococcal vaccine in children aged 5 to 15 years. Am J Dis Child 140:135 (1986).
17. Austrian R, Howie VM, Ploussard JH. The bacteriology of pneumococcal otitis media. Johns Hopkins Med J 141:104 (1977).
18. Wald ER, Haemophilus influenzae as a cause of acute otitis media. Pediatr Infect Dis J 8 (Suppl):S28 (1989).
19. Kilian M. A taxonomic study of the genus Haemophilus. J Gen Microbiol 93:9 (1976).
20. Musser JM, Granoff DM, Pattison PE, Selander RK. A population genetic framework for the study of invasive diseases caused by serotype b strains of Haemophilus influenzae. Proc Natl Acad Sci USA 82:5078 (1985).
21. Murphy TF, Dudas KC, Mylotte JM, Apicella MA. A subtyping system for nontypable Haemophilus influenzae based on outer membrane proteins. J Infect Dis 147:838 (1983).
22. Campagnari AA, Gupta MR, Dudas KC, et al. Antigenic diversity of the lipooligosaccharides of nontypable Haemophilus influenzae. Infect Immun 55:882 (1987).
23. Barenkamp SJ, Shurin PA, Marchant CD, et al. Do children with recurrent Hemophilus influenzae otitis media become infected with a new organism or reacquire the original strain? J Pediatr 105:533 (1984).
24. Sanyal MA, Henderson FW, Stempel EC, et al. Effect of upper respiratory tract infections on eustachian tube ventilatory function in the preschool child. J Pediatr 97:11 (1980).
25. Straker E, Hill AB, Lovell R. A study of the nasopharyngeal flora of different groups of persons observed in London and South-east England during the years 1930 to 1937. London, England: His Majesty's Stationery Office, p.7-51 (1939).
26. Gwaltney JM, Sande MA, Austrian R, Hendley OJ. Spread of Streptococcus pneumoniae in families. II. Relation of transfer of S. pneumonia to incidence of colds and serum antibody. J Infect Dis 132:62 (1975).

27. Henderson FW, Collier AM, Sanyal MA, et al. A longitudinal study of respiratory viruses and bacteria in the etiology of acute otitis media with effusion. N Engl J Med 306:1377 (1982).
28. Sarkkinen H, Ruuskanen I, Meurman O, et al. Identification of respiratory virus antigens in middle ear fluids of children with acute otitis media. J Infect Dis 151:444 (1985).
29. Giebink GS, Berzins IK, Marker, SC, et al. Experimental otitis media after nasal inoculation of Streptococcus pneumoniae and influenza A virus in chinchillas. Infect Immun 30:445 (1980).
30. Giebink GS, Wright PF. Different virulence of influenza A virus strains and susceptibility to pneumococcal otitis media in chinchillas. Infect Immun 41:913 (1983).
31. Giebink GS. Studies of Streptococcus pneumoniae and influenza virus vaccines in the chinchilla otitis media model. Pediatr Infect Dis J 8 (Suppl):S42 (1989).
32. Barenkamp SJ. Protection by serum antibodies in experimental nontypable Haemophilus influenzae otitis media. Infect Immun 52:572 (1986).
33. Murphy TF, Campagnari AA, Nelson B, Apicella MA. Somatic antigens of Haemophilus influenzae as vaccine components. Pediatr Infect Dis J 8 (Suppl):S66 (1989).
34. Loeb MR, Phillips E. Evaluating vaccine candidates for the prevention of otitis media. Pediatr Infect Dis J 8 (Suppl):S48 (1989).
35. Brinton CC Jr, Fusco P, Wood S, et al. A complete vaccine for neonatal swine colibacillosis and the prevalence of Escherichia coli pili on swine isolates. Vet Med Sm Animal Clin J 78:962 (1983).
36. Vilella D. Moraxella bovis somatic pili: purification, characterization, role in infectious bovine keratoconjunctivitis and use in the prophylaxis of the disease. M.S. Thesis, University of Pittsburgh, Pittsburgh, PA, 1981.
37. Brinton CC, Carter MJ, Derber DB, et al. Design and development of pilus vaccines for Haemophilus influenzae diseases. Pediatr Infect Dis J 8 (Suppl):S54 (1989).
38. Karasic RB, Beste DJ, To SC.-M, et al. Evaluation of pilus vaccines for prevention of experimental otitis media caused by nontypable Haemophilus influenzae. Pediatr Infect Dis J 8 (Suppl):S62 (1989).
39. Gray BM, Dillon HC Jr. Epidemiological studies of Streptococcus pneumoniae in infants: antibody in types 3, 6, 14, and 23 in the first two years of life. J Infect Dis 158:948 (1988).
40. Sloyer JL, Howie VM, Ploussard JH, et al. Immune response to acute otitis media: association between middle ear fluid antibody and the clearing of clinical infection. J Clin Microbiol 4:306 (1976).
41. Giebink GS. The pathogenesis of pneumococcal otitis media in chinchillas and the efficacy of vaccination in prophylaxis. Rev Infect Dis 3(Suppl 2):42 (1981).
42. Shurin PA, Giebink GS, Wegman DL, et al. Prevention of pneumococcal otitis media in chinchillas with human bacterial polysaccharide immune globulin. J Clin Micro 26:755 (1988).
43. Toews GB, Hart DA, Hansen EJ. Effect of systemic immunization in pulmonary clearance of Haemophilus influenzae type b. Infect Immun 48:343 (1985).
44. Giebink GS, Schiffman G. Humoral immune response in chinchillas to the

capsular polysaccharides of Streptococcus pneumoniae. Infect Immun 39:638 (1983).

45. Giebink GS, Berzins IK, Schiffman G, et al. Experimental otitis media in chinchillas following nasal colonization with type 7F Streptococcus pneumoniae: Prevention after vaccination with pneumococcal capsular polysaccharide. J Infect Dis 140:716 (1979).
46. Giebink GS. Discussion. Efficacy studies of pneumococcal vaccine in infants and children. Rev Infect Dis 3:S131 (1981).
47. Robbins JR, Schneerson R, Pittman M. Haemophilus influenzae type b infections. In: Germanier R, ed. Bacterial vaccines. New York: Academic Press, p. 289 (1984).
48. Schneerson R, Robbins JB, Parke JC Jr, et al. Quantitative and qualitative analyses of serum antibodies elicited in adults by Haemophilus influenzae type b and pneumococcus type 6A capsular polysaccharide-tetanus toxoid conjugates. Infect Immun 52:519 (1986).
49. Anderson P, Betts R. Human adult immunogenicity of protein-coupled pneumococcal capsular antigens of serotypes prevalent in otitis media. Pediatr Infect Dis J 8 (Suppl):S50 (1989).

PROTECTION AGAINST STREPTOCOCCAL PHARYNGEAL COLONIZATION WITH VACCINES COMPOSED OF M PROTEIN CONSERVED REGIONS

V.A. Fischetti[1], D.E. Bessen[1], O. Schneewind[1] and D.E. Hruby[2]

The Rockefeller University, 1230 York Avenue, New York, NY 10021[1]
and Center for Gene Research and Biotechnology, Department of
Microbiology, Oregon State University, Corvallis, OR 97331[2]

INTRODUCTION

Group A streptococci are widespread human pathogens which are responsible for nasopharyngeal infections and impetigo. There are over 25 million cases of group A streptococcal infections each year in the United States alone, at a cost of over $1 billion to the public. Despite the fact that streptococcal infection can be successfully treated with antibiotics, in the interim it often causes significant discomfort and loss in productivity. Furthermore, about 3% of infected individuals afflicted with streptococcal infection develop a more serious illness such as rheumatic fever and rheumatic heart disease. Because rheumatic heart disease occurs in as many as 1% of school-age children in developing countries (Agarwal, 1981; Dodu and Bothig, 1989), there is a strong impetus to develop a safe and effective vaccine against group A streptococcal pharyngitis.

The mechanisms by which protective immunity to group A streptococcal infection is achieved are detailed in recent reviews (Fischetti, 1989b; Bessen and Fischetti, 1988b). A major virulence factor present on the surface of virtually all clinical isolates is M protein, a fibrillar structure which is antigenically variable and provides the basis for the serological typing scheme (Fischetti, 1989b). Serum IgG directed to the antigenically variable, NH_2-terminal portion of M protein leads to complement fixation and opsonophagocytosis of the homologous serotype by polymorphonuclear leukocytes, however, this protective mechanism is limited to type-specific IgG (Jones and Fischetti, 1988; Lancefield, 1962).

M PROTEIN STRUCTURE

Analysis of the complete sequence of the M6 molecule (derived from strain D471) reveals that the N-terminal two-thirds of the mature protein is composed of three major regions containing tandem repeated segments (Hollingshead et al., 1986). Region A is composed of five direct tandem repeats of 14 amino acids each, region B also has five tandem repeats of 25 amino acids each, and region C contains two and a half repeats of 42 amino acids apiece. Recent reports indicate that extensive amino acid sequence repeats also occur in the M24 (Mouw et al., 1988), M5 (Miller et al., 1988b), M12 (Robbins et al., 1987), M49 (Haanes and Cleary, 1989) and M2 (Bessen and Fischetti, 1990) molecules. Detailed analysis of the structure and antigenic determinants of the M molecule reveals that it forms an α-helical coiled-coil extending nearly 60 nm from the cell surface (Phillips et al., 1981).

Immunobiology of Proteins and Peptides VI
Edited by M.Z. Atassi, Plenum Press, New York, 1991

Based on sequence studies of M proteins from three different M serotypes (M6, M5, M24) it was observed that all have a seven residue periodicity in the placement of hydrophobic residues within the amino acid sequence (Manjula and Fischetti, 1980). This type of periodicity is characteristic of molecules whose tertiary structure is either completely or partially coiled-coil (e.g. myosin, tropomyosin and fibrinogen) (Cohen and Parry, 1986; Cohen and Parry, 1990). Fiber X-ray diffraction data substantiates the fact that the M6 protein is a coiled-coil structure (Phillips et al., 1981). DNA sequence analysis of the C-terminal end of the M6 molecule reveals a stretch of 20 hydrophobic amino acids which could act as a membrane anchor, and an adjacent proline- and glycine-rich region (Hollingshead et al., 1986). It has been proposed that the proline and glycine residues allow this region of the M molecule to assume a conformation necessary to traverse the cross-linked region of the peptidoglycan, thereby stabilizing the molecule within the cell wall (Fischetti et al., 1988). Recent evidence indicates that the M molecule is not covalently linked to the cell wall matrix but is bound through the cell membrane (Pancholi and Fischetti, 1988; Pancholi and Fischetti, 1989).

IMMUNOCHEMISTRY

It is clear from the early work of Lancefield that type-specific antibodies directed to the M molecule are opsonic and allow for effective phagocytosis of streptococci of the homologous serotype (Lancefield, 1962; Lancefield, 1959). Beachey et al. (Beachey et al., 1979) used pepsin extracted M24 protein (the N-terminal half of the native M molecule) to immunize human volunteers. Unlike earlier acid-extracted products (Fox, 1974), skin delayed-type hypersensitivity tests in 37 adults proved to be negative with this PepM protein. Immunization with alum-precipitated PepM24 protein led to the development of type-specific opsonic antibodies in 10 of 12 volunteers and the formation of positive delayed-cutaneous hypersensitivity to PepM24 in 11 individuals. None of the volunteers developed heart-reactive antibodies as determined by immunofluorescence. These studies clearly indicated that M protein vaccines free of sensitizing antigens could be produced, but also illustrated the type-specificity of the immune response. Since there are over 85 distinct serotypes of M protein, a vaccine based on type-specific epitopes may not be practical in spite of its high efficacy.

In limited M protein vaccine trials, humans immunized at an intranasal site displayed lower rates of both nasopharyngeal colonization and clinical illness following streptococcal challenge as compared to the placebo group (Polly et al., 1975). In contrast, individuals immunized subcutaneously exhibited a decrease in clinical illness only, and had no change in the level of nasopharyngeal colonization. These studies suggest that local immune factors play a key role in providing protection against initial infection by group A streptococci.

TWO CLASSES OF STREPTOCOCCAL M PROTEIN

In a recent antigenic analysis of surface-exposed M protein epitopes, it became apparent that there exists two basic M protein structures (Bessen et al., 1989). A panel of seven antibodies directed to well-defined epitopes located in the B-repeat, pepsin site, and C-repeat regions of M6 protein were tested for immunoreactivity to over 130 streptococcal isolates representing 50 different serotypes. Isolates of only four serotypes (M5, M14, M19, M36) shared antigenic determinants located in the B-repeat and pepsin site regions of M6 protein. In contrast, about half of the serotypes tested displayed strong immunoreactivity with antibody probes directed to C-repeat region epitopes of M6 protein. Nearly all isolates which produce opacity factor (an apolipoproteinase) lack immunological crossreactivity with antibodies to the C-repeat region. Thus, we have identified two major structural classes of M protein molecules, termed class I and II. Streptococci bearing class I M proteins share surface-exposed antigenic epitopes with the C-repeat region of M6 protein and fail to produce OF, whereas class II serotypes produce OF but are deficient in epitopes cross-reactive with the M6-like C repeat region (Bessen et al., 1989). The close proximity of the highly conserved C-repeat region to the poorly homologous B-repeat and pepsin site, strongly suggests that the C-repeat region forms a distinct antigenic domain (Bessen et al., 1989). In support of the immunochemical analysis, structural data suggest that the C-repeat antigenic domain may have unique conformational characteristics as well.

The degree of homology between M protein genes of different serotypes has been addressed in nucleic hybridization studies. Using DNA probes corresponding in positions to the C-repeat region and to the cell wall and membrane anchor regions of M6, a class I M protein (Scott et al., 1986; Miller et al., 1988a), distinctions can be made between class I and II serotypes (Bessen et al., 1989). Hybridization to chromosomal DNA of most class II serotypes is observed only under conditions of low stringency, whereas class I serotypes hybridize under high stringency conditions. These findings suggest that there are fundamental differences between class I and II M proteins at the DNA level (Bessen et al., 1989). There is a strong correlation between certain serological M types and outbreaks of acute rheumatic fever (Bisno, 1980). Of 16 streptococcal serotypes which have been associated with one or more major outbreaks of rheumatic fever, all but one type express the C-repeat region epitope which defines class I M protein (Bessen et al., 1989).

NON-TYPE-SPECIFIC PROTECTION

The frequency of group A streptococcal respiratory infection rises sharply at age four, peaks at age six, and rapidly declines above age 10, reaching adult levels by 18 years (Breese and Hall, 1978). At its peak incidence, 50% of children between the ages of five and seven suffer from streptococcal infection each year. The decreased occurrence of streptococcal pharyngitis in adults might be explained by an age-related host factor. Alternatively, protective antibodies directed to antigens common to all group A streptococcal serotypes might arise as a consequence of multiple infections experienced during childhood, resulting in an elevated response to conserved protective epitopes of the M molecule. In exploring the second hypothesis, it was found that sera of most adults examined have a strong antibody response to the conserved regions of M6 protein (Bessen and Fischetti, 1988d; Bessen and Fischetti, 1988c; Bessen and Fischetti, 1990b). While antibodies directed to conserved epitopes fail to neutralize the antiphagocytic property of M protein (Jones and Fischetti, 1988), they might afford protection by an alternative mechanism.

The C-repeat region of type 6 streptococci is highly conserved among organisms of many distinct serological types (Jones et al., 1985; Jones et al., 1986; Bessen and Fischetti, 1988d; Bessen et al., 1989; Jones and Fischetti, 1988). Monoclonal antibodies directed to determinants localized to the C-repeat region of M6 protein bind to the surface of whole streptococci which represent more than half of the serotypes examined (Jones et al., 1985; Jones et al., 1986). In view of the strong evidence for the presence of exposed conserved epitopes on the native M molecule, it should be possible to generate antibodies reactive to the majority of serological types using only a few distinct antigens for immunization. If a conserved region vaccine is to be effective in preventing nasopharyngeal infection by group A streptococci, it may require stimulation of the secretory immune response.

Passive immunization at the mucosa allowed for a more precise evaluation of the role played by secretory IgA (sIgA) in preventing streptococcal infection at a mucosal site (Bessen and Fischetti, 1988a). It was found that affinity-purified human anti-M protein sIgA administered intranasally protected mice against systemic infection after intranasal challenge with group A streptococci. In contrast, anti-M protein specific IgG administered intranasally had no protective effect at this site, although it was highly opsonic and promoted streptococcal phagocytosis in whole blood.

To determine whether antibodies directed to the conserved exposed epitopes of M protein influence the course of mucosal colonization by group A streptococci, peptides corresponding to these regions were used as immunogens in a mouse model (Bessen and Fischetti, 1988c; Bessen and Fischetti, 1990b). Synthetic peptides corresponding to conserved region sequences of the M6 sequence were covalently linked to the mucosal adjuvant cholera toxin B subunit (CTB) and administered intranasally to the mice. About one month later, animals were then challenged intranasally with live streptococci (either homologous M6 or heterologous M14) and pharyngeal colonization monitored for 10-15 days. It was found that mice immunized with the peptide-CTB complex showed a significant reduction in colonization after challenge with either M6 or M14 streptococci compared to mice receiving CTB alone (Bessen and Fischetti, 1988c; Bessen and Fischetti, 1990), (Figure 1). Thus, despite the fact

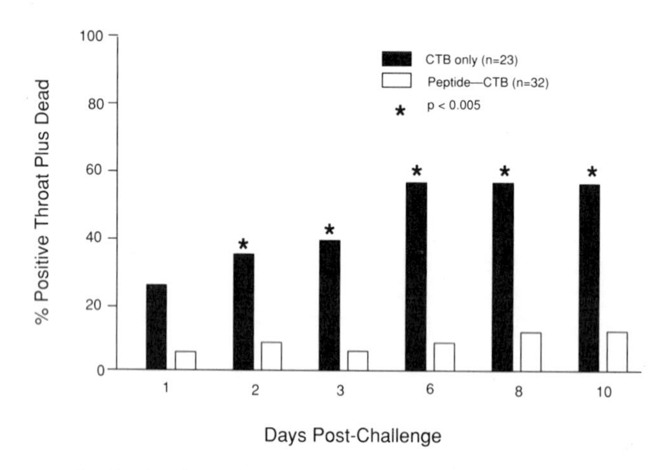

Figure 1. Pharyngeal colonization of immunized mice challenged with group A streptococci. Mice were immunized intranasally with either cholera toxin B-subunit (CTB) or synthetic peptides derived from sequences in the conserved region of the M6 molecule coupled to CTB. Pharyngeal cultures were taken up to 10 days following intranasal challenge with 5×10^5 streptococci. Results are expressed as the number of culture positive mice plus dead at each time point. Statistical difference was significant according to chi-square analysis.

that conserved region peptides are unable to evoke an opsonic antibody response (Jones and Fischetti, 1988), these peptides have the capacity to induce an immune response capable of influencing the colonization of group A streptococci at the nasopharyngeal mucosa in this model system. These findings suggest for the first time that protection against multiple serotypes of group A streptococci can be achieved with a vaccine consisting of the widely shared C-repeat region of M6 protein. This approach may perhaps form the basis of a broadly protective vaccine for the prevention of streptococcal pharyngitis.

VACCINIA VIRUS AS A VECTOR TO DELIVER M PROTEIN CONSERVED REGIONS

The successful cloning of the *emm6.1* gene into vaccinia virus and its high expression in mammalian cells infected with the recombinant vaccinia may prove to be a powerful vector for delivery of the M molecule to mucosal surfaces (Hruby et al., 1988). In an effort to express only the conserved region epitopes located in the C-terminal half of the molecule, genetic engineering methods were used to remove the N-terminal half of the M6 gene and recombine the C-terminal gene fragment into the vaccinia genome (VV:M6'). Western blot analysis revealed that the cells infected with the VV:M6' recombinant produce a molecule of about 30 kDa that is reactive with a monoclonal specific for the C-terminal half of the M protein (Fischetti et al., 1989a). Mice were immunized intranasally with either the VV:M6' recombinant or wild-type vaccinia and challenged intranasally and orally one month later with 5 x 10^6 streptococci. Pharyngeal cultures taken up to 10 days post challenge revealed that the VV:M6' immunized animals differed significantly from controls. Of the VV:M6' immunized animals, only 16% of 152 total swabs taken were streptococcal positive with 10 (6%) yielding >100 CFU, whereas 69% of 115 swabs were positive in the wild-type group and 40 (35%) displaying >100 CFU. On average >70% of the animals immunized with wild-type virus were culture positive for group A streptococci at every swab day up to 10 days after challenge. This is compared with <30% colonization of mice immunized with the VV:M6' recombinant. These results confirm and extend the previous studies using synthetic peptides from conserved regions of the M molecule to protect against streptococcal colonization (Bessen and Fischetti, 1988c; Bessen and Fischetti, 1990b).

Because the VV:M6' recombinant virus exhibits obvious potential as an effective anti-streptococcal immunization vehicle, it was of interest to further develop this vector system to enhance the antigenicity of the protective M protein epitopes. In the course of these experiments it was found that when the C-repeat region was genetically repeated 3 times and inserted into the vaccinia virus genome (VV:CRR3X), it exhibited some surprising and unusual recombinant properties which may enhance the effectiveness of the vaccinia virus as a vector for similarly constructed antigens (Hruby et al., 1990). The recombinant VV:CRR3X virus which was isolated appeared to represent not an individual recombinant virus, but a complex mixture of variants which contained from 1 to >20 tandem copies of the CRR region at the insertion site. This complexity was also seen at the transcription level in which VV:CRR3X infected cells revealed transcripts which increased in size from 1,400 to 6,600 bases by increments of about 300 bases. All were functional by Western blot using M protein specific antibody.

It might be anticipated that high levels of recombination may induce mutations into the target sequence. Some mutations will likely be silent or induce stop codons, while others will introduce missense mutations. It has been found that recombination events within the repeat regions of the M molecule result in changes in both sequence (Hollingshead et al., 1987) and antigenic epitopes (Jones et al., 1988). Thus, the changes that may occur within the repeats in VV:CRR3X may mimic those occurring within the M molecule. For vaccine purposes, the vaccinia virus recombinants such as the VV:CRR3X may offer an opportunity to present the host with an antigenic mosaic for the induction of a broad array of antibodies against a variable epitope. Utilizing the recombinant properties of the VV vectors may be of particular relevance with regard to developing effective vaccination strategies against infection by serotypically diverse pathogens.

There is great potential for the vaccinia virus as a vector for delivering foreign antigens. It is, however, not without concern. For instance, safety issues of vaccinia virus-based vaccines have been raised based on rare complications resulting from VV immunizations, specifically in immunocompromised individuals. It will be several years before these viruses are engineered such that they retain their infectivity and immunogenicity without pathogenicity. Vectors such as fowl pox are currently being considered as viable alternatives to the vaccinia virus. These viruses have all the advantages of vaccinia virus without apparent side effects. To date, there have been no recorded human infection as a result of a fowl pox virus. These viruses, which have a host range that is limited to avian species, have been used successfully to express the G glycoprotein of rabies virus in non-avian species (Taylor et al., 1988b; Taylor et al., 1988a). Animals immunized with the recombinant fowl pox were protected from lethal challenge with rabies virus. Thus, such vectors may serve as the choice to existing methods for delivering foreign antigens to protect man from bacterial as well as viral diseases.

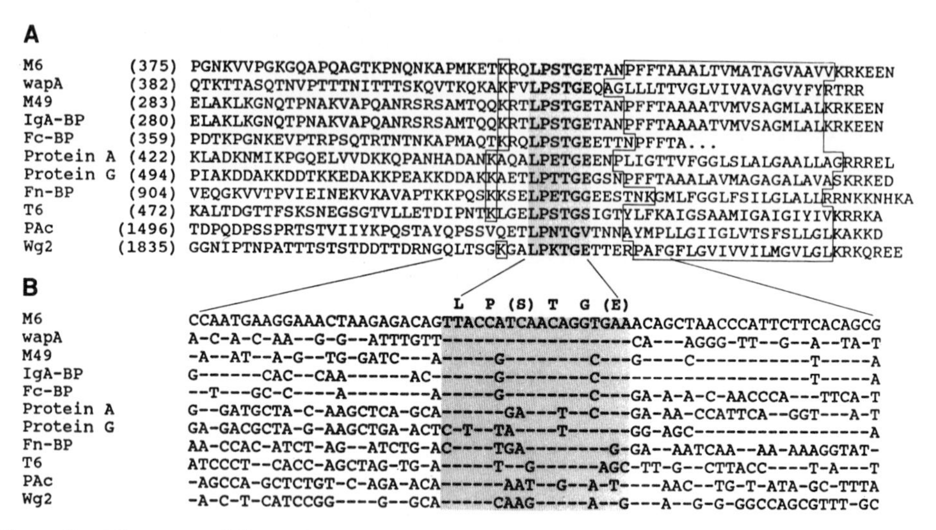

Figure 2. Alignment of C-terminal amino acid sequences from 11 surface proteins from gram-positive cocci (A) and their coding sequences (B). The LPSTGE sequence is shaded and the proteins were aligned along this consensus sequence. In 10 out of 11 proteins, a conserved K residue was found 2 or 3 residues preceding the consensus sequence (boxed). The C-terminal hydrophobic regions are also boxed. The amino acid number of the first residue of each sequence is in parenthesis. Abbreviations: M6: M protein (*S. pyogenes*) (Hollingshead et al., 1986); IgA-BP: IgA binding protein (from an M4 *S. pyogenes*) (Frithz et al., 1989); wapA: wall-associated protein A (*S. mutans*) (Ferretti et al., 1989); Fc-BP: Fc binding protein from *S. pyogenes* (Heath and Cleary, 1989); Protein G: IgG binding protein (group G streptococi) (Olsson et al., 1987); PAc: cell surface protein (*S. mutans*) (Okahashi et al., 1989); Protein A: IgG binding protein (*S. aureus*) (Guss et al., 1984); wg2: cell wall protease (*S. cremoris*) (Kok et al., 1988); Fn-BP: fibronectin binding protein (*S. aureus*) (Signas et al., 1989); T6: surface protein (*S. pyogenes*) (Schneewind et al., 1990).

BLOCKING THE ATTACHMENT OF SURFACE PROTEIN IN BACTERIAL CELLS

Proteins on bacterial surfaces perform a variety of functions for the organism from enabling resistance to phagocytosis as seen with the M protein to adherence to specific receptors on human tissue cells. We have initiated a study to understand the mechanism of attachment of the M molecule within the streptococcal cell in order to devise strategies to block the attachment process. In doing so we may be able to circumvent many problems associated with vaccines directed to surface exposed epitopes.

The complete sequence of group A streptococcal M6 protein (Hollingshead et al. 1986) has revealed a 20 hydrophobic amino acid segment at the C-terminal end predicted to be a membrane anchor followed by a 6 amino acid charged tail acting as a stop-transfer segment (Blobel, 1980). Sequence analyses of M proteins from other streptococcal serotypes and other surface molecules from gram-positive cocci (Olsson et al., 1987; Signas et al. 1989; Guss et al., 1984; Kok et al., 1988; Frithz et al., 1989; Ferretti et al., 1989; Okahashi et al., 1989; Schneewind et al., 1990; Haanes and Cleary, 1989; Heath and Cleary, 1989) have revealed a similar arrangement of hydrophobic amino acids and charged tail despite sequence variation within this region among the M molecules and the other surface proteins. This common motif suggests a conventional mechanism of attachment for these different surface molecules. Examination of the C-terminal region of surface molecules from Gram-positive cocci has revealed a hexapeptide sequence which begins 9 residues N-terminal to the putative membrane anchor region. This sequence is found in virtually all sequenced surface molecules from Gram-positive cocci (particularly streptococcal and staphylococcal species) (Figure 2). The hexapeptide, with the consensus sequence LPSTGE, has not as yet been reported in proteins from gram-positive bacilli or gram-negative organisms.

We chose the cytoplasmic membrane of *E. coli* as a model system to examine the properties of the M protein LPSTGE motif and C-terminal hydrophobic domain with specifically constructed mutants. The *emm*6.1 gene and its derivatives were selectively expressed with T polymerase in *E. coli* strain HMS174(DE3) (Studier and Moffatt, 1986). Plasmid pM6.1 contains the complete *emm*6.1 reading frame. The LPSTGE motif and C-terminal hydrophobic domain from position 407 onwards is deleted in $pM6.1_{1\text{-}406}$. In $pM6.1_{\Delta LPSTGE}$, only the LPSTGE sequence is deleted and replaced by a two amino acid substitution (GS). All three constructs were expressed in *E. coli* HMS174(DE3), labeled with [^{35}S] methionine, subjected to cell fractionation (Davis et al., 1985), and immunoprecipitated with M6-specific antibodies. Results of this analysis revealed that the wild-type M6 protein was found only in the membrane fraction and exhibited its typical multiple banding pattern on SDS-PAGE. The truncated M6 protein ($pM6.1_{1\text{-}406}$) and the deletion mutant ($pM6.1_{\Delta LPSTGE}$) were both secreted into the periplasmic space. These results indicate that the highly conserved LPSTGE motif is crucial for anchoring the M protein within the bacterial cell. A further understanding of the anchor mechanism may allow us to a means by which to block the surface translocation of these molecules. In doing so, the ability of these organisms to cause disease could be markedly affected.

References

Agarwal, B.L., 1981, Rheumatic heart disease unabated in developing countries, Lancet i:910.

Beachey, E.H., Stollerman, G.H., Johnson, R.H., Ofek, I. and Bisno, A.L., 1979, Human immune response to immunization with a structurally defined polypeptide fragment of streptococcal M protein, J. Exp. Med. 150:862.

Bessen, D. and Fischetti, V.A., 1988a, Passive acquired mucosal immunity to group A streptococci by secretory immunoglobulin A, J. Exp. Med. 167:1945.

Bessen, D. & Fischetti, V.A. 1988b, Vaccination against Streptococcus pyogenes infection, in: New Generation Vaccines, Levine, M.M. & Wood, G., Eds., Marcel Dekker,Inc., New York. p 599.

Bessen, D. and Fischetti, V.A., 1988c, Influence of intranasal immunization with synthetic peptides corresponding to conserved epitopes of M protein on mucosal colonization by group A streptococci, Infect. Immun. 56:2666.

Bessen, D. & Fischetti, V.A. 1988d, Role of nonopsonic antibody in protection against group A streptococcal infection, in: Technological Advances in Vaccine Development, Lasky, L., Ed., Alan R. Liss Inc., New York. p 493.

Bessen, D. and Fischetti, V.A., 1990, Synthetic peptide vaccine against mucosal colonization by group A streptococci. I. Protection against a heterologous M serotype with shared C repeat region epitopes, J. Immunol. 145:1251.

Bessen, D., Jones, K.F. and Fischetti, V.A., 1989, Evidence for two distinct classes of streptococcal M protein and their relationship to rheumatic fever, J. Exp. Med. 169:269.

Bessen, D. and Fischetti, V.A., 1990, Differentiation between two biologically distinct classes of group A streptococci by limited substitutions of amino acids within the shared region of M protein-like molecules, J. Exp. Med. In press:

Bisno, A.L. 1980, The concept of rheumatogenic and nonrheumatogenic group A streptococci, in: Streptococcal diseases and the immune response, Read, S.E. & Zabriskie, J.B., Eds., Academic Press, New York. p 789.

Blobel, G., 1980, Intracellular protein topogenesis, Proc. Natl. Acad. Sci. U. S. A. 77:1496.

Breese, B.B. & Hall, C.B. (1978), Beta hemolytic streptococcal diseases. Houghton Mifflin, Boston.

Cohen, C. and Parry, D.A.D., 1986, Alpha-helical coiled coils-a widespread motif in proteins, T. I. B. S. 11:245.

Cohen, C. and Parry, D.A.D., 1990, Alpha-helical coiled coils and bundles: How to design an alpha-helical protein, Proteins Struct. Funct. Genet. 7:1.

Davis, N.G., Boeke, J.D. and Model, P., 1985, Fine structure of a membrane anchor domain, J. Mol. Biol. 181:111.

Dodu, S.R.A. and Bothig, S., 1989, Rheumatic fever and rheumatic heart disease in developing countries, World Health Forum 10:203.

Ferretti, J.J., Russell, R.R.B. and Dao, M.L., 1989, Sequence analysis of the wall-associated protein precursor of Streptococcus mutans antigen A, Molec. Microbiol. 3:469.

Fischetti, V.A., Parry, D.A.D., Trus, B.L., Hollingshead, S.K., Scott, J.R. and Manjula, B.N., 1988, Conformational characteristics of the complete sequence of group A streptococcal M6 protein, Proteins: Struct. Func. Genet. 3:60.

Fischetti, V.A., Hodges, W.M. and Hruby, D.E., 1989a, Protection against streptococcal pharyngeal colonization with a vaccinia:M protein recombinant, Science 244:1487.

Fischetti, V.A., 1989b, Streptococcal M protein: Molecular design and biological behavior, Clin. Microbiol. Rev. 2:285.

Fox, E.N., 1974, M proteins of group A streptococci, Bacteriol. Rev. 38:57.

Frithz, E., Heden, L-O. and Lindahl, G., 1989, Extensive sequence homology between IgA receptor and M protein in Streptococcus pyogenes, Molec. Microbiol. 3:1111.

Guss, B., Uhlen, M., Nilsson, B., Lindberg, M., Sjoquist, J. and Sjodahl, J., 1984, Region X, the cell-wall-attachment part of staphylococcal protein A, Eur. J. Biochem. 138:413.

Haanes, E.J. and Cleary, P.P., 1989, Identification of a divergent M protein gene and an M protein related gene family in serotype 49 Streptococcus pyogenes, J. Bacteriol. 171:6397.

Heath, D.G. and Cleary, P.P., 1989, Fc-receptor and M protein genes of group A streptococci are products of gene duplication, Proc. Natl. Acad. Sci. U. S. A. 86:4741.

Hollingshead, S.K., Fischetti, V.A. and Scott, J.R., 1986, Complete nucleotide sequence of type 6 M protein of the group A streptococcus: repetitive structure and membrane anchor, J. Biol. Chem. 261:1677.

Hollingshead, S.K., Fischetti, V.A. and Scott, J.R., 1987, Size variation in group A streptococcal M protein is generated by homologous recombination between intragenic repeats, Mol. Gen. Genet. 207:196.

Hruby, D.E., Hodges, W.M., Wilson, E.M., Franke, C.A. and Fischetti, V.A., 1988, Expression of streptococcal M protein in mammalian cells, Proc. Natl. Acad. Sci. USA 85:5714.

Hruby, D.E., Schneewind, O., Wilson, E.M. and Fischetti, V.A., 1990, Assembly and analysis of a functional vaccinia virus amplicon containing the C-repeat region from the M protein of Streptococcus pyogenes, Submitted

Jones, K.F., Manjula, B.N., Johnston, K.H., Hollingshead, S.K., Scott, J.R. and Fischetti, V.A., 1985, Location of variable and conserved epitopes among the multiple serotypes of streptococcal M protein, J. Exp. Med. 161:623.

Jones, K.F., Khan, S.A., Erickson, B.W., Hollingshead, S.K., Scott, J.R. and Fischetti, V.A., 1986, Immunochemical localization and amino acid sequence of cross-reactive epitopes within the group A streptococcal M6 protein, J. Exp. Med. 164:1226.

Jones, K.F. and Fischetti, V.A., 1988, The importance of the location of antibody binding on the M6 protein for opsonization and phagocytosis of group A M6 streptococci, J. Exp. Med. 167:1114.

Jones, K.F., Hollingshead, S.K., Scott, J.R. and Fischetti, V.A., 1988, Spontaneous M6 protein size mutants of group A streptococci display variation in antigenic and opsonogenic epitopes, Proc. Natl. Acad. Sci. USA 85:8271.

Kok, J., Leenhouts, K.J., Haandrikman, A.J., Ledeboer, A.M. and Venema, G., 1988, Nucleotide sequence of the cell wall proteinase gene of Streptococcus cremoris Wg2, Appl. Environ. Microbiol. 54:231.

Lancefield, R.C., 1959, Persistence of type specific antibodies in man following infecion with group A streptococci, J. Exp. Med. 110:271.

Lancefield, R.C., 1962, Current knowledge of the type specific M antigens of group A streptococci, J. Immunol. 89:307.

Manjula, B.N. and Fischetti, V.A., 1980, Tropomyosin-like seven residue periodicity in three immunologically distinct streptococcal M proteins and its implication for the antiphagocytic property of the molecule, J. Exp. Med. 151:695.

Miller, L., Burdett, V., Poirier, T.P., Gray, L.D., Beachey, E.H. and Kehoe, M.A., 1988a, Conservation of protective and nonprotective epitopes in M proteins of group A streptococci, Infect. Immun. 56:2198.

Miller, L., Gray, L., Beachey, E.H. and Kehoe, M.A., 1988b, Antigenic variation among group A streptococcal M proteins: Nucleotide sequence of the serotype 5 M protein gene and its relationship with genes encoding types 6 and 24 M proteins, J. Biol. Chem. 263:5668.

Mouw, A.R., Beachey, E.H. and Burdett, V., 1988, Molecular evolution of streptococcal M protein: Cloning and nucleotide sequence of type 24 M protein gene and relation to other genes of Streptococcus pyogenes, J. Bacteriol. 170:676.

Okahashi, N., Sasakawa, C., Yoshikawa, S., Hamada, S. and Koga, T., 1989, Molecular characterization of a surface protein antigen gene from serotype c Streptococcus mutans implicated in dental caries, Molec. Microbiol. 3:673.

Olsson, A., Eliasson, M., Guss, B., Nilsson, B., Hellman, U., Lindberg, M. and Uhlen, M., 1987, Structure and evolution of the repetitive gene encoding streptococcal protein G, Eur. J. Biochem. 168:319.

Pancholi, V. and Fischetti, V.A., 1988, Isolation and characterization of the cell-associated region of group A streptococcal M6 protein, J. Bacteriol. 170:2618.

Pancholi, V. and Fischetti, V.A., 1989, Identification of an endogeneous membrane anchor-cleaving enzyme for group A streptococcal M protein, J. Exp. Med. 170:2119.

Phillips, G.N., Flicker, P.F., Cohen, C., Manjula, B.N. and Fischetti, V.A., 1981, Streptococcal M protein: alpha-helical coiled-coil structure and arrangement on the cell surface, Proc. Natl. Acad. Sci. USA. 78:4689.

Polly, S.M., Waldman, R.H., High, P., Wittner, M.K., Dorfman, A. and Fox, E.N., 1975, Protective studies with a group A streptococcal M protein vaccine. II. Challenge of volunteers after local immunization in the upper respiratory tract, J. Infect. Dis. 131:217.

Robbins, J.C., Spanier, J.G., Jones, S.J., Simpson, W.J. and Cleary, P.P., 1987, Streptococcus pyogenes type 12 M protein regulation by upstream sequences, J. Bacteriol. 169:5633.

Schneewind, O., Jones, K.F. and Fischetti, V.A., 1990, Sequence and structural characterization of the trypsin-resistant T6 surface protein of group A streptococci, J. Bacteriol. 172:3310.

Scott, J.R., Hollingshead, S.K. and Fischetti, V.A., 1986, Homologous regions within M protein genes in group A streptococci of different serotypes, Infect. Immun. 52:609.

Signas, C., Raucci, G., Jonsson, K., Lindgren, P., Anantharamaiah, G.M., Hook, M. and Lindberg, M., 1989, Nucleotide sequence of the gene for fibronectin-binding protein from Staphylococcus aureus: Use of this peptide sequence in the synthesis of biologically active peptides, Proc. Natl. Acad. Sci. USA 86:699.

Studier, F.W. and Moffatt, B.A., 1986, Use of bacteriophage T7 RNA polymerase to direct selective high-level expression of cloned genes, J. Mol. Biol. 189:113.

Taylor, J., Weinberg, R., Languet, B., Desmettre, P. and Paoletti, E., 1988a, Recombinant fowlpox virus inducing protective immunity in non-avian species, Vaccine 6:497.

Taylor, J., Weinberg, R., Kawoaka, Y., Webster, R.G. and Paoletti, E., 1988b, Protective immunity against avian influenza induced by a fowlpox virus recombinant, Vaccine 6:504.

ORAL DELIVERY OF ANTIGENS IN LIVE BACTERIAL VECTORS

Robert N. Brey, Garvin S. Bixler, James P. Fulginiti,
Deborah A. Dilts, and Marta I. J. Sabara

Praxis Biologics
300 East River Road
Rochester, New York 14623

INTRODUCTION

Most available vaccines for enteric diseases are inactivated whole cell preparations which provide incomplete protection against infection and have undesirable side effects. As alternative vaccines for enteric infection, several live oral vaccines have been recently developed. These vaccines are derived from virulent pathogens by introduction of stable attenuating mutations in metabolic pathway genes. Ideally, these mutations are non-reverting and genetically well defined. The *Salmonella typhi* live oral vaccine Ty21a has been tested in several field trials, first in Egypt and more recently in four separate trials in Chilean schoolchildren (Ferreccio et al, 1989; Levine et al 1987a; Wahdan et al, 1982). The results of those trials indicate that Ty21a is safe, well tolerated, and provides 66% protection in a typhoid fever endemic area. Protection against typhoid fever was at least that achieved with parenterally administered whole cell vaccine preparations and was retained for at least three years.

However, Ty21a suffers from several drawbacks. It contains a number of uncharacterized point mutations in genes for galactose metabolism (including *galE*). It is difficult to maintain *in vitro* and an effective course of oral vaccination involves three or four doses taken in enteric coated capsules on different days. Finally, because precise mutations engineered into the *galE* locus of *S. typhi* induced typhoid fever symptoms in 2 of 4 volunteers, it is unlikely that *galE* is the sole attenuating locus of Ty21a (Hone et al, 1988a).

Salmonella typhi strains attenuated at other loci have been under development as successors of Ty21a. Most promising of these approaches to date has been the development of strains requiring aromatic compounds which are attenuated *in vivo* because of nutritional requirement for p-aminobenzoic acid. A strain containing a Tn*10*-generated deletion of *aroA* locus and a second deletion in *purA* was tested in human volunteers (Levine et al. 1987b). In these experiments, volunteers showed poor seroconversion to O-specific somatic antigen, yet a significant number of them manifested a cellular immune response specific for S. typhi. By introducing these mutations into S. typhimurium, it was later found that the second attenuating mutation, purA, "hyperattenuated" the strains in mice (O'Callaghan, D.O. et al, 1988). An effective live oral S. typhi vaccine based on aromatic deficiency will likely contain two non-reverting mutations in the same aromatic biosynthetic pathway. Other enteric organisms, such as Shigella flexneri and Yersinia enterocolitica, have been attenuated by introduction of similar deletions in the aromatic pathway and are promising live oral vaccine candidates for those enteric diseases (Sory et al, 1990). Other chromosomal loci have been used successfully to attenuate virulent S. typhimurium including cya crp, ompR, and phoP. S. typhi strains carrying some of these biochemical pathway mutations or mutations affecting virulence are being developed for human vaccines (Dorman et al. 1989; Miller and Mekalanos, 1990; Curtiss and Kelly, 1987).

ATTENUATED RECOMBINANT SALMONELLA TO DELIVER FOREIGN ANTIGENS

Because of the potent secretory and cellular immune responses induced by ingestion of attenuated Salmonella and the ease of introducing foreign genes using common plasmid cloning vectors, the use of Salmonella to deliver antigens of other enteric pathogens has been explored. In addition, Salmonella has been used as an oral delivery system for a number of antigens derived from both viral and parasitic sources. In some instances, the ability of the recombinants to protect against disease has been tested in mouse models. The Salmonella recombinant system has special appeal because vaccines can be ingested orally (ideally in a single dose); production of the vaccines would involve a minimum of downstream processing, unlike subunit vaccines for parenteral injection, and the vaccines could ultimately bypass the cold chain. These features make potential bivalent or polyvalent vaccines for a number of enteric, viral and parasitic diseases feasible. Further, it is likely that Salmonella vaccines containing a biochemical attenuation which causes a self-limiting infection could be used in undernourished or immune-deficient individuals.

A number of heterologous antigens have been expressed in a variety of attenuated Salmonella strains, which in some cases elcited a protective immune response in animals. Oral vaccination of mice with S.typhimurium aroA expressing the Streptococcus pyogenes M5 fimbrial antigen resulted in significant IgA titers in salivary secretions and protected mice against homologous challenge (Poirier et al, 1988). Recently, Sadoff and his co-workers demonstrated that a laboratory attenuated strain of S. typhimurium (M206) expressing the P.berghei CS protein induced partial protection against the sporozoite stage of malaria in the absence of detectable antibody response (Sadoff et al, 1988), implying that effector T cell mechanisms could be induced by the recombinant. In fact, Class I MHC-restricted CTL's, associated with protection against the sporozoite stage of malaria, recognizing specifically a peptide of the P. berghei CS protein, were induced as a consequence of immunization with the recombinant Salmonella.

Similarly, CTL's directed to a peptide of P. falciparum CS protein were also induced by oral vaccination of mice with Salmonella expressing either full length or repeatless CS protein constructs (Agarwal et al, 1990; Romero et al, 1989). Oral vaccination of mice with a S. typhimurium expressing a blood stage antigen of P. vinckei along with parenteral injection of parasite extract resulted in complete protection against lethal challenge with blood stage parasites (Kumar et al., 1990). In protecting against influenza A infection, only Class II MHC-restricted CTL's were induced following oral ingestion of Salmonella carrying the influenza A nucleoprotein (Tite et al, 1990). Partial protection against Leishmania infection was induced by oral vaccination of mice with a Salmonella recombinant expressing gp63 (Yang et al., 1990).

Several recent clinical trials with Salmonella recombinants based on Ty21a host strain also demonstrate the promise of live attenuated organisms as an approach to antigen delivery in humans. Ty21a expressing Form I antigen of S. sonnei has induced protection against homologous Shigella challenge, but the vaccine seemed to vary from lot to lot, drastically affecting its efficacy (Herrington et al, 1990). Although no significant protection against V. cholerae challenge was seen in a recent clinical trial using Ty21a expressing V. cholerae Inaba LPS antigen, significant titers of intestinal IgA were achieved (Tacket et al., 1990).

RESULTS AND DISCUSSION

LOCALIZATION OF EPITOPES CELL SURFACE ORGANELLE HYBRIDS

A number of groups have reported the construction of genetic hybrid proteins containing inserted sequences which can be directed to the external surface of E. coli cells as hybrid outer membrane proteins. For example, epitopes inserted in the coding sequence of the E. coli lamB protein, which is involved with the uptake of maltose and maltodextrins and also is the receptor for bacteriophage lambda, have been localized on the cell surface (van der Werf et al, 1990). Extracellular organelles such as pilins, flagella, or bacteriophage-encoded coat proteins, have been adapted to accept varying amounts of DNA sequence. These approaches to localizing epitope sequences to the cell surface have been explored in the development of live attenuated Salmonella vaccines. Such live vaccines in which critical T and B cell epitopes derived from heterologous antigens are expressed on the cell surface may be more effective in inducing desired humoral and cellular immune responses than epitopes or proteins expressed cytoplasmically.

The flagellar filaments of Salmonella are composed of a single protein, flagellin, which is exported by a flagellum specific pathway to the cell surface. A single flagellum may be composed of several thousand flagellin molecules and there may be 10-30 flagella per cell. The flagellins from a number of different Salmonella species have been cloned and sequenced (Wei and Joys, 1985). These data have shown that flagellins are conserved at the amino and carboxy ends of the molecules, while the central region is hypervariable. Conservation of the terminal regions spans species and genera, indicating that the

Table 1. Heterologous epitopes inserted in the coding region of fliC of Salmonella muenchen

CONSTRUCT	INSERTED EPITOPE SEQUENCE[1]		size	motil.[2]
Plasmodium CS protein				
P.berghei				
	$(DPAPPNAN)_4$	repeats x 4	32	+
	$(DPAPPNAN)_8$	repeats x 8	64	+
	DIIDGS NDDSYIPSAEKIL GS	CTL epitope	21	+
P.falciparum				
	$(NPNANPNANPNANPNA)_1N$	repeats x 3	17	+
	$(NPNANPNANPNANPNA)_3N$	repeats x 10	49	+
Rotavirus				
VP7				
278-295	DIIDGS APQTERMMRINWKKWWQV GS		26	-
	DIIDGS $(APQTERMMRINWKKWWQV\ GS)_2$		46	-
VP4				
232-255	D NIVPVSIVSRNIVYTRAQPNQDIV GS		27	-
220-255	D INNGGLPPIQQNTRNIVVPVSIVSRNIVYTRAQPNQDIV GS		42	-
C.dipththeriae				
crm197				
366-383	DIIDGS NLFQVVHNSYNRPAYSP GS		25	-
N. meningiditis class I OMP				
"VR1" series	DIIDGS $(AANGGASGQV\ GS)_2$	VR1 x 2	30	+
	DIIDGS $(AANGGASGQV\ GS)_4$	VR1 x 4	54	+
	DIIDGS $(AANGGASGQV\ GS)_6$	VR1 x 6	78	+
	DIIDGS $(AANGGASGQV\ GS)_8$	VR1 x 8	102	+
	DIIDGS $(AANGGASGQV\ GS)_{10}$	VR1 x 10	126	+
"VR2" series	DIIDGS $(TKDTNNNLTL\ GS)_3$	VR2 x 3	42	+
VR1 VR2 hybrid	DIIDGS $(AANGGASGQV\ GS)_4$ $(TKDTNNNLTL\ GS)_3$		90	+

1. Sequences are contained in the EcoRV site of pPX1650, which contains a 48 bp deletion of DNA from the hypervariable region of H1d. Underlined sequences are derived from linker DNA.
2. Motility was assessed by restoration of swarming of S. dublin SL5927 (aroA fliC)

structural constraints that control export and assembly are generally confined to the termini of the molecules. This suggests the possibility of inserting large segments of heterologous antigenic proteins or precisely defined epitopes into the hypervariable region of flagella for vaccine use without disruption of flagellar localization or function (Newton et al, 1989; Majarian et al, 1989). Because flagella are surface localized and flagellin comprises a major protein of

Salmonella, it is possible to purify easily and economically large amounts of hybrid protein for use as subunit vaccines, which can be used as an alternative to a synthetic peptide-based product. As shown in Table 1, we have expressed a number of defined epitopes from various pathogenic organisms in the hypervariable region of the fliC (formerly H1) gene of Salmonella muenchen. In most cases, hybrid molecules encode functional, surface-localized flagella, in which the inserted epitope is accessible to antibodies that recognize the inserted epitope. Oral or parenteral administration of the live Salmonella expressing the hybrids induces antibodies which reacted with the cognate protein immunogen. A hybrid flagellin containing hepatitis B virus preS2 121-145 and S122-137 epitopes does not yield functional flagella, although oral administration of the recombinant bacteria induced a low level of antibodies capable of reacting with whole virus as well as to the encoded peptides (Wu et al, 1989). Although these hepatitis B-flagellin hybrid molecules were expressed, they may not have been directed to the cell surface, or may have been exported but not assembled into filamentous structures. Cloning the immunodominant B cell epitope repeat regions of P. berghei and P. falciparum yielded hybrid molecules containing up to 64 amino acids. These hybrid molecules encoded functional flagella in which the inserted epitopes were recognized on the cell surface by specific monoclonal antibodies directed against the corresponding repeat regions (Majarian et al, 1989). Further, partially purified flagellin preparations were extremely immunogenic in Balb/c mice, inducing high level of antibodies directed against the inserted epitope (data not shown). A hybrid flagellin encoding a mapped CTL epitope of P. berghei (Romero et al, 1989) also encoded a functional flagellin in which the inserted epitope was recognized specifically by anti-peptide antibodies. Likewise, a hybrid which expresses a characterized T-helper cell epitope encompassing residues 366-383 of C. diphtheriae CRM197 (Bixler et al, 1989) is recognized by specific antibodies, capable of priming T-cells in vivo following parenteral immunization, and is capable of stimulating T cell clones specific for the peptide 366-383 (Bixler, G.S., and Brey, R. N., unpublished experiments).

ROTAVIRUS/SALMONELLA BIVALENT LIVE OR SUBUNIT VACCINES BASED ON VP7 AND VP4 PROTEINS

To develop a potential vaccine against rotavirus diarrhea which could be employed as a purified subunit or delivered in attenuated Salmonella, several hybrid flagellin molecules encoding a rotavirus neutralizing epitope derived from a composite sequence of the VP7 protein from both human and bovine rotavirus were constructed and tested for immunogenicity in mice. Previously, a peptide conjugate vaccine containing the VP7 275-295 sequence has been shown to induce high level of virus neutralizing antibodies in mice (Frenchick et al, 1989). Plasmids encoding hybrid flagellins containing either a single insert or a tandem insert of VP7 278-295 were isolated and the inserts were confirmed by DNA sequencing. Neither of these hybrid flagellin molecules specified a functional flagellum, since the plasmids did not restore motility to non-motile S. dublin SL5927 (fliC aroA). However, the hybrids did specify surface localized molecules since molecules of the predicted

Table 2. Immunogenicity of rotavirus hybrid flagellins as subunits and delivered in live attenuated *Salmonella dublin*

Endpoint ELISA titers, pooled serum

			Anti VP7 275-295 at week			Anti-flagellin week 6
Immunogen[1,2]	Dose	Adjuv.	0	4	6	
1650 (wt flagellin)	10 mg IM	alum	25	30	10	141,800
	10 "	none	25	30	10	128,700
1592 fla 278-295	1 "	alum	20	100	100	44,300
	10 "	alum	20	126	126	71,700
	10 "	CFA	20	300	151	298,200
	10 "	none	20	490	490	53,800
1594 fla(278-295)x2	1 "	alum	20	125	125	2,900
	10 "	alum	16	100	85	1800
	10 "	CFA	16	120	120	28,100
	10 "	none	15	100	100	1,363
SL5927	10^{10}p.o.	none	10	nd	30	<1,000
pPX1596/SL5927	"	"	10	nd	700	1,500
pPX1598/SL5927	"	"	10	nd	260	<1000
SL5927	10^{7}i.p.	none	20	nd	63	1,900
pPX1596/SL5927	"	"	25	nd	1200	238,000
pPX1598/SL5927	"	"	18	nd	100	28,500

1. Swiss Webster mice were immunized at weeks 0 and 2 with partially purified flagellin preparations; for immunization with live bacteria, mice were immunized with 10^{10} organisms orally at week 0 and 4 or intraperitoneally with 10^{7} organisms on week 0 and 4.
2. pPX1596 is a single insert VP7 278-295 hybrid expressed in a pSC101-based vector; pPX1598 is a tandem insert in a pSC101-based vector. SL5927 is *S. dublin aroA fliC*.

molecular weights which reacted with anti-VP7 polyclonal antibodies were easily recovered from culture supernatants and were the predominant species present. These materials were partially purified from culture supernatants of the recombinant *S. dublin* by precipitation with 50% ammonium sulfate and used to test for biological activity. Both partially pure hybrids were capable of competing with live virus in inhibiting virus plaquing versus control flagellin, confirming the suggestion that this region of VP7 is involved in cell attachment (Frenchick et al, 1989). This result suggests that the epitope is presentedin the context of the hybrid flagellin in a biologically relevant manner. These partially purified materials were also used to immunize Swiss Webster mice in either alum or complete Freund's adjuvant. As shown in Table

2, potent anti-flagellar responses were observed (by ELISA using highly purified S. muenchen flagellin as the capture antigen) with either the wild type flagellin (1650) or the hybrid containing a single insert of the 278-295 epitope. Using the 275-295 peptide as the capture antigen in ELISA, a response in the serum approximately 20-50 fold over background levels was obtained with the hybrid containing a single insert. The hybrid containing a double insert elicited poor responses to the flagellin moiety and marginal response to the 275-295 peptide. Both these hybrid molecules react poorly on western blots with several virus neutralizing monoclonal antibodies, suggesting reduced antigenicity. These results suggest that the inserted peptide may destabilize the flagellar structure, and further, that crucial conformational aspects involved in antigenicity and immunogenicity of the epitope may not be necessarily retained in the hybrid constructs. The VP7 275-295 region contains several tryptophan residues, residues absent in flagellin. The presence of these amino acids in the hybrids may interfere with assembly of intact filaments.

Several regions of VP4 encoding a trypsin cleavage site involved in uptake of the rotavirus particle by enterocytes have also been identified (Arias et al, 1989) and inserted into flagellin (Table 1). These molecules reacted very well with a series of neutralizing monoclonal antibodies and polyclonal anti-peptide sera, suggesting retention of antigenicity in these hybrid molecules (data not shown).

In developing a hybrid flagellin approach to a live Salmonella vaccine, the strategy has been to evaluate the hybrid molecules first as subunit components, followed by further evaluation in Aro$^-$ S. dublin. For delivery of the VP7 278-295 epitope as a live vaccine, Swiss Webster mice were immunized orally with several doses of S. dublin SL5927 recombinant bacteria at weeks 0 and 4 or intraperitoneally with 10^7 organisms at week 0 and 4. In this experiment the hybrid flagellins were subcloned into a genetically stable vector. The results are shown in the lower half of Table 2. A low but significant titer against both flagellin moiety and the inserted peptide was observed when the single insert hybrid was used to immunize orally. However, a more potent response was seen to both components following two doses of parenterally administered bacteria, as high against the peptide as any of the paratially purified flagellin subunits. These results show that it may be feasible to develop a Salmonella/rotavirus bivalent vaccine, but that the effort has to be accompanied by an increased understanding of the conformational factors involved in expressing these particular rotavirus epitopes in bacteria.

HYBRID FLAGELLIN MOLECULES EXPRESSING EPITOPES OF CLASS IN OUTER MEMBRANE PROTEINS OF N. MENINGIDITIS

To develop a vaccine effective against most of the clinical types of group B N. meningiditis infections, a subunit strategy based on the Class I outer membrane proteins (OMP) has been pursued. Class I OMP's, along with other membrane components, have been implicated as protective immunogens in group B disease. A series of monoclonal antibodies which define type B subtypes have been shown to confer passive protection in the infant rat model. Further,

Table 3. Immunogenicity of hybrid N. meningiditis flagellin molecules

Endpoint serum ELISA titers at week 10

Immunogen[1]	Dose (mg)	Adjuvant	Coating antigens[2]			
			Flagellin	M21-BSA (VR1)	M20-BSA (VR2)	PI.7,16 OMP
1650 (wt flagellin)	1	alum	427,800	170	<100	350
	10	alum	468,400	150	<100	250
pVR1 x 4	1	alum	787,100	4525	260	38,600
	10	alum	887,800	17,600	<100	100,300
pVR2 x 3	1	alum	263,100	300	500	1800
	10	alum	1,493,200	750	5,476	4,700
p1210-6	1	alum	299,900	4,650	800	10,500
	10	alum	497,600	3,900	2,250	47,500
pVR1 x 4	10	CFA	1,840,000	19,950	350	17,500
pVR2 x 3	10	CFA	1,217,000	700	17,750	17,860

1. Swiss Webster mice were immunized IM with partially purified flagellin preparations at weeks 0, 2, and 8.
2. Flagellin was purified H1d; M20-BSA is VR2 epitope peptide-conjugate: AQAANGGSGQVK; M21-BSA is VR1 epitope peptide-conjugate: YYTKDTNNNLTLVP; PI.7,16 is purified outer membrane protein

epitopes to which these monoclonal antibodies bind have been defined by peptide mapping. DNA sequencing data has elucidated two variable regions within the class I OMP's, termed VR1 and VR2 (McGuiness et al, 1990). The possibility that a vaccine composed of a number of such VR1 or VR2 epitopes, constructed in flagellin, may successfully address the bulk of the type B disease was considered. A series of hybrid flagellin molecules containing arrays of reiterated VR1, VR2, or composite epitopes, based on the primary amino acid sequence of VR1 and VR2 regions from N. meningiditis PI:7,16 was initially constructed and tested for immunogenicity. Three constructs were initially chosen: one having 4 tandem inserts of VR1, one having 3 tandem inserts of VR2, and one being a composite having 4 VR1 inserts followed by 3 inserts of VR2. Each of the VR1 or VR2 hybrids encoded a flagellin which conferred motility on S. dublin SL5927. Anti-VR1 or anti-VR2 monoclonal antibodies reacted with the appropriate constructs on western blots and inhibited motility of the recombinant S. dublin in the predicted fashion. This suggests that the epitopes expressed in these molecules were freely accessible to the monoclonal antibodies. Recombinant flagellins were partially purified from culture supernatants by precipitation with 50% ammonium sulfate. In these crude culture supernatants, the hybrid flagellin molecules comprised greater than 90% of the material and were produced at yields in excess of 100 mgs/L. The hybrid materials were injected in alum, or in some cases with CFA, at weeks 0, 2, and 8. The hybrid VR1 flagellin induced potent anti-flagellar antibodies as well as high titer antibodies against the cognate peptide (Table 3). Likewise, the VR2 hybrid induced anti-flagellar antibodies and antibodies directed to the inserted peptide, though at apparently lower magnitude. The composite VR1 VR2 hybrid (p12-10-6) induced responses against both peptides. Significantly, each of the hybrids induced antibody capable of recognizing the isolated outer membrane protein. The efficacy of the hybrid remains to be determined.

SIZE LIMITATIONS OF INSERTED DNA ON THE FUNCTION OF HYBRID FLAGELLIN MOLECULES

From the nature of the kinds of heterologous peptide sequence inserted in flagellin *fliC*, it seems that the constraints on the export of monomeric flagellin and subsequent filament formation are not strictly related to the size of the insert. To test this hypothesis more fully, a "tolerated" epitope, VR1, was reiterated in flagellin to yield a series of hybrid molecules containing up to 10 tandem inserts of VR1. The VR1 x 10 insert encoded 126 amino acids, was produced in amounts similar to the wild type, and encoded motile flagella which were inhibited by the cognate monoclonal antibody. Similarly, segments of genes containing up to 330 amino acids were inserted within the coding sequence. An insert of 330 amino acids derived from the *P. berghei* CS protein was inserted; in this case, the resultant hybrid molecule was produced in miniscule amounts in the bacterium, although the detectable protein was surface localized. In contrast, a segment of VP7 (aa 180-320) was also inserted but was localized internally. A 250 amino acid segment derived from respiratory syncytial virus F protein was also inserted, normally exported and yielded functional flagella. Clearly, the size *per se* is not the determining factor in localization, assembly, or motility.

It is obvious that *fliC* of *S. muenchen* can accommodate large amounts of heterologous peptide sequence of varying amino acid composition without disrupting function. It is also clear that inserted peptides can be immunogenic in the context of the recombinant hybrid molecules. What is less clear, is what role expression of epitopes on the surface of live bacteria will have upon the presentation of antigens to the host immune system.

GENETIC STABILITY OF *SALMONELLA* FOR VACCINE USE

It is thought that constant presentation of endogenous antigens synthesized intracellularly during the course of the transient infection established by recombinant attenuated *Salmonella* is advantageous in inducing a desired immune response. Segregation of expression plasmids and subsequent loss of antigen could result in variation of vaccine preparation, which in turn may influence efficacy. A number of experimental modes to address genetic stability have been developed for *Salmonella* vaccines. Most common cloning and expression vectors are deficient in partitioning functions which control segregation of plasmids into daughter cells. Stability has to be reintroduced into expression plasmids either by complementation of a lethal chromosomal defect or by introduction of plasmid stability functions. Alternatively, antigen-expressing genes can be stably incorporated into the chromosome of the vaccine Salmonella, usually by homologous recombination at a particular locus (Hone et al, 1988b).

STABLE GENE INSERTION BY MODIFIED TRANSPOSABLE ELEMENTS

A generalized means to introduce heterologous genes into the chromosome of Salmonella sp. has been developed. Gene expression cassettes contained within the inverted repeats of Tn10, also containing a selectable marker, can be transposed to random locations in the chromosome (Brey, R.N., and Deich, R.A., unpublished observations). In these constructs the specific Tn10 transposase is located outside of the inverted repeats and the inserted gene expression cassette and selectable marker, so that only DNA contained within the inverted repeats transposes. Because the transposase protein does not itself insert into the chromosome, DNA contained within the inverted repeats is essentially stabilized on the chromosome.

To develop a means for directly comparing the in vivo immunogenicity of stable versus unstable recombinant S. typhimurium strains, an antigen known to elicit high serum (and intestinal mucosal) antibody titers upon oral vaccination with Aro$^-$ Salmonella, LT-B, was expressed either as a chromosomal integrant, on a stable plasmid vector, or on an unstable cloning vector. LT-B is the non-toxic B subunit of enterotoxigenic E. coli labile toxin (LT); it induces antibodies that cross react with and neutralize cholera toxin (Clements and El-Morshidy, 1984). LT-B expression was stabilized by random insertion into the chromosome of S. typhimurium aroA SL3261. A segment of a plasmid containing LT-B coding sequence controlled by the lac promoter was cloned into a site of a plasmid vector located between the transposable ends of Tn10, which also contained the kanamycin-resistance determinant of Tn5. To create a suicide vector for delivery of the modified transposable element in Salmonella, an E. coli strain carrying the transposable LT-B expression cassette was lysogenized with a lambda containing homology with Tn10. By inducing the culture for phage production, it was possible to isolate hybrid λ phage particles containing the modified LT-B transposon. These events were detected by selecting for kanamycin-resistant chloramphenicol-sensitive lysogens in E. coli. These clones arose at a frequency of approximately 1/1000 by either homologous recombination or a transposase-mediated event between the inverted repeats located either on the plasmid or on the lysogen. Lambda lysogens containing the LT-B transposon were induced and the resulting phage were infected into S. typhimurium LB5010 (r-m+) or 523Ty (S. typhi aroA), each expressing the lamB receptor of E. coli. Kanamycin-resistant clones arose in both S. typhimurium and S. typhi strains. Each of the clones from both strains of Salmonella expressed LT-B antigen, and Southern blotting of DNA from 10 independent LB5010 isolates revealed random localization of LT-B. Interestingly, some of the isolates expressed more antigen than others, suggesting positional effects in the chromosome. Those isolates that expressed well were transduced into SL3261 to construct appropriate vaccine strains. Several isolates were tested for

Table 4. Persistence of plasmid-containing S. typhimurium aroA Peyer's patches of individual mice following oral vaccination[1]

Immunizing bacteria, 10^9 p.o.	Animal #	Day 10 total	Day 10 drugR	Day 10 %	Day 15 total	Day 15 drugR	Day 15 %	Day 29 total	Day 29 drugR	Day 29 %
pUC8/SL3261 (vector control)										
	#1	2	2	100	1	1	100	0	0	na
	#2	0	0	na	0	0	na	6	0	0
	#3	1	1	100	158	3	2	0	0	na
pPX100/SL3261 (pUC LT-B)										
	#1	3	3	100	59	0	0	1	0	0
	#2	36	0	0	44	0	0	1	0	0
	#3	360	0	0	0	0	na	1	0	0
pGD103/SL3261										
	#1	612	612	100	0	0	na	8	8	100
	#2	370	370	100	103	103	100	0	0	100
	#3	456	456	100	102	102	100	8	8	100
pPX3005/3261 (pGD103 LT-B)										
	#1	600	600	100	40	40	100	1	1	100
	#2	218	218	100	31	31	100	0	0	na
	#3	436	436	100	30	30	100	0	0	na
3261λ3 (LT-B integrant)										
	#1	28	28	100	18	18	100	1	1	100
	#2	300	300	100	43	43	100	1	1	100
	#3	480	480	100	29	29	100	1	1	100

1. Total bacterial counts in tissue homogenates were obtained by enumerating colonies appearing after overnight growth on SS agar without antibiotic. Colonies were replicated to SS agar with antibiotic to assess retention of drug-resistance.

stability *in vitro* by subculturing in the absence of drug-selection for up to 80 generations. None of the gene-insert isolates of SL3261 showed any loss of drug resistance and each of the colonies recovered expressed the LT-B antigen. Two of these isolates were finally chosen to examine *in vivo* immunogenicity and stability.

A stable plasmid vector expressing the *lac* promoter-controlled LT-B gene was constructed by cloning the expression cassette into a low copy number vector (approximately 6-7 copies

Table 5. Effect of genetic stability on immunogenicity of live Salmonella

Immunizing strain	Immunized at week[1]	Endpoint α-LT-B ELISA titers, pooled sera reciprocal dilution (Week)					
		0	2	4	6	10	13
SL3261	0, 10	<50	<50	<50	<50	<50	<50
	0, 4, 10	<50	<50	<50	<50	<50	<50
pPX100/SL3261	0, 10	<50	400	200	200	200	5,800
(pUC LT-B)	0, 4, 10	<50	50	400	3200	1615	8,600
pPX3005	0, 10	<50	50	200	200	200	26,600
(pGD103 LT-B)	0, 4, 10	<50	50	200	800	600	7,570
3261λ3	0, 10	<50	50	50	50	50	312
(integrant)	0, 4, 10	<50	50	50	200	350	1,100
3261λ14	0, 10	<50	50	100	800	100	4,070
(integrant)	0, 4, 10	<50	50	100	800	800	4,000

1. Balb/c mice were immunized orally with 10^{10} bacteria at weeks 0 and 10 or 0, 4, and 10.

per cell), pGD103, derived from pSC101. Plasmid pSC101 retains partition functions and is known to be stable in E. coli. The pGD103-LT-B construct was stable in vitro upon culturing for up to 80 generations in the absence of drug selection. Expression of lac promoter controlled LT-B was somewhat proportional to gene dosage with high copy number > low copy number > gene integrant (data not shown).

A number of these LT-B expressing strains and plasmid vector control strains were used to immunize Balb/c mice orally with 10^{10} organisms. From the data presented in Table 4, it is clear that either the gene integrant or the low copy number stable expression plasmid were stable for up to 29 days following oral inoculation. On the other hand, both the puC8 vector and its LT-B expression plasmids, pPX100, were drastically unstable, with none of the organisms recovered from mesenteric lymph nodes on day 15 showing retention of the expression plasmid. Further, the high copy vector itself and the LT-B expression vector affected the in vivo growth of the vaccine strain, suggesting that these high copy number plasmids are unsuitable for in vivo gene expression.

As shown in Table 5, although the level of gene expression encoded by the low copy number pGD103 vector is significantly lower than that encoded by the high copy number pUC vector, pPX100, serum ELISA response to the LT-B antigen was higher for the stable plasmid in mice immunized twice at weeks 0 and 10. This suggests in vivo expression from stabilized genes is beneficial in design of live Salmonella vaccines. In addition, two vaccine strains, SL326λ3 and SL3261λ14, each bearing a single copy gene stabilized on a modified transposon, were capable of inducing serum antibodies close to the level of high copy number unstable plasmids. It may be possible to enhance gene regulatory signals, such as promoters, and overcome possible gene copy number barriers. Using the modified transposon system, it has been possible to engineer gene insertions of a number of antigens, including the *P. berghei* CS protein. In that case, expression from a stable single copy gene insert controlled by the strong bacteriophage lambda P_L promoter encodes expression at levels identical to or higher than multicopy plasmids.

CONCLUSIONS

It is clear that heterologous genes can be expressed in a large variety of *Salmonella* species, including attenuated vaccine candidate strains of *S. typhi*, *S. typhimurium*, and *S. dublin*. Heterologous genes can sometimes be expressed to high levels either intracellularly, in the periplasmic space, or surface localized. Genetic stability can be achieved by either stabilizing plasmids or by integrating gene expression cassettes into the *Salmonella* chromosome. High level gene expression may not be a prerequisite for efficacy of antigen delivery, since a number of poorly expressed antigens, such as the *P.berghei* or *P. falciparum* CS protein, can induce either protective immunity or a correlate thereof. It seems that high level gene expression will also influence the *in vivo* growth parameters of the vaccine strains, further altering mechanisms of delivery. In the ideal *Salmonella* vaccine, it will be desirable to achieve the proper balance of gene expression so that the desired immune responses can be induced. An effective delivery system is primarily designed to induce the kind of immunity not achieved by other modes of vaccination. For any vaccine based on a live attenuated *Salmonella* vector, whether for human or veterinary use, oral delivery, ease of manufacture and cost effectiveness are primary issues.

ACKNOWLEDGMENTS

We would like to thank Jan Poolman for providing purified *Neisseiria meningiditis* outer membrane proteins and monoclonal antibodies and Robert A. Deich for construction of some of the bacteriophage lambda transposition vectors. Special thanks are due to Bruce Stocker for kindly supplying attenuated *Salmonella* strains and the flagellin clone.

REFERENCES

Agarwal, A., Kumar, S., Jaffe, R., Hone, D., Gross, M., and Sadoff, J. 1990. Oral *Salmonella*: malaria circumsporozoite recombinants induce specific CD8+ cytotoxic T cells. J. Exp. Med. 172:1083-1090.

Arias, C.F., Garcia, G., and Lopez, S. 1989. Priming for rotavirus neutralizing antibodies by a VP4 protein-derived synthetic peptide. J. Virol. 63:5393-5398.

Bixler, G.S., Eby, R., Dermody, K.M., Woods, R.M., Seid, R.C., and Pillai, S. 1989. Synthetic peptide representing a T-cell epitope of CRM197 substitutes as carrier molecule in a *Haemophilus influenzae* type B (HIB) conjugate vaccine. In

"Immunobiology of proteins and peptides V" (Atassi, M.Z., ed.) pp.175-180, Plenum Press, New York.

Clements, J.D., and El-Morshidy, S. 1984. Construction of a potential oral live bivalent vaccine for typhoid fever and cholera-Escherichia coli-related diarrheas. Infect. Immun.. 466:564-569.

Curtiss, R. III, and Kelly, S.M. 1987. Salmonella typhimurium deletion mutants lacking adenylate cyclase and cyclic AMP receptor protein are avirulent and immunogenic. Infect. Immun. 55:3035-3043.

Dorman, C.J., Chatfield, S., Higgins, C.F., Hayward, C., and Dougan, G. 1989. Characterization of porin and ompR mutants of a virulent strain of Salmonella typhimurium: ompR mutants are attenuated in vivo. Infect. Immun. 57:2136-2140.

Ferreccio, C., Levine, M.M., Rodriguez, H., Contreras, R., and Chilean Typhoid Committee. 1989. Comparative efficacy of two, three, or four dose of Ty21a live oral typhoid vaccine in enteric-coated capsules: a field trial in an endemic area. J. Infect. Dis. 159:766-769.

Frenchick, P.J., Sabara, M.I.J., and Babiuk, L.A. 1989. Use of a viral nucleocapsid particle as a carrier for synthetic peptides. In Vaccine 89: modern approaches to new vaccines including prevention of AIDS" (Lerner, R.A. et al, eds.) pp. 479-483, Cold Spring Harbor Laboratory, New York.

Herrington, D.A., Van De Verg, L., Formal, S.B., Hale, T.L., Tall, B.D., Cryz, S.J., Tramont, E.C., and Levine, M.M. 1990. Studies in volunteers to evaluate candidate Shigella vaccines: further experience with a bivalent Salmonella typhi-Shigella sonnei vaccine and protection conferred by previous Shigella sonnei disease. Vaccine 8:353-357.

Hone, D.M., Attridge, S.R., Forrest, B., Morona, R., Daniels, D., LaBrooy, J.T., Bartholomeusz, R.C.A., Shearman, D.J.C., and Hackett, J. 1988a. A galE via (Vi antigen-negative) mutant of Salmonella typhi Ty2 retains virulence in humans. Infect. Immun. 56:1326-1333.

Hone, D., Attridge, S., van den Bosch, L., and Hackett, J. 1988b. A chromosomal integration system for stabilisation of heterologous genes in Salmonella based vaccine strains. Microb. Path. 5:407-418.

Kumar, S., Gorden, J., Flynn, J.L., Berzofsky, J.A., and Miller, L.H. 1990. Immunization of mice against Plasmodium vinckei with a combination of attenuated Salmonella typhimurium and malarial antigen. Infect. Immun. 58:3425-3429.

Levine, M.M., Ferreccio, C., Black, R.E., Germanier, R, and Chilean Typhoid Committee. 1987a. Large-scale field trial of Ty21a live oral vaccine in enteric-coated capsule formulation. Lancet i:1049-1052.

Levine, M.M., Herrington, D., Murphy, J.R., Morris, J.G., Losonsky, G., Tall, B., Lindberg, A.A., Svenson, S., Baqar, S., Edwards, M.F., and B. Stocker. 1987b. Safety, infectivity, immunogenicity, and in vivo stability of two attenuated auxotrophic mutant strains of Salmonella typhi, 541Ty and 543Ty, as live oral vaccines in humans. J. Clin. Invest. 79:888-902.

Majarian, W.R., Kasper, S.J., and Brey, R.N. 1989. Expression of heterologous epitopes as recombinant flagella on the surface of attenuated Salmonella. In "Vaccine 89:modern approaches to new vaccines including prevention of AIDS"

(Lerner, R.A.,et al, eds.) pp. 277-281, Cold Spring Harbor Laboratory, New York.
McGuiness, B., Barlow, A.K., Charles, I.N., Farley, J.E., Anilionis, A., Poolman, J.T., and Heckles, J.E. 1990. Deduced amino acid sequence of class I protein (porA) from three strains of Neisseria meningiditis. J. Exp. Med. 171:1871-1882.
Miller, S., and Mekalanos, J.J. 1990. Constitutive expression of the phoP regulon attenuates Salmonella virulence and survival within macrophages. J. Bacteriol. 172:2485-2490.
Newton, S.M.C., Jacob, C.O., and Stocker, B.A.D. 1989. Immune response to cholera toxin epitope inserted in Salmonella flagellin. Science 244:70-72.
O'Callaghan, D.O., Maskell, D., Liew, F.Y., Easmon, C.S.F., and Dougan, G. 1988. Characterization of aromatic- and purine-dependent Salmonella typhimurium: attenuation, persistence, and ability to induce protective immunity in BALB/c mice. Infect. Immun. 56:419-423.
Poirier, T.P., Kehoe, M.A., and Beachey, E.H. 1988. Protective immunity evoked by oral administration of attenuated aroA Salmonella typhimurium expressing cloned streptococcal M protein. J. Exp. Med. 168:25-32.
Romero, P., Maryanski, J.L., Corradin, G., Nussenzweig, R.S., Nussenzweig, V., and Zavala, F. 1989. Cloned cytotoxic T cells recognize an epitope in the circumsporozoite protein and protects against malaria. Nature (Lond.). 341:323-326
Sadoff, J.C., Ballou, W.R., Baron, L.S., Majarian, W.R., Brey, R.N., Hockmeyer, W.T., Young, J.F., Cryz, S.J., Ou, J., Lowell, G.H., and Chulay, J.D. 1988. Salmonella typhimurium vaccine expressing circumsporozoite protects against malaria. Science 240:336-338.
Sory, M., Hermand, P., Vaerman, J.P., and Cornelis, G.R. 1990. Oral immunization of mice with a live recombinant Yersinia enterocolitica O:9 strain that produces the cholera toxin B subunit. Infect. Immun. 58:2420-2428.
Tacket, C.O., Forrest, B., Morona, R., Attridge, S.R., LaBrooy, J., Tall, B.D., Reymann, M., Rowley, D., and Levine, M.M. 1990. Safety, immunogenicity, and efficacy against cholera challenge in humans of a typhoid-cholera hybrid vaccine derived from Salmonella typhi Ty21a. Infect. Immun. 58:1620-1627.
Tite, J.P., Gao, X.-M., Hughes-Jenkins, C.M., Lipscombe, M., O'Callaghan, D., and Dougan, G. 1990. Antiviral immunity induced by recombinant nucleoprotein of influenza A virus III. delivery of recombinant nucleoprotein to the immune system using attenuated Salmonella typhimurium as a live carrier. Immunology. 70:540-546.
van der Werf, S., Charbit, A., Leclerc, C., Mimic, V., Ronco, J., Girard, M., and Hofnung, M. 1990. Critical role of neighboring sequences on the immunogenicity of the C3 poliovirus neutralization epitope expressed at the surface of recombinant bacteria. Vaccine 8:269-277.
Wahdan, M.H., Serie, C., Cerisier, Y., Sallam, S., and Germanier, R. 1982. A controlled field trial of live Salmonella typhi strain Ty21a oral vaccine against typhoid: three-year results. J. Infect. Dis. 145:292-295.
Wei, L.-N., and Joys, T.M. 1985. Covalent structure of three phase-1 flagellar filament proteins of Salmonella. J. Mol. Biol. 186:791-803.

Wu., J., Newton, S., Judd, A., Stocker, B., and Robinson, W.S. 1989. Expression of immunogenic epitopes of hepatitis B surface antigen with hybrid flagellin proteins by a vaccine strain of Salmonella. Proc. Nat. Acad. Sci.(USA) 86:4726-4730.

Yang, D.M., Fairweather, N., Button, L.L., McMaster, W.R., Kahl, L.P., and Liew, F.Y. 1990. Oral Salmonella typhimurium (aroA) vaccine expressing a major leishmanial surface protein (gp63) preferentially induces T helper cells and protective immunity against leishmaniasis. J. Immunol. 145:2281-2285.

AUGMENTATION BY INTERLEUKINS OF THE ANTIBODY RESPONSE TO A CONJUGATE VACCINE AGAINST *Haemophilus influenzae* B

Garvin S. Bixler, Jr. and Subramonia Pillai

Praxis Biologics, Inc.
300 East River Road
Rochester, New York 14623

ABSTRACT

Interleukins have been recognized as potential adjuvants for use during vaccination. The immunogenicity of some poorly immunogenic bacterial capsular polysaccharides have been improved by conjugation to a protein carrier. Augmentation of the immune response to these glycoconjugates, however, may be realized in the presence of interleukins. The antibody response to one such vaccine which comprises a oligosaccharide derived from the capsule of *Haemophilus influenzae* type b coupled to CRM_{197} (HbOC) can be augmented in this manner. A suboptimal dose (0.1 μg) of HbOC and varying concentrations of IL-1α or IL-1β (10^2 - 5 x 10^5 U) were injected intramuscularly at 0 and 2 weeks into Swiss Webster mice. Vaccines were also formulated with and without aluminum phosphate. Antibody to the oligosaccharide was determined by Farr assay. In 3/3 experiments, IL-1α enhanced primary and secondary antibody responses whereas with IL-1β, only a slight increase in the primary antibody response was seen but enhanced secondary responses were observed. Thus, IL-1α and to some extent IL-1β enhanced the primary and secondary antibody responses to a glycoconjugate vaccine.

INTRODUCTION

Interleukin 1 (IL-1) is a macrophage-derived cytokine which mediates a broad range of physiological effects (For review, see Dinarello, 1988). Of particular interest for vaccine development is its capacity to regulate the immune response. Previous studies have shown that IL-1 can influence both T and B-cell activities in vitro. In the presence of IL-1, augmented T-cell proliferative responses to antigens (Mizel and Ben-Zui, 1980) or mitogens (Rosenwasser and Dinarello, 1981) have been observed. IL-1 is also known to directly influence B cell growth and differentiation (Hoffmann, 1979; Booth et al., 1983). IL-1 has been shown to enhance primary antibody responses in vitro (Schnader, 1973; Hoffman, et al., 1987; Reed et al., 1989) as well as in vivo (Staruch and Wood, 1983; Reed et al., 1989).

IL-1 exists in two forms (IL-1α and IL-1β) which despite biochemical similarities share a relatively small degree of sequence homology (March et al., 1985). Immunologically, both forms have been shown to augment immune responses in vivo (Reed et al., 1989; Nencioni et al., 1987). In the present study, we have directly compared the capacity of IL-1α and IL-1β to augment the antibody response to a suboptimal concentration of a glycoconjugate vaccine, HbOC, which is comprised of the bacterial capsular polysaccharide from H. influenzae type b covalently coupled to CRM_{197}, a non-toxic form of diphtheria toxin (Eby et al., 1986; Anderson and Eby, 1987, 1990).

RESULTS AND DISCUSSION

The capsular polysaccharide of H. influenzae type b is a T-independent antigen which is poorly immunogenic in young children (Anderson et al, 1972). Through covalently coupling of the polysaccharide to a suitable carrier protein such as CRM_{197}, significant improvement in the antibody response to the polysaccharide in this age group was obtained (Anderson et al., 1985; Insel and Anderson, 1986; Madore et al, 1990a,b). The conversion of the polysaccharide to a T-dependent antigen circumvents many of the problems associated with immunization of an immature host. Further augmentation of the immune response, however, may be realized by taking advantage of the immunomodulating capacities of interleukins.

To investigate this possibility, the antibody response to a suboptimal dose of HbOC was examined. Previous studies have shown that in outbred Swiss Webster mice 2.5 μg of HbOC is an optimal immunizing dose (data not shown). As shown in Figure 1, when a dose of 0.1 μg was employed to immunize mice intramuscularly at 0 and 2 weeks a substantially lower anti-polysaccharide antibody response to the glycoconjugate was observed as determined by the standard Farr assay (Farr, 1958). Adsorption of the HbOC to aluminum phosphate improves the response to the suboptimal dose, but it was still substantially below that obtained with an optimal antigen challenge. In subsequent studies of the adjuvant effects of IL-1, the suboptimal dose of 0.1 μg administered at 0 and 2 weeks was routinely used.

To assess the effect of recombinant human IL-1α and IL-1β (Immunex Corporation, Seattle, Washington), on the immune response to the glycoconjugate, varying concentrations of the interleukins were mixed directly with the vaccine formulation shortly before injection. Swiss Webster mice were immunized as above with the freshly formulated mixture. As shown in Figure 2A, an increase in the antibody response to the glycoconjugate was observed which correlated directly to the concentration (10^2 - 10^5 U/mouse) of IL-1α injected. In 3/3 experiments, an enhanced response was most clearly reflected in the higher secondary titers although increases in the primary responses were also observed. For example, the antibody response to PRP at week 2 was 0.30 μg/ml without IL-1α but reached 1.47 μg/ml with 10^4 U of IL-1α. Interestingly, the kinetics of the antibody response were shifted slightly since a peak response in the IL-1α groups was achieved at week four whereas the response of the control group which received 0.1 μg HbOC alone peak at week four and remained at that level at week six.

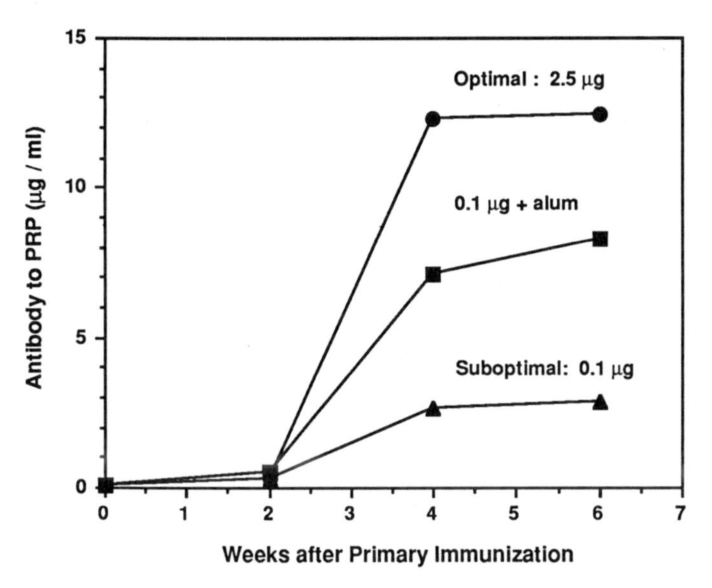

Fig.1. Kinetic of antibody response to a suboptimal dose of HbOC in Swiss Webster mice.

Similarly, augmented secondary antibody responses were also observed in the presence of IL-1β (Figure 2B). However, the relationship between degree of enhancement and concentration was not as clear. In the 3 experiments performed, maximal enhancement of the secondary response occurred in those groups receiving 1×10^3 - 5×10^4 units/mouse. In the experiment shown, at 10^2 units/mouse a decrease in the response relative to controls was observed. Depression of the secondary responses, however, was only observed in this one experiment. As seen here, IL-1β had minimal effect on the primary response to this antigen.

Mineral suspensions are frequently used as adjuvants in vaccine formulations. IL-1α and IL-1β may also derive benefit by adsorption to aluminum phosphate through improved stability or through a longer half-life. To explore the possibility that IL-1α and IL-1β act synergistically with aluminum phosphate, mice were immunized with a suboptimal dose of HbOC formulated with varying concentrations of IL-1α or β and a constant concentration of aluminum phosphate (100 μg/mouse). At all concentrations of IL-1α examined in the experiment shown in Figure 3, enhanced secondary antibody responses to a suboptimal dose of HbOC was observed. In 2/3 experiments, a synergistic effect between aluminum phosphate and IL-1α was observed. However, in contrast with the results obtained with IL-1α alone (Figure 2A), as shown in Figure 3, an inverse relationship between IL-1α concentration and the antibody response to the glycoconjugate adsorbed to aluminum phosphate was found. It is possible that this inverse relationship may be attributable to either the slow release of IL-1α from the aluminum phosphate or to a toxic effect from high localized concentrations of IL-1α at the injection site.

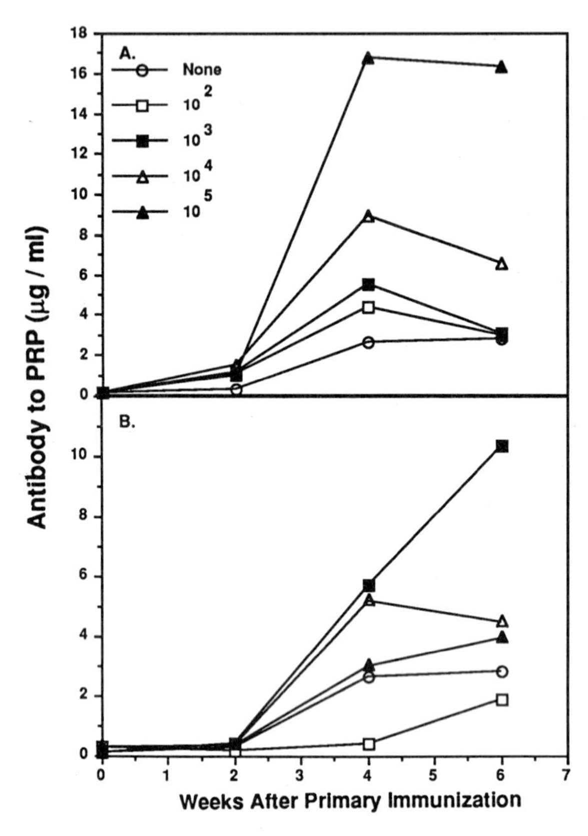

Fig.2. Augmentation of the antibody response to HbOC in the presence of (A) IL-1α or (B) IL-1β.

Similarly, mice were also immunized with HbOC formulated with both IL-1β and aluminum phosphate. Although a broad range of doses of IL-1β (10^2 - 5 x 10^5 units/mouse) was examined in three experiments, the mixture of IL-1β and aluminum phosphate slightly reduced or had no significant effect relative to controls on the antibody response to the glycoconjugate. In one experiment, the highest concentration of IL-1β examined (5 x 10^5 units/mouse) was found to have an enhanced response to a suboptimal dose of HbOC (data not shown).

These studies have compared the capacity of IL-1α and β to enhance the antibody response to a glycoconjugate vaccine in Swiss Webster mice. It has been shown that the response to a suboptimal concentration of HbOC formulated with IL-1α or β can be augmented. Inclusion of aluminum phosphate in addition to IL-1 had in some cases synergistic effects. The capability to augment the immune response to a poorly immunogenic antigen or to a immunogen administered to an immature host has significant implications for the future of vaccine development.

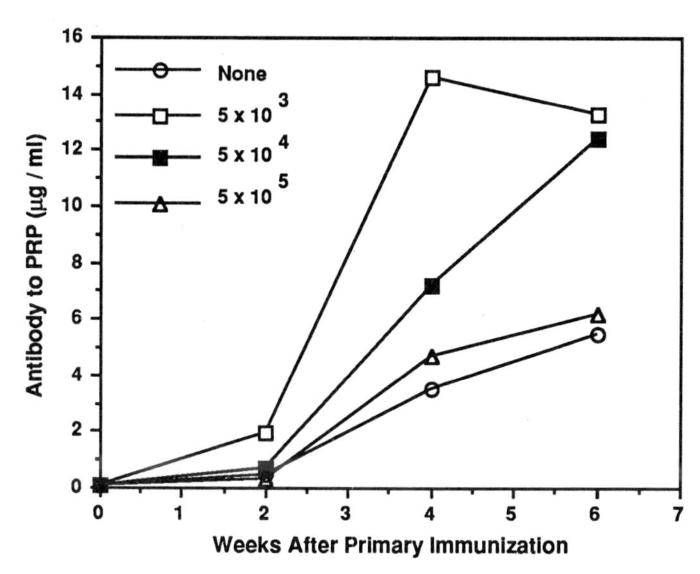

Fig.3. Augmentation of the antibody response to HbOC in the presence of IL-1α and alum.

REFERENCES

Anderson, P. W. and Eby, R.J., 1987, European Patent 0-245-045-A2, Praxis Biologics, Inc.

Anderson, P. W. and Eby, R. J., 1990, U. S. Patent 4,902,506, University of Rochester.

Anderson, P., Peter, G, Johnston, R. B. J., Wetterlow, L.H., and Smith, D. H., 1972, Immunization of humans with polyribophsophate, the capsular antigen of Hemophilus influenzae type b, J. Clin. Invest., 51:39.

Anderson, P., Pichichero, M., and Insel, R.,.1985, Immunogens consisting of oligosaccharides from the capsule of Haemophilus influenzae type b coupled to diphtheria toxoid or the toxin protein CRM_{197}, J. Clin. Invest., 76:52.

Booth, R. J., Prestige, R. L., and Waston, J. D., 1983, Constitutive production by the WEHI-3 cell line of a B cell growth and differentiation factor that co-purifies with interleukin 1, J. Immunol., 131:577.

Dinarello, C.A., 1988, Biology of interleukin 1, FASEB J., 2:108.

Eby, R. J., Madore, D., Johnson, C., Smith, D.H., Connelly, B., McHenry, C. L., and Meyers. M., 1986, A new stable vaccine for Haemophilus influenzae b highly immunogenic for human infants composed of oligosaccharides of capsular polymer (PRP) linked to CRM_{197}, Pediatr. Res., 20:308A.

Farr, R.S., 1958, A quantitative immunochemical measure of the primary interaction between I*BSA and antibody, J. Infect. Dis., 103:239.

Hoffman, M. K., 1979, Control of B cell differentiation by macrophages, Ann. N. Y. Acad. Sci., 322:577.

Hoffman, M.K., Gilbert, K.M., Hirst, J. A., and Scheid, M., 1987, An essential role for interleukin 1 and a dual function for interleukin 2 in the immune response of murine B lymphocytes to sheep erythrocytes, J. Mol. Cell. Immunol., 3:29.
Insel, R.A., and Anderson, P.W., 1986, Oligosaccharide-protein conjugate vaccines induce and prime for oligoclonal IgG antibody responses to the Haemophilus influenzae b capsular polysaccharide in human infants, J. Exp. Med., 163:262.
Madore, D. V., Johnson, C. L., Phipps, D. C., Pennridge Pediatric Associates, Popejoy, L. A., Eby, R., and Smith, D. H., 1990a, Safety and immunologic response to Haemophilus U type b oligosaccharide-CRM_{197} conjugate vaccine in 1- to 6-month-old infants, Pediatrics, 85:331.
Madore, D. V., Johnson, C. L., Phipps, D. C., Pennridge Pediatric Associates, Meyers, M. G., Eby, R., and Smith, D. H., 1990b, Safety and immunogenicity of Haemophilus influenzae type b oligosaccharide-CRM_{197} conjugate vaccine in infants aged 15 to 23 months, Pediatrics, 86:527.
March, C. J., Mosley, B., Larsen, A., Cerretti, P., Braedt, G., Price, V., Gillis, S., Henny, C.S., Kronheim, S.R., Grabstein, K., Conlon, P. J., Hopp, T. P., and Cosman, D., 1985, Cloning, sequence and expression of two distinct human interleukin-1 complementary DNAs, Nature, 315:641.
Mizel, S. B., and Ben-Zui, A., 1980, Studies on the role of LAF (interleukin 1) in antigen induced lymph node lymphocyte proliferation, Cell. Immunol., 54:382.
Nencioni, L., Villa, L., Tagliabue, A., Antoni, G., Presentini, F., Perin, R., Siverstri S., and Bonaschi, D., 1987, In vivo immunostimulating activity of the 163-171 peptide of human IL-1β. J. Immunol., 139:800.
Reed, S.G., Pihl, D. L., Conlon, P.J., and Grabstein K. H., 1989, IL-1 as adjuvant. Role of T cells in the augmentation of specific antibody production by recombinant human IL-1α, J. Immunol., 142:3129.
Rosenwasser, L.J., and Dinarello, C. A., 1981, Ability of human leukocytic pyrogen to enhance phytohemagglutinin induced murine thymocyte proliferation, Cell. Immunol., 63:134.
Schnader, J. W., 1973, Mechanism of activation of the bone-marrow-derived lymphocyte. III. A distinction between a macrophage-produced triggering signal and the amplifying effect on trigger B lymphocytes of allogeneic interactions, J. Exp. Med., 138:1446.
Staruch, M. J., and Wood, D.D., 1983, The adjuvanticity of interleukin 1 in vivo, J. Immunol., 130:2191.
Wood, D. D., 1979, Mechanism of action of human B cell-activating factor. I. Comparison of plaque-stimulating activity with thymocyte-stimulating activity, J. Immunol., 123:2400.

FACTORS PRODUCED BY STROMAL CELLS INVOLVED IN B-CELL DEVELOPMENT

A. Cumano, A. Narendran and C. J. Paige

The Ontario Cancer Institute
500 Sherbourne Street
Toronto, Canada M4X 1K9

INTRODUCTION

B lymphocytes are produced in fetal liver and bone marrow from multipotent precursor cells. The possibility of manipulating the expansion and differentiation of those precursors would allow the identification of the different steps involved in this process, and lead to the identification of growth factors required for the transition along the B-cell differentiation pathway. Major efforts have, therefore, been concentrated on the definition of such culture systems. Immature B-cells which lack detectable amounts of immunoglobulin on the surface and that are unresponsive to mitogens require particular culture conditions. Until recently, the only way to support growth and differentiation of B-cell precursors "in vitro" was to use adherent stromal cells from lymphoid organs. We will, therefore, concentrate our discussion on the role of stromal cells and soluble factors they produce in the generation of B lymphocytes from committed B-cell precursors.

COLONY ASSAYS USED TO STUDY FACTORS INVOLVED IN B-CELL DEVELOPMENT

We have utilized colony assays that allow not only the study of mature lymphocytes but also provide the culture requirements which induce the proliferation and differentiation of B-cell progenitors.

The B-cell colony assay is based on a two layer agar culture where mitogenic agents such as LPS and/or sheep red blood cells are added in the bottom agar layer of the cultures. The agar is then allowed to solidify and overlaid with another agar layer containing the cells. Growth factors, mitogens and other stimulants diffuse readily across the two agar layers. Cells proliferate and form colonies (50-100 such colonies can be easily analyzed in a single culture) that can be studied in different ways, including; 1) enumeration and morphological assertion under microscope; (Paige and Skarvall, 1982) 2) evaluation of secreted antibody by plaque assay or nitrocellulose protein blot assay; (Sauter and Paige, 1987) and 3) evaluation of RNA transcripts using the RNA colony blot technique (Wu and Paige, 1986)

The pre-B cell colony assay was developed as a tool to analyze large number of clones originating from B-cell precursors in hematopoietic organs that can differentiate to mitogen responsive and immunoglobulin-secreting plasma cells (Paige, 1983). As in the mature B-cell colony assay the addition of mitogen is necessary to attain optimal responses. However, B-cell precursors require additional stimuli which can be provided by fresh adherent cell layers from fetal liver. This requirement was not fullfiled by recombinant growth factors tested alone or in combination. This result suggests that the molecules produced by the adherent cell layer required to support B-cell differentiation in this system, have not yet been isolated.

STROMAL CELLS

Stromal cells are adherent cells isolated from fetal liver, bone marrow or spleen. They cannot differentiate into any of the hematopoietic cell types but they are able to support lymphocyte development "in vitro". There has been much controversy as to where they originate. A common precursor for hematopoietic cells and stromal cells has been suggested but there is no experimental evidence supporting it. Although their morphology is distinct from typical fibroblasts it is not clear whether they represent a distinct cell type or simply the heterogeneity in morphology and function selected by the physiological environment or by "in vitro" culture conditions.

Several stromal cell lines have been derived. Some express adhesion molecules on their surface (N-CAM)(Pietrangeli et al., 1988) as well as surface antigens present in different stages of pre-B-cell differentiation (Bp-1/6C3, CALLA) (Welch et al., 1990). S17 and S10 stromal cell line were obtained from a long-term bone marrow culture. They express class I histocompatibility antigens but not the B-cell differentiation antigen Bp-1/6C3 (Collins and Dorshkind, 1987).

Fetal liver derived adherent cell layers are effective in providing the stimuli in the pre-B cell colony assay necessary for B-cell precursors to become mitogen responsive cells. These cells are a heterogeneous mixture of stromal cells, macrophages and other adherent cell types. Using the stromal cell line S17 as an example (Collins and Dorshkind, 1987), we have investigated whether stromal cell lines can substitute for heterogeneous adherent layers, in this assay. We showed that S17 cells can completely replace the adherent cell layer in this assay. By decreasing the number of responding cells in the top layer of the agar culture, we also found that they are likely to have a direct effect on the B-cell precursor. Consistent with this idea, a highly enriched population of B-cell progenitors, selected for expression of the B220 form of the CD45 antigen, also formed colonies when grown over S17 stromal cells (Cumano et al. 1990).

INTERLEUKIN-7 (IL-7) AND OTHER FACTORS PRODUCED BY STROMAL CELLS

Recently a growth factor designated interleukin-7 (Namen et al. 1988) was isolated from bone marrow stromal cells. It was shown to act directly on B-cell precursors from bone marrow and fetal liver, and on T-cell precursors in the thymus as well as mature activated T-cells.

We used a modified pre-B cell colony assay to determine the effects of IL-7 and S17 cells in B-cell development (Cumano et al. 1990) (Table I). A number of previous studies have confirmed that IL-7 is not produced by S17 cells.

For this purpose we have used B220$^+$ cells isolated from fetal liver at day 15 of gestation. These cells constitute ~2% of the total population and ~100% of the colony-forming-pre-

Table II. Maturation of B-cell precursors recovered from liquid culture

Liquid culture[a)]	Cells recovered ($x10^3$)[b)]	Plaque-forming colonies/10^4 cells[c)].
Day 3		
Medium	10	100
S17 cells	15	110
IL-7	50	1,100
Day 4		
Medium	8	80
S17 cells	23	420
IL-7	100	1,500
Day 5		
Medium	2	20
S17 cells	25	2,200
IL-7	210	1,600

a) $B220^+$ 15-day fetal liver cells ($4x10^4$) were cultured as described in Table I
b) After 3, 4 or 5 days in culture with the different stimuli cells were harvested and counted. Half the cell content of two individual wells was used corresponding to $4x10^4$ cells.
c) Cells were replated in agar with S17 cells and LPS as stimuli. Cells were at the same concentrations as mentioned in Table I except for D 5 of cells cultured with S17 cells which were plated at $1.25x10^3$, 625 and 312 cells per culture. Plaque-forming colonies were counted after 6 days and the results are expressed as plaque-forming colonies in 10^4 cells seeded in the culture.

Table I. Plaque-forming colonies generated from fetal liver cells in a two step culture assay

Liquid culture[a)]	Agar culture[b)]	Plaque forming colonies/5.000 cells[c)]
	IL-7+LPS	22.4
IL-7	S17+LPS	515.2
	LPS	3.2
	IL-7+LPS	19.2
S17	S17+LPS	62.4
	LPS	8
	IL-7+LPS	3.2
Medium	S17+LPS	42.4
	LPS	2.4

a) $B220^+$ 15-day fetal liver cells ($4x10^4$) were cultured for 3 days with 250 U IL-7, $2x10^4$ S17 adherent cells, irradiated with 2000 rads or medium alone in 2 ml medium in 24-well Costar plates. After this period cells were counted and plated in agar; cell counts were $4x10^4$ for pre-B cells in culture with S17 cells, $2x10^5$ for cells in IL-7 and $2.8x10^4$ for cells in medium alone.
b) Cells harvested from the liquid culture were plated in agar IL-7, S17 cells or LPS alone. Cells growing with IL-7 were replated at starting concentrations of 625 and 312 cells per culture. Cells in either S17 cells or medium alone were replated at concentrations of $2.5x10^3$ and $1.25x10^3$ cells/culture.
c) After 6 days colonies were enumerated and tested for the presence of plaque-forming colonies. The numbers of colonies and plaque-forming colonies coincided, showing that virtually all colonies were secreting Ig. Results are expressed as plaque-forming colonies in $5x10^3$ cells seeded in the liquid culture.

B cells. We incubated these cells in liquid culture with IL-7, S17 cells or with medium alone. After 3 days in culture, cells were harvested, counted and plated in a agar colony assay containing LPS alone, S17+LPS or IL-7+LPS. After the liquid culture period only the cells cultured with IL-7, showed a significant increase in proliferation (5-10 fold). Cells cultured with S17 cells or medium alone did not show an increase in the number of cells recovered. These results indicate that whatever stimuli are provided by S17 stromal cells in support of pre-B cell colony formation it is not via expansion of the precursor cell as it is the case with IL-7. We also found that the cells which responded to IL-7 in the liquid culture phase failed to continue to expand when subsequently cloned in agar with LPS+IL-7. In contrast, mitogen responsiveness was demonstrated in the presence of S17. The results suggest a novel function of stromal cells in the pre-B colony assay.

These findings suggest that S17 cells play a significant role in "in vitro" differentiation of B-cell precursors at a stage after cells can react to IL-7 stimulus (mature B cells do not respond to IL-7). This was further investigated in two additional experiments. In the first one we reasoned that if S17 cells have a role in maturation of B-cell precursors, incubation of $B220^+$ enriched fetal liver cells for prolonged periods of time would lead to elevated levels of LPS responsive cells independent of cell division. The results are shown in Table II.

After enrichment cells were incubated with medium alone, S17 adherent cells or IL-7. After 3, 4 and 5 days cells were harvested, counted and used in a colony assay stimulated by LPS and S17 cells. These results show the cell recovery after the liquid incubation period and the number of plaque-forming-colonies in 10^4 $B220^+$ cells. After 3 days cells incubated with medium alone or with S17 cells behaved similarly, whereas in IL-7 expanded the cells a proportional increase in PFCs was observed. S17 cells induced a marginal proliferation (2 fold in 5 days) but at day 5 there is a dramatic increase in the PFCs recovered which is higher than cells incubated with IL-7 alone. Results of this experiment support our conclusion that S17 cell products have an important role in the maturation in B-cell precursors.

The ability of S17 cells to promote maturation of responsivness to LPS is not the only effect on B-cell precursors. Our results also show a significant enhancement in cell proliferation in response to IL-7. Exposure of B-cell precursors at low cell concentrations (100-500 cells /well), to S17 adherent cells along with IL-7 results in a 10 fold increase in the amount of thymidine incorporation. However, in the absence of IL-7, S17 cells do not induce a substantial proliferation (Cumano et al. 1990). The major implication of this result is that S17 factors deliver a complementary signal to IL-7 in lymphopoiesis. The nature of this stimulus can be to support the survival of cells and therefore to provide them the opportunity to complete a pre-determined developmental program. Alternatively, accessory molecules responsible for signal transduction have been described both for T and B lymphocytes which are associated to the antigen receptor. It is possible that at certain stages triggering one specific receptor alone is not sufficient to induce cell division but two or more combined stimuli are required. The protein produced by stromal cells might be the ligand of one such receptor.

STROMAL CELLS PRODUCE FACTORS THAT INCREASE THE CLONING EFFICIENCY OF SPLENIC B-CELLS IN RESPONSE TO MITOGENS

Polyclonal activators (agar mitogen, LPS) can induce B-cell proliferation and differentiation to plasma cells. In C57BL/6 mice the frequency of splenic cells stimulated with LPS, in a colony assay is 1:9. The presence of S17 cells as an adherent layer can increase this frequency to 1:2 which means that essentially all splenic B-cells respond to LPS (Table III).

The X linked genetic defect in the CBA/N strain of mice results in a wide spectrum of immune abnormalities (reviewed by Scher 1982, Kincade 1988). These include the inability of their B lymphocytes to respond to certain thymus independent antigens and B cell mitogens (Mosier et al. 1976, Huber and Melchers 1979). One intriguing "in vitro" manifestation of this condition is the virtual inability of CBA/N B cells to form colonies in LPS containing semi-solid agar cultures (Kincade 1977). As a step towards understanding the mechanism by which the genetic defect of CBA/N animals may relate to the observed immunological abnormalities, we are studying the conditions that could overcome the lack of colony growth in semisolid media. It was found that the colony forming ability of CBA/N spleen cells can be efficiently restored by feeder layers of S17. This confirms that LPS sensitive clonable B cells are present in these animals and the factor(s) provided by S17 either helps to overcome the deficiency of a critical accessory cell or an autocrine factor, or helps bypass a defect in the relevant biochemical pathways imposed by the genetic abnormality (Narendran et al. 1991, manuscript in preparation).

Table III. Number of plaque-forming colonies generated from unseparated spleen cells in C57BL/6 mice

Culture conditions	Plaque-forming colonies/100 cells[a)]	Frequency
AGAR	2.5	1:40
AGAR+S17	20	1:5
AGAR+LPS	11	1:9
AGAR+S17+LPS	48	1:2
AGAR+IL-7	1.5	1:63
AGAR+IL-7+LPS	9	1:10
AGAROSE+LPS	1	1:90
AGAROSE+S17+LPS	25	1:4

a) Cells were plated at concentrations of 250 and 125 nucleated cells per culture, except when agarose was used in which case 1000 and 500 cells were plated. Plaqu-forming colonies were tested at day 6 from four to six individual cultures.

ADDITIONAL REMARKS

It has been recently reported (Hayashi et al., 1990) that prior to a stage where cells can respond to IL-7, B-cell precursors require the presence of stromal cells "in vitro" in conjunction with IL-7 to proliferate. These cells can subsequently become stromal cell independent. Those results suggest that also before cells can respond to IL-7, stromal cells are important in lymphopoiesis and we have preliminary evidence that this effect can also be mediated by stromal cell conditioned medium.

We have no evidence that the latter effect is mediated by the same factor as the ability to support differentiation into mitogen responsiveness or increased proliferative capacity of mature B-cells. We have, however, evidence that the same molecule produced by S17 stromal cells can mediate CBA/N spleen-cell colony formation, increase LPS responsiveness in normal splenic cells and support pre-B cell differentiation to mature lymphocytes.

CONCLUSIONS

Here we propose that stromal cells isolated from hematopoietic and lymphoid organs produce factors, yet undefined, that can have an effect in a wide spectrum of B-cell developmental stages. Rather than different factors acting at different stages of development we hypothesize that a single molecule is able to support the differentiation from pre-B cell to mature B-cell and to increase the proliferation of activated peripheral B-cells. In Figure 1 we show a schematic representation of this concept. Cells that are committed to the B-cell lineage are identified by having the B220 surface marker. These cells have started to rearrange the heavy-chain immunoglobulin locus and have the ability to respond to a stimulus delivered by stromal cells alone or in combination with IL-7. Later, at the pre-B cell stage, such cells can still respond to stromal cells although they now have the ability to respond to IL-7 alone. Stromal cells alone can support differentiation to mitogen responsiveness although little proliferation takes place. Mature B-cells increase their proliferative capacity to mitogens in the presence of stromal cells. Possibly, the signal delivered by S17 stromal cell line triggers an accessory molecule involved in signal transduction which is required to induce mitosis in some stages of B-cell differentiation. Alternatively, it may lead to an increase in the concentration of receptors on the cell surface. This effect may be associated with a specific molecule and its effects are not specific for a particular stage in cell differentiation.

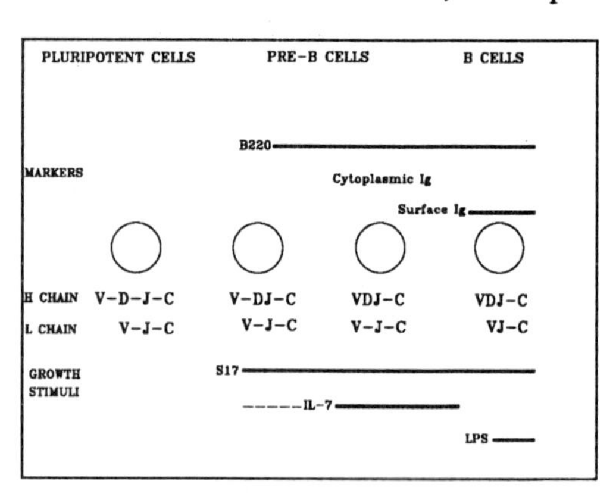

Figure 1

Conditioned media from stromal cell lines as well as from fresh adherent cells are particularly ineffective in duplicating the activities observed when compared to the stromal cell themselves. This observation indicates that the amount of active protein produced by those cells in culture is either extremely low or unstable. If we associate low production of factors with the presence of adhesion molecules on the surface of stromal cells as well as shared surface markers with the B-cell precursors we might conclude that cell to cell interaction is important in vivo signaling between B-cell precursors and stromal cells and that contact between those cell types is the physiological way growth factors are delivered in a regulated manner in hematopoietic organs to target precursor cells.

Identification of the molecules involved in this developmental process and the function of stromal cells in the interaction with their target precursors is essential towards understanding B-cell development and hematopoiesis in general.

This work was supported by the Medical Research Council and the National Cancer Institute of Canada. A.C. is a fellow of the Medical Research Council of Canada and A.N. is a fellow of the Arthritis Society of Canada.

REFERENCES

Collins, L. S. and Dorshkind, K., 1987. A stromal cell line from myeloid long-term bone marrow cultures can support myelopoiesis and B lymphopoiesis. J. Immunol. 138:1082.

Cumano, A., Dorshkind, K., Gillis, S. and Paige, C. J., 1990. The influence of S17 stromal cells and interleukin 7 on B cell development. Eur. J. Immunol. 20:2183.

Hayashi, S.-I., Kunisada, T., Ogawa, M., Sudo, T., Kodama, H., Suda, T., Nishikawa, S. and Nishikawa, S.-I., 1990. Stepwise progression of B lineage differentiation supported by interleukin 7 and other stromal cell molecules. J. Exp. Med. 171:1683.

Huber, B. and Melchers, F., 1979. Frequencies of mitogen-reactive B cells in the mouse. Lipopolyssacharide-, lipoprotein- and Nocardia mitogen-reactive B cells in CBA/N mice. Eur. J. Immunol. 9:827.

Kincade, P. W., 1977. Defective colony formation by lymphocytes from CBA/N and C3H/HeJ mice. J. Exp. Med. 145:249.

Kincade, P. W., 1988. Experimental models for understanding B lymphocyte formation. Adv. Immunol. 41:181.

Mosier, D. E., Scher, I. and Paul, W. E., 1976. In vitro responses of CBA/N mice: spleen cells of mice with an X-linked defect that precludes immune responses to several thymus-independent antigens can respond to TNP-lipopolyssacharide. J. Immunol. 117:1363.

Namen, A. E., Lupton, S., Hjerrild, K., Wignall, J., Mochizuki, D. Y., Schmierer, A., Mosley, B., March, C. J., Urdal, D., Gillis, S. and Goodwin, R. G. 1988. Stimulation of B-cell progenitors by cloned murine interleukin 7. Nature 333:571.

Paige, C. J. and Skarvall, H. J., 1982. Plaque formation by B cell colonies. J. Immunol. Methods. 52:51.

Paige, C. J., 1983. Surface immunoglobulin-negative B-cell precursors detected by formation of antibody-secreting colonies in agar. Nature 302:711.

Pietrangeli, C. E., Hayshy, S.-I. and Kincade, P. W., 1988. Stromal cell lines which support lymphocyte growtth: characterization, sensitivity to radiation and responsivness to growth factors. Eur. J. Immunol. 18:863.

Sauter, H. and Paige, C. J., 1987. Detection of normal B-cell precursors that give rise to colonies producing both k and l immunoglobulin chains. Proc. Natl. Acad. Sci. USA 84:4989.

Scher, I. 1982. The CBA/N mouse strain: An experimental model illustrating the influence of the X-chromossome on immunity. Adv. Immulogy 33:1.

Welch, P. A., Burrows, P. D., Namen, A., Gillis, S. and Cooper, M. D., 1990. Bone marrow stromal cells and interleukin-7 induce coordinate expression of the BP-1/6C3 antigen and pre-B cell growth. International Immunology 2:697.

Wu, G. E. and Paige, C. J., 1988. V_H gene family utilization is regulated by a locus outside of the V_H region. EMBO J. 5:3475.

SUPPRESSOR T CELLS INDUCED IN VIVO BY TOLEROGENIC CONJUGATES OF A GIVEN ANTIGEN AND MONOMETHOXYPOLYETHYLENE GLYCOL DOWNREGULATE ANTIBODY FORMATION ALSO TO A SECOND ANTIGEN, IF THE LATTER IS PRESENTED AS A COVALENT ADDUCT WITH THE FORMER

Alec H. Sehon

MRC Group for Allergy Research
Department of Immunology
The University of Manitoba
Winnipeg, MB
Canada R3E OW3

ABSTRACT

Recent results of studies employing tolerogenic monomethoxypolyethylene glycol (mPEG) conjugates of antigens are briefly reviewed. Administration of antigen$(mPEG)_n$ conjugates into mice induced antigen-specific suppressor T (Ts) cells, from which a suppressor factor (TsF) was extracted. These Ts cells were cloned and shown to be Thy1.2$^+$, CD3$^+$, CD4$^-$, CD5$^-$, CD8$^+$ and to express the $\alpha\beta$ heterodimer of conventional T cell receptors (TCR). The TsF of a clone of OVA-specific Ts cells shared the epitopes of the α and β chains of TCR, whereas the TsF of T cells of an HIgG-specific clone shared only the epitope of the α chains of TCR; OVA and HIgG represent ovalbumin and human monoclonal (myeloma) IgG. These studies have provided evidence for the phenomenon of "linked immunological suppression" which may be summarized by the statement "Ts cells specific for an epitope of a given antigen, Ag_A, suppress the antibody response to an unrelated antigen, Ag_B, only if the latter is presented in the form of a covalent adduct, Ag_A-Ag_B, to the immune system of the animal pretolerized with $Ag_A(mPEG)_n$, but not if Ag_B is presented as a mixture with Ag_A.

INTRODUCTION

The possibility of *selectively* downregulating the host's immune response to a given antigen represents one of the most formidable challenges of modern immunology in relation to the development of new therapies for IgE-mediated allergies, autoimmune diseases, and the prevention of the immune rejection of organ transplants. Similar considerations apply to an increasing number of promising therapeutic modalities for a broad spectrum of diseases, which would involve the use of foreign biologically active agents potentially capable of modulating the immune response, provided they were not also immunogenic. Among these agents, one may cite (i) xenogeneic monoclonal or polyclonal antibodies

Immunobiology of Proteins and Peptides VI
Edited by M.Z. Atassi, Plenum Press, New York, 1991

(collectively referred to here as xIg) against different epitopes of the patients' $CD4^+$ cells (Cruse and Lewis, 1989; Diamantstein and Osawa, 1986), administered alone or in combination with immunosuppressive drugs for treatment of rheumatoid arthritis and other autoimmune diseases, or for the suppression of graft versus host reactions and of the immune rejection of organ transplants (Cruse and Lewis, 1989), (ii) "magic bullets" for the destruction of tumour cells (Vogel, 1987; Frankel, 1988; Sedlacek et al., 1988), which consist of anti-tumour xIg to which are coupled toxins (Tx), or radionuclides, or chemotherapeutic drugs. However, in most cases the patients produce antibodies to the injected xIg and to the even more immunogenic immunotoxins (xIg-Tx); consequently, the therapeutic effectiveness of these immunological strategies is undermined by the patients' antibodies which prevent these "bullets" from reaching their target cells. Similar limitations apply to the use of other biologically active substances synthesized by recombinant DNA technology such as hormones (Buzi et al., 1989) or growth factors and lymphokines (Katre, 1990), which are often immunogenic as a result of small differences in their conformational characteristics or in their glycosidic constituents in relation to their natural counterparts.

However, as shown in a series of papers from the author's laboratory, it is possible to convert an antigen, including xIg, to the corresponding tolerogenic derivatives by coupling an optimal number (n) of molecules of monomethoxypolyethylene glycol (mPEG) onto the antigen molecule. Thus, as substantiated by recently published results (Wilkinson et al., 1987; Maiti et al., 1988), it was possible to administer multiple immunizing injections of unmodified xIg into mice, which had been pretolerized by treatment with the corresponding $xIg(mPEG)_n$ conjugates, without inducing a significant antibody response in the recipients, i.e., the immunosuppression was in excess of 90% and the immunological unresponsiveness could be maintained for periods longer than one year.

The present paper represents an update of the progress made in the author's laboratory since the review of his studies published in the last volume of this series (Sehon, 1989). It provides additional evidence for the activation of antigen-specific Ts cells by mPEG conjugates of the respective antigens and the generation of the corresponding Ts cell clones (Takata et al., 1990). Moreover, evidence has been adduced (unpublished data) that mice pretolerized to a given antigen Ag_A by injection of $Ag_A(mPEG)_n$ were co-suppressed to a second antigen (Ag_B), on condition that Ag_B was administered to these mice as a conjugate, Ag_A-Ag_B, but not as a mixture of the two antigens; this phenomenon will be referred to as "linked immunological suppression".

GENERATION OF ANTIGEN-SPECIFIC Ts CELL CLONES

In earlier experiments (Sehon, 1982) it had been shown that transfer of spleen cells from mice tolerized by mPEG conjugates of ovalbumin (OVA) into naive syngeneic recipients led to significant suppression of anti-OVA antibody responses in the latter to subsequent injections of unmodified OVA. It was, therefore, concluded that the downregulation of the primary antibody responses was due, at least in part, to the activation of antigen-specific Ts cells in immunosuppressed animals. In more recent experiments (Wilkinson et al., 1987), injection of spleen cells from mice tolerized with mPEG conjugates of HIgG induced HIgG-specific tolerance in syngeneic recipients. The extracts of the spleen cells of these immunosuppressed animals, which were obtained by freezing and thawing of the cells (referred to as F/T extracts), were also capable of downregulating the *specific immune response to HIgG*; hence, this extract was deemed to contain the suppressor T cell factor (TsF). The characteristics of OVA- and HIgG-specific Ts cells, induced *in vivo* by mPEG conjugates of these antigens, were studied with the aid of cell culture systems designed for the induction of antibody responses *in vitro* (Mokashi et al., 1989; Takata et al., 1990). Thus, it was established that injection of $OVA(mPEG)_{13}$ into mice induced splenic Ts cells which could suppress secondary *in vitro* IgG responses to $OVA(DNP)_3$ in an antigen-specific manner.

In very recent experiments *non-hybridized HIgG-specific Ts clones* were generated from spleen cells of B6D2F1 mice tolerized with $HIgG(mPEG)_{26}$ (Takata et al., 1990). One of these clones (clone 23.32), possessing the highest immunosuppressive activity for *in vitro* antibody formation, exhibited the following phenotypic characteristics: $Thy1.2^+$, $CD3^+$, $CD4^-$, $CD5^-$, $CD8^+$, and expressed the $\alpha\beta$ heterodimer of the T cell receptor (TCR). Moreover, the suppression exerted by the cells of this clone and by their F/T extract was HIgG-specific and MHC class I ($H\text{-}2K^d$) restricted. Similarly, the TsF of clone 23.32 was able to inactivate *in vitro* the function of HIgG-primed Th cells via a mechanism requiring the participation of accessory cells of the $H\text{-}2^d$ haplotype. On the basis of immunochemical results, it was also concluded that the TsF produced by clone 23.32 had a determinant which was serologically related to that of the α chain of TCR. However, because the TsF did not react with two monoclonal antibodies specific for epitopes of the β chain of TCR, it was not possible to conclude that the TsF of this clone was identical to TCR. On the other hand, in very recent unpublished experiments, OVA-specific Ts cell clones -- with phenotypic and functional characteristics similar to those of clone 23.32 -- which were generated from spleen cells of mice tolerized with $OVA(mPEG)_{12}$ were shown to be devoid of cytotoxic activity. Moreover, the TsF of these cloned Ts cells shared the epitopes of both α and β chains of TCR. All these results taken together support the view that the specific immunosuppression induced by tolerogenic mPEG conjugates involves the activation of the corresponding antigen-specific Ts cells. Experiments designed to clarify the precise relationship between the TCR of these Ts cells and their suppressor factors at the molecular genetic level are in progress.

CO-SUPPRESSION OF THE IMMUNE RESPONSE TO TWO COVALENTLY LINKED ANTIGENS IN MICE PRETOLERIZED BY mPEG CONJUGATES OF ONE OF THE ANTIGENS

The unexpected discovery was recently made that pretreatment of recipient mice with a tolerogenic mPEG conjugate suppressed the immune response not only to the antigen incorporated in the tolerogen, but also to a covalent compound consisting of that antigen and another unrelated antigen (unpublished data). This principle of "linked immunological suppression" is illustrated by the results in Table 1. For this experiment, the test and control mice received first one injection of the tolerogen, $HIgG(mPEG)_{21}$, and PBS, respectively. All the mice were thereafter immunized on days 7 and 37 with 20 μg of heat aggregated HIgG (haHIgG) and on days 49 and 59 with a conjugate of OVA and HIgG. All mice were bled on day 21 for the determination of their anti-HIgG antibody titers and on days 66 and 84 for the assay of their anti-OVA and anti-HIgG antibody titers; these titers were determined by enzyme linked immunoassays. It is obvious from the data presented in this Table that the antibody response of the animals which had been pretreated with $HIgG(mPEG)_{21}$ was suppressed not only to HIgG, but also to OVA, when the latter was presented to mice in the form of a *covalent copolymer* with HIgG. By contrast (data not shown), the anti-OVA response in mice pretolerized with $HIgG(mPEG)_{21}$ was not affected when these mice were immunized with a *mixture* of HIgG and OVA; however, as expected, the anti-HIgG antibody response was strongly suppressed.

TABLE. 1. CO-SUPPRESSION OF ANTIBODY RESPONSES TO OVA AND HIgG ON IMMUNIZATION, WITH COVALENT OVA-HIgG ADDUCTS, OF B6D2F1 MICE PRETOLERIZED WITH HIgG(mPEG)$_{21}$ CONJUGATES

Treatment of mice[a]	Immunizations[b]				Anti-HIgG ELISA titres[c]						
						IgG1		IgG+IgM+IgA		IgG2a	
0	d7	d37	d49	d59	d21	d66	d84	d66	d84	d66	d84
PBS	haHIgG (20 μg)		OVA-HIgG (50 μg)		28,000	>100,000	>100,000	>100,000	>100,000	1,200	700
HIgG(mPEG)$_{21}$	"		"		3,500	3,000	2,500	2,000	1,500	<200	<200

	Anti-OVA ELISA titres					
	IgG1		IgG+IgM+IgA		IgG2a	
	d66	d84	d66	d84	d66	d84
	88,000	45,000	25,000	11,000	900	800
	6,000	11,000	2,500	3,500	<200	<200

(a) The test mice received one injection of 100 μg of tolerogenic HIgG(mPEG)$_{21}$ conjugate on day 0, and the control mice received PBS.

(b) All mice received two injections of 20 μg haHIgG on days 7 and 37, and two injections of 50 μg OVA-HIgG on days 49 and 59; the OVA-HIgG conjugate was prepared by crosslinking with EDCI.

(c) ELISA titres were determined using conjugates of alkaline phosphatase and rabbit antibodies to mouse IgG1; IgG + IgM + IgA; Ig2a; these reagents were purchased from ZYMED, San Francisco, CA.

DISCUSSION

Since the study relating to the generation of antigen-specific Ts cell clones and to the relationship between the TCR of the cloned Ts cells and their TsF was recently reported elsewhere (Takata et al., 1990), this discussion will be limited to an analysis of the possible mechanism underlying the phenomenon of "linked immunological suppression" observed in the experiment described above (and in other similar systems developed in this laboratory). In very recent unpublished *in vitro* experiments, it was established that the cloned Ts cells and their TsF exerted their antigen-specific downregulating effect on antibody formation by inactivation (but not by cytolytic elimination) of the corresponding carrier-specific helper T (Th) cells. Moreover, as illustrated in Table 1, the animals pretolerized to a given antigen exhibited co-suppression to the *covalent conjugate* of that antigen and an unrelated antigen.

Taking into account all these facts, the author believes that the hypothetical, simplistic model, illustrated in Figure 1, encompasses the main features of the observed phenomenon. Accordingly, it is visualized that: (i) the covalently linked heteropolymer Ag_A-Ag_B is first processed by an antigen presenting cell (APC), capable of presenting simultaneously epitopes of an exogeneous antigen in association with both MHC class I and class II molecules (Rock et al., 1990), (ii) the epitopes of Ag_A (i.e., α_1, α_2, ... α_n) and of Ag_B (i.e., β_1, β_2, ... β_n) reappear subsequently on the membrane of the APC in the form of complexes with appropriate MHC class I and class II molecules, (iii) these complexes interact on the one hand with the corresponding Ts_α and on the other with Th_α and Th_β cells.

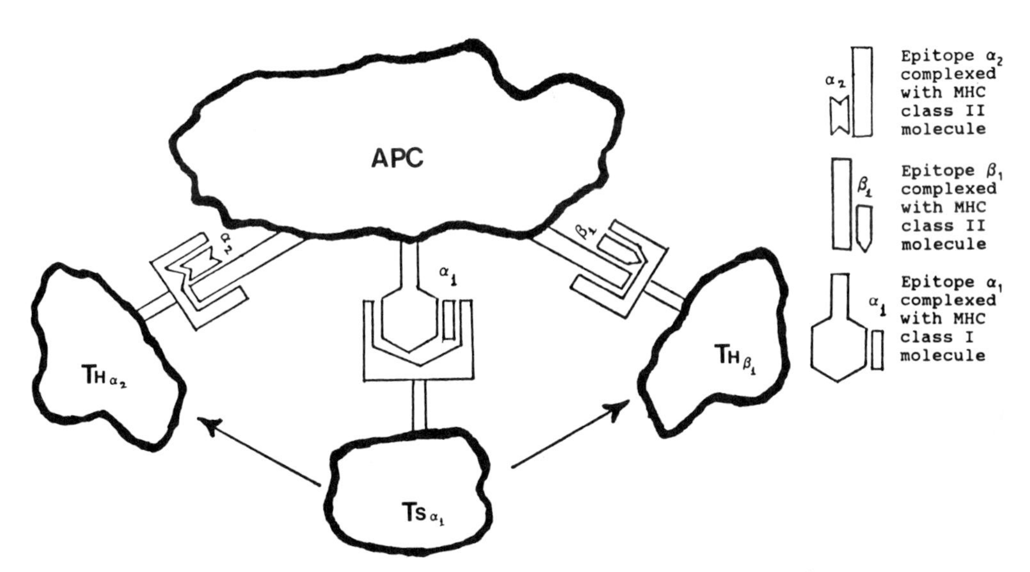

Fig. 1. Co-suppression of Th_α and Th_β cells by Ts_α cells. In this model the APC is assumed to have processed the antigenic adduct Ag_A-Ag_B and re-exposed on its membrane (i) the epitopes α, and α_2 of Ag_A in association with the APC's MHC class I and class II molecules, respectively, and (ii) the epitopes β_1 of Ag_B in association with the APC's MHC class II molecule. These complexes are shown as interacting with the corresponding Th and Ts cells leading to a multicellular cluster within a given sector of the APC. The arrows emanating from the Ts_α cell represent lymphokine molecules which downregulate closely neighbouring Th cells in an antigen-nonspecific manner.

This multicellular cluster model is in essence a variation on Mitchison's theme proposed for silencing the appropriate Th cell by an idiotype-specific cytolytic Ts cell (Mitchison, 1989). This model incorporates the essential prerequisites for the linked suppression to occur, viz., (i) it explains the co-existence of the Th_{α} and Th_{β} cells -- recognizing, respectively, the different α and β epitopes complexed with MHC class II molecules -- on the membrane of the same APC, and allows also for the simultaneous accommodation of the Ts_{α} cell recognizing one of the α epitopes which is associated with a class I molecule on the same APC, (ii) it allows for short range interactions between Th and Ts cells on the membrane of the APC, which is essential for the effective transfer of downregulating signals (TsF and possibly also nonspecific negative lymphokines) from the carrier-specific Ts cell to the Th cells.

This hypothetical model implies that the *apparently antigen-specific* negative effect of a Ts_{α} cell on a Th_{α} cell, which is observed when both cells are linked to the same APC -- as would be the case under normal circumstances involving a monomolecular antigen -- may be the result of the release of specific and/or nonspecific cytokines by the Ts_{α} cell. As a corollary, the encounter between a Ts_{α} cell and a Th_{β} cell on the membrane of the same APC may result in the inactivation of the Th_{β} cell by the presumed nonspecific suppressor cytokine(s) released by the Ts_{α} cell. Obviously, as substantiated by the results listed in Table 1, if Ag_A and Ag_B were not covalently linked and were injected as a mixture, the probability that both antigens would encounter the same APC and would be processed by it would be minimal.

On the other hand, if the antigenic copolymer, Ag_A-Ag_B, were not degraded *in vivo*, i.e., the bimolecular antigen were not processed by the APC, these results could be explained by an alternative model involving the simultaneous interaction of the Th_{α} and Th_{β} cells with Ts_{α} cells, as a result of their association with a nondegraded form of the duplex antigen. Although this concept is currently not *en vogue*, it is to be noted that convincing experimental evidence has been recently provided (Atassi et al., 1989) supporting this alternative view, i.e., that T cell recognition of some antigens may take place by interaction with the respective, whole (i.e., not processed) protein antigen molecules.

Obviously, for a more detailed understanding of the mechanisms underlying the phenomenon of "linked immunological suppression", additional experimental systems will have to be designed with the aid of chemically precisely defined conjugates of the type $(Ag_A)_x$-$(Ag_B)_y$, i.e., it will be interesting to establish the effect of the molecular composition of the bimolecular antigenic adducts (differing in the numbers x and y of the constituent molecules of each antigen) on the linked suppression induced *in vivo* as a result of pretolerizing the animals with tolerogenic mPEG conjugates of either Ag_A or Ag_B.

CONCLUSIONS

Studies involving tolerogenic antigen-mPEG conjugates have established that (i) presentation of a given antigen covalently linked to an optimal number of mPEG molecules results in abrogation of the antibody response in an antigen-specific manner, (ii) this downregulating effect of the immune response is due to the activation of antigen-specific, non-cytolytic Ts cells, and (iii) Ts cells specific for Ag_A can suppress the immune response to Ag_B, when Ag_B is presented in the form of a covalent Ag_A-Ag_B adduct to an animal pretolerized to Ag_A.

ACKNOWLEDGEMENTS

Valuable discussions with Dr. Eli Sercarz are gratefully acknowledged. The studies reported in this paper have been conducted in collaboration with Drs. Soji Bitoh, Youhai Chen, Valerie Holford-Strevens, Danuta Kierek-Jaszczuk, Glen M. Lang, P.K. Maiti, E. Rector and Masaru Takata. The skillful secretarial assistance of Y. Hein and K. Risk is also greatly appreciated. These studies were supported by generous grants from the Medical Research Council of Canada and the National Institute of Allergy and Infectious Diseases of the US National Institutes of Health.

REFERENCES

Atassi, M. Z., Zouhair, M., Yoshioka, M., and Bixler, Y. S., Jr., 1989, T cells specific for α-β interface regions of hemoglobin recognize the isolated subunit but not the tetramer and indicate presentation without processing, Proc. Natl. Acad. Sci. U.S.A., 86:6729.

Buzi, F., Buchanan, C.R., Morrell, D.J., and Preece, M.A., 1989, Antigenicity and efficacy of authentic sequence recombinant human growth hormone (somatropin): First-year experience in the United Kingdom, Clin. Endocrinology, 30:531.

Cruse, J. M. and Lewis, R. E., Jr., eds., 1989, "Therapy of autoimunne diseases," Karger, Basel.

Diamantstein, T. and Osawa, H., 1986, The interleukin-2 receptor, its physiology and a new approach to a selective immunosuppressive therapy by anti-interleukin-2 receptor monoclonal antibodies, in: "Immunol. Rev.," 92:5, G. Möller. ed., Munksgaard, Copenhagen.

Frankel, A. E., 1988, "Immunotoxins," Kluwer Academic Publishers, Boston.

Katre, N. V., 1990, Immunogenicity of recombinant IL-2 modified by covalent attachment of polyethylene glycol, J. Immunol., 144:209.

Maiti, P. K., Lang, G. M., and Sehon, A. H., 1988, Tolerogenic conjugates of xenogeneic monoclonal antibodies with monomethoxypolyethylene glycol. I. Induction of long-lasting tolerance to xenogeneic monoclonal antibodies, Int. J. Cancer, 3:17.

Mitchison, N. A., 1989, Suppression of the response to murine alloantigens: Four-cell-type clusters, function-flipping and idiosyncratic responses, in: "Antigenic determinants and immune regulation", Chem. Immunol., 46:157, E. Sercarz, ed., Karger, Basel.

Mokashi, S., Holford-Strevens, V., Sterrantino, G., Jackson, C-J. C., and Sehon, A.H., 1989, Down-regulation of secondary *in vitro* antibody responses by suppressor T cells of mice treated with a tolerogenic conjugate of ovalbumin and monomethoxypolyethylene glycol, $OVA(mPEG)_{13}$, Immunol. Lett., 23:95.

Rock, K.L., Gamble, S., and Rothstein, L., 1990, Presentation of exogeneous antigen with class I major histocompatibility complex molecules, Science, 249:918.

Sedlacek, H. H., Schulz, G., Steinstraesser, A., Kuhlmann, L., Schwarz, A., Seidel, L., Seemann, G., Kraemer, H. P., and Bosslet, K., 1988, "Monoclonal antibodies in tumor therapy - Present stage, chances and limitations," Karger, Basel.

Sehon, A. H., 1982, Suppression of IgE antibody responses with tolerogenic conjugates of allergens and haptens, in: "Regulation of the IgE antibody response," Prog. Allergy, 32:161, K. Ishizaka, ed., Karger, Basel.

Sehon, A., 1989, Modulation of antibody responses by conjugates of antigens with monomethoxypolyethylene glycol, in: "Immunobiology of proteins and peptides V", M. Z. Atassi, ed., Plenum Publishing Corporation, New York.

Takata, M., Maiti, P. K., Kubo, R. T., Chen, Y., Holford-Strevens, V., Rector, E., and Sehon, A. H., 1990, Cloned suppressor T cells derived from mice tolerized with conjugates of antigen and monomethoxypolyethylene glycol, J. Immunol., 145:2846.

Vogel, C-W., 1987, "Immunoconjugates: Antibody Conjugates in Radioimaging and Therapy of Cancer", Oxford University Press, Oxford.

Wilkinson, I., Jackson, C-J. C., Lang, G. M., Holford-Strevens, V., and Sehon, A. H., 1987, Tolerance induction in mice by conjugates of monoclonal immunoglobulins and monomethoxypolyethylene glycol. Transfer of tolerance by T cells and by T cell extracts, J. Immunol., 139:326.

A PURIFIED SAPONIN ACTS AS AN ADJUVANT FOR A T-INDEPENDENT ANTIGEN

A. C. White, P. Cloutier, and R. T. Coughlin

Cambridge Biotech Corporation, 365 Plantation St, Worcester, MA 01605

ABSTRACT

Three strains of mice were injected with a T-independent antigen, Escherichia coli 055:B5 polysaccharide (PS) combined with purified saponin, QS-21, isolated from Quillaja saponaria bark. PS was prepared by hydrolysis of lipopolysaccharide (LPS). Nine week old mice were injected intradermally with 60 ug PS, as determined by an anthrone assay, with or without 15 ug QS-21 on days 0 and 14. On day 22 sera were assayed by EIA for PS specific antibodies. Titers were 11-fold higher in CD-1 mice with QS-21. C3H/HeJ (lps^d) and C3H/HeSnJ (lps^r) mice also showed an adjuvant associated increase in titer with saponin. Therefore, LPS responsiveness was not required for the adjuvant effect. PS vaccinated C3H and CD-1 mice with and without QS-21 had similar antibody isotype profiles. IgG2b titers accounted for more than half of the total Ig response. IgG2a was next highest followed by IgG3, IgM, IgG1, and IgA. In comparison, CD-1 mice injected with 0.1 ug intact LPS had a different LPS specific isotype profile. IgG3 was the highest followed by IgG1, IgG2b, IgM, IgG2a, and IgA.

INTRODUCTION

Bacterial PS vaccines can stimulate B-cells directly to produce a rapid antibody response. These protective antibodies can opsonize and assist complement mediated lysis of bacteria. However, without T-cell help the antibodies produced are of low affinity and have a restricted isotype profile (Coughlin and Bogard, 1987). Furthermore, young children are frequently unable to respond to T-independent antigens. Researchers are currently exploring various methods to make polysaccharide vaccines more effective. Conjugation of PS to proteins is one method which has proven successful (Cryz, Sadoff, and Furer, 1988).

Our approach was to test purified saponins isolated from the bark of the South American tree Q. saponaria, as potential adjuvants for PS vaccines. We focused on a purified saponin fraction, QS21, alone and in combination with an oil and water emulsion. QS21 is significantly purer than crude bark extracts such as Quil-A, which have been used to formulate ISCOMs; in addition QS21 is less toxic in mice than these crude extracts (Kensil, Patel, Lennick, and Marciani, in press).

MATERIALS AND METHODS

Preparation of polysaccharide and saponin

Phenol extracted LPS from E. coli 055:B5 (Sigma, St. Louis, MO) was hydrolyzed at

Immunobiology of Proteins and Peptides VI
Edited by M.Z. Atassi, Plenum Press, New York, 1991

100C in 0.1 N HCl for one hour. The PS solution was neutralized with 5 N NaOH, centrifuged at 4C, and dialyzed against deionized water. An anthrone assay was used to determine the PS concentration. Hydrolysis of 136 mg LPS yielded 35 mg of pure PS. A neocuproine assay (Chaplin, 1986) adapted to a microplate format was used to determined the concentration of reducing sugar. Titration curves showed the PS reducing sugar level to be 3.6 times greater than LPS, but to be 61 times less than glucose. A LAL test (Whittaker Bioproducts, Inc.) determined the endotoxin level to be less than one percent by weight. Thus, the PS preparation was free of contaminating lipid A and LPS and was not denatured. QS21, a purified saponin, was purified from a water extract of Q. saponaria bark followed by silica chromatography (Kensil, ibid). The resulting product consisted of one major peak when analyzed by reverse phase HPLC.

Immunizations

Nine week old female CD-1 (Charles River Laboratories, Kingston, NY), C3H/HeSnJ, and C3H/HeJ mice (Jackson Laboratory, Bar Harbor, ME) were injected i.d. with 60 ug PS with or without 15 ug of QS21 in 0.2 ml PBS on days 0 and 14. Earlier mouse experiments determined 60 ug PS with 15 ug QS21 to be the optimal vaccine formulation. For the oil and water emulsion (OW) doses, the PS was homogenized with 0.2% Tween 80 in PBS and 1% squalane. On day 22 mice were bled out by cardiac puncture. Nine week old female CD-1 mice received either 1 or 0.1 ug LPS, i.d., in 0.2 ml PBS on days 0, 15, and 29. Tail bleeds were done on day 35.

Solid Phase EIA

Sera were assayed by enzyme linked immunoassay using LPS coated 96-well plates. Plates were coated overnight with 200 ul/well of 10 ug/ml E. coli 055:B5 LPS in PBS with 0.05% triethylamine and then blocked with 4% bovine serum albumin (BSA) in PBS overnight. Serially diluted immune mouse sera in 1% BSA/PBS were incubated for one hour and washed with 0.05% Tween 20. For total Ig determination, horseradish peroxidase labelled (HRP) affinity purified goat $F(ab')_2$ anti-mouse IgG + IgM (Tago, Burlingame, CA) was added for one hour. HRP labelled subclass specific reagents were also used (Southern Biotech Assoc., Birmingham, AL). A chromogenic substrate containing tetramethylbenzidine was used and absorbance were measured at 450nm. Serum antibody titers were expressed as endpoint dilutions producing an absorbance of 1.0 for total Ig assays and 0.5 for Ig subclass assays. Differences between antibody titers were determined to be significant by the z-test.

RESULTS AND DISCUSSION

The influence of QS21 and a lipophilic carrier on humoral immunity to E. coli 055:B5 PS in groups of CD-1 mice are shown in figure 1. An oil and water emulsion had no influence on serum titers ($P<0.05$) of PS immunized mice. QS21 enhanced the antibody titer of PS immunized mice 11 fold over PS alone.

The PS specific antibody subclass responses were measured on LPS coated plates. As shown in Table 1, normal mouse sera and PBS injected CD-1 mouse sera were not reactive. Although, PS alone produced very low LPS specific titers, QS21 showed an adjuvant effect for most isotypes. For PS + QS21 immunized CD-1 mice, titers were IgG2b>>IgG3>IgG2a>IgM >IgG1. Although LPS was at least a 600 fold more potent immunogen than PS with QS21 a very different antigen specific isotype profile was produced. For LPS immunized mice the order was IgG3>IgM>IgG1>>IgG2b>IgG2a. Antigen specific IgA was not detected.

The serum antibody responses to PS in C3H/HeJ (lps-nonresponder) and C3H/HeSnJ (lps-responder) mice are shown in Table 2. Both strains of mice responded well to PS vaccination with or without QS21. Both strains of mice responded better when given QS21 ($P<0.01$). Therefore, LPS responsiveness is not required for the QS21 adjuvant effect on PS.

Table 1. LPS specific subclass responses in CD-1 mice to PS and LPS

Immunogen	Adjuvant	Number of mice	IgM	IgG3	IgG1	IgG2b	IgG2a	IgA
None	none	10	<50	<50	<50	<50	<50	<25
PBS	none	10	<50	<50	<50	<50	<50	<25
PS, 60 ug	none	20	170	360	<150	950	<170	<25
PS, 60 ug	QS21, 15 ug	20	650	2200	<220	12000	1300	<25
LPS, 0.1 ug	none	10	3200	9500	1600	670	190	<25

Table 2. E. coli 055:B5 LPS specific titers for sera from individual C3H/HeJ (lps-nonresponder) and C3H/HeSnJ (lps-responder) mice immunized with PS

Vaccine	Number of mice	C3H/HeJ	C3H/HeSnJ
PBS	5	58 +/- 29	41 +/- 42
PS	10	1277 +/- 2029	2468 +/- 1989
PS + QS21	10	4805 +/- 5861	8140 +/- 10943

Table 3. LPS specific subclass response of C3H/HeJ and C3H/HeSnJ female mice

Strain	Vaccine	Number of mice	IgM	IgG3	IgG1	IgG2b	IgG2a	IgA
C3H/HeJ	PBS	5	<50	<50	<50	<50	<50	<25
	PS	10	105	150	285	2300	3200	<25
	PS+QS21	10	860	2300	250	13000	920	<25
C3H/HeSnJ	PBS	5	<50	<50	<50	<50	<50	<25
	PS	10	580	520	100	3200	620	<25
	PS+QS21	10	521	510	1050	6000	4150	<25

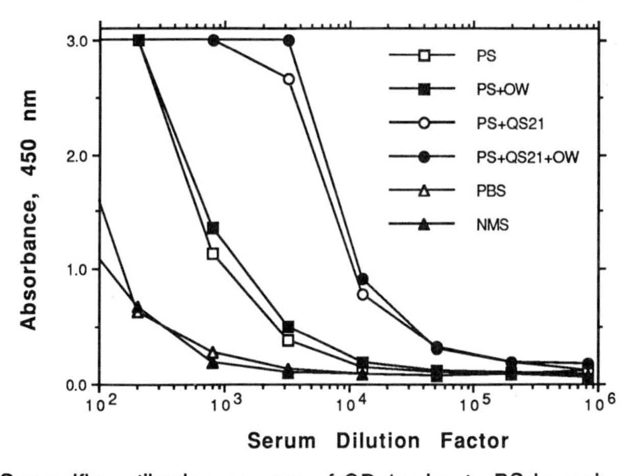

Figure 1. LPS specific antibody response of CD-1 mice to PS in various formulations.

Like PS vaccinated CD-1 mice, IgG2b was the dominant subclass present in serum for both C3H/HeJ and C3H/HeSnJ female mice (Table 3) and QS21 significantly enhanced specific IgG2b titers. C3H mice were unique in having a relatively high IgG2a titer.

The data presented suggests that QS21 is an effective adjuvant for a PS antigen. Rudbach et al. (1990) have shown that monophosphoryl lipid A is also an effective adjuvant for a T-independent antigen. Our experiments demonstrate that CD-1, C3H/HeJ, and C3H/HeSnJ mice have significantly enhanced serum titers to PS when QS21 is used. The titer increase was largely a result of increases in IgG2b subclass antibodies. The response in C3H/HeJ mice shows that contaminating LPS or lipid A in the PS preparation did not account for the vigorous response to PS.

Research is ongoing in our laboratory to determine whether QS21 also augments the immune response to PS protein conjugates.

ACKNOWLEDGEMENTS

We would like to thank Dr. Charlotte Kensil for supplying QS21 and Cindy Greer for her skillful technical assistance.

REFERENCES

Chaplin, M. F., 1986, Monosaccharides, in: "Carbohydrate Analysis, A Practical Approach," M. F. Chaplin and J. F. Kennedy, eds., IRL Press, Oxford.

Coughlin, R. T. and Bogard, W. C. (1987) Immunoprotective murine monoclonal antibodies specific for the outer-core polysaccharide and for the O-antigen of *Escherichia coli* 0111:B4 lipopolysaccharide (LPS), *J. Immunol.* 139:557.

Cryz, S., Sadoff, J., and Furer, E. (1988) Immunization with a *Pseudomonas aeruginosa* immunotype 5 O polysaccharide-Toxin A conjugate vaccine: Effect of a booster dose on antibody levels in humans, *Infect. Immun*, 56:1829.

Kensil, C., Patel, U., Lennick, M., and Marciani, D. J. (in press: *J. Immunol*) Separation and characterization of saponins with adjuvant activity from *Quillaja saponaria* Molina.

Rudbach, J. A., Cantrell, J. L., Ulrich, J. T., and Mitchell, M. S., 1990, Immunotherapy with bacterial endotoxins, in: "Endotoxin," H. Friedman, T.W. Klein, M. Nakano, and A. Nowotny, eds., Plenum Press, New York

MODIFIED-LIVE INFECTIOUS BOVINE RHINOTRACHEITIS VIRUS (IBRV) VACCINE EXPRESSING FOOT-AND-MOUTH DISEASE VIRUS (FMDV) CAPSID PROTEIN EPITOPES ON SURFACE OF HYBRID VIRUS PARTICLES

Saul Kit[1], Malon Kit[2], Richard DiMarchi[3], Sheila Little[3], and Charles Gale[3]

Baylor College of Medicine[1], and NovaGene, Inc.[2], Houston, TX; Eli Lilly Research Laboratories[3], Indianapolis, IN U.S.A.

INTRODUCTION

Foot-and-Mouth Disease (FMD) is a highly contagious disease of cattle and other cloven hoofed animals caused by a Picornavirus. It is enzootic in most South American countries, where hundreds of outbreaks are recorded each year, and endemic in many Asian, Middle East, African and European countries. The disease can spread rapidly due to its short incubation period and contagious nature and has a dramatic impact on livestock production and trade. In countries where FMD is endemic, binary ethylenimine-inactivated vaccines have been widely used to control disease. Annual production of FMD vaccine worldwide is probably between 700 and 800 million doses. Nearly 500 million doses are applied to cattle annually (Bahnemann, 1990). Yet, despite intensive vaccination programs, outbreaks of FMD occurred in 22 countries in the last quarter of 1989 (Foreign Animal Disease Report of USDA-APHIS,No. 18-2. Summer 1990).

Unfortunately, inactivated vaccines do not confer long-lasting immunity, possibly because they are less competent than live virus in inducing T cell responses and T cell memory. In endemic areas, vaccination consists of 2 inoculations of killed vaccine two to four weeks apart, and revaccination two or three times each year to maintain protective levels of immunity. Further, the inactivated vaccines are not highly purified and require adjuvants to enhance the immune response. Also, production of inactivated vaccines is complex and expensive.

Another problem with present control measures for FMD involves the remarkable antigenic variability of FMDV (Carrillo et al, 1990). Like other RNA viruses, FMDV has a high mutation rate. Seven serotypes and over 65 subtypes are known, necessitating either expensive polyvalent vaccines or vaccines appropriate to particular areas which may have to be changed rapidly to meet new strains arising in the field. The evolution of antigenic variants has been mimicked _in vitro_ in studies where FMDV populations under antibody pressure have acquired electrophoretically altered VP1 and resistance to neutralization by antiviral sera. Viruses isolated from the field are often composed of mixed populations of antigenically distinguishable variants. Another potential source of antigenic variation is the selection involved in adaptation of virus field strains to growth in cultured cells. Finally, in areas where the incidence of FMD is low, vaccine production plants and/or procedures of virus inactivation have constituted a source of infection (Carrillo et al, 1990). Thus, although much has been achieved with available vaccines, there is still a real need for improved FMD vaccines.

In recent years, a great deal of research has focused on the development of safer, more stable, better defined, and more efficacious FMD vaccines. This research has been facilitated by the findings that the major immunogenic regions providing protective immunity are located in the viral capsid protein, VP1, at amino acid residues 140 to 160 and 200-213 (Strohmaier, et al, 1982), and this has made it possible to mimic these epitopes by preparing synthetic peptide and biosynthetic fusion protein vaccines.

SYNTHETIC PEPTIDES

The conventional method for preparing synthetic peptide vaccines is to conjugate the peptides to proteins or synthetic polymer carriers, and then to administer them together with adjuvants. Methods designed to avoid the use of carriers by polymerizing or cyclizing the peptides have also been reported, as has the approach in which a small peptidyl core matrix is covalently bound to radially branching synthetic peptides as dendridic arms. In 1982, Bittle and coworkers synthesized peptides corresponding to 4 different regions of the FMDV VP1 protein and coupled them to keyhole limpet hemocyanin. The conjugated peptides induced neutralizing antibodies in guinea pigs, rabbits, and cattle, and protected guinea pigs against subsequent challenge with virulent FMDV (Bittle et al, 1982). However, the synthetic peptides exhibited only one-tenth the immunogenicity of killed virus particles. Later, DiMarchi et al (1986) prepared a peptide in which amino acids 141 to 158 of the 01K serotype of VP1, or the equivalent sequence of other serotypes, was linked through proline-proline-serine to the carboxy-terminal amino acids 200 to 213, and on which cysteines were present at or near the amino and carboxy-termini of the peptides so as to permit polymerization and/or cyclization. The DiMarchi peptide elicited a markedly en-hanced neutralizing antibody response against FMDV, protected cattle after a single vaccination, and provided complete protection after repeated immunization. While the DiMarchi synthetic peptide offered many advantages, it required adjuvants and was costly to produce. As an approach to improving the immunogenicity of FMDV peptides, Francis and Clarke (1989) constructed peptides in which the FMDV B-cell epitopes were linked to T-helper cell epitopes. Also, Wiesmuller et al (1989) produced a novel synthetic peptide vaccine, consisting of a synthetic activator of T cells and macrophage, covalently linked to an amphophilic alpha-helical T cell epitope. The vaccine was composed of the VP1 amino acid sequence of 135-154 and the adjuvant, tripalmitoyl-S-glyceryl-cysteinyl-serine, the synthetic analogue of the N-terminal portion of an E. coli lipoprotein. Wiesmuller's vaccine was found to induce long-lasting high protection against FMD in guinea pigs after a single administration without additional protein or carrier, but there have been no reports so far on the evaluation of the vaccine in cattle.

In summary, carrier free polypeptide vaccines are anticipated to be safer than inactivated virus vaccines because they are chemically defined, do not contain nucleic acids or extraneous proteins, and are unlikely to have any of the pathogens that might be present in serum or tissue extracts. They are stable and withstand thermal extremes, an important consideration where access to refrigeration equipment may be limited. Synthetic peptides, although currently expensive, laborious to produce, and less effective than natural viral proteins in eliciting immune responses, may become more feasable as T-helper and cytotoxic T cell epitopes are elucidated and their sequences linked to the B-cell epitope sequences.

BIOSYNTHETIC PEPTIDE VACCINES

A major contribution by Kleid et al (1981) consisted of the cloning of a DNA sequence coding for the VP1 of the FMDV (A_{12}) strain and the ligation of this sequence to a plasmid designed to express a chimeric protein from the E. coli tryptophane promoter-operator system. Two injections of this protein in an oil adjuvant elicited virus neutralizing antibodies in cattle and provided protection against challenge with FMDV.

Proteins consisting of one, two, or four copies of the amino acids 137 to 162 of FMDV attached to the N-terminus of E. coli beta galactosidase have also been expressed in E. coli (Broekhuijsen et al, 1987). In guinea pigs, the fusion protein emulsified in incomplete Freund's adjuvant and containing one copy of the FMDV determinant elicited only low levels of neutralizing antibodies, but protective levels were found by emulsified fusion proteins consisting of two or four copies of the FMDV determinant. A single injection of the fusion protein with two or four FMDV determinants was sufficient to protect guinea pigs and Landrace pigs against challenge infection.

Other systems for producing FMDV capsid proteins have included insect cells and baculovirus expression vectors (Roosien et al, 1990) and vaccinia virus recombinants (Newton et al, 1986). Enhanced immunogenicity of the FMDV VP1 peptide has also been described by Clarke et al (1990), when the peptide epitope was presented as a fusion protein linked to the N-terminus of the core protein of hepatitis B virus. When the fusion protein was expressed in E. coli, it retained the ability to self-assemble into virus-like particles with regular arrays of the novel FMDV sequence on their surfaces. The FMDV peptide presented in this way approached the immunogenicity of the original virus (Rowlands, 1989; Clarke et al, 1987). The studies of Clark et al (1990) also showned that the nature of the immune response was a reflection of T cell recognition sites within hepatitis B virus core proteins and the particulate nature of the immunogen. Physical disruption of the particles decreased immunogenicity considerably.

RECOMBINANT IBRV-FMDV

We now describe a new approach to biosynthetic FMD vaccines which utilizes infectious bovine rhinotracheitis virus [(IBRV); bovine herpesvirus-1] as a vector to express FMDV epitopes as part of a fusion protein with a major IBRV glycoprotein, that is glycoprotein gIII (U.S. patent pending). The IBRV-FMDV recombinants here described optimize antigen presentation and utilize some of the characteristics of the synthetic peptide and subunit vaccines which provide improvements in safety. The following important advantages for mass vaccination purposes may be noted. First, as attenuated modified-live virus vaccines, the IBRV-FMDV recombinants can protect cattle against both IBR and FMD. Second, administration of the IBRV-FMDV vaccines allows *in vivo* amplification of the same chemically defined FMDV VP1 epitopes that are used in synthetic peptide vaccines and, as with biosynthetic subunit vaccines, there is no infectious FMDV present to cause disease. Third, the vaccine can be produced at low cost and delivered without adjuvants either by the convenient intramuscular route, or, if desired, by the more labor intensive intranasal route which results in the induction of IgA immunoglobulins. Fourth, the vaccine is by definition a "marker" vaccine. That is, cattle vaccinated with the IBRV-FMDV recombinant can be distinguished serologically from cattle infected with FMDV field strains.

The genomic organization of the IBRV-FMDV vector is shown schematically in Figure 1. The parental virus is the attenuated vaccine strain, IBRV(NG)dltk, which has a thymidine kinase (TK) gene deletion and an insertion of a NovaGene (NG) signature sequence at the site of the TK gene deletion (about 0.40 map units). The foreign FMDV dexyribonucleotide sequence is inserted at the NH_2-terminal end of the IBRV glycoprotein gIII gene (about 0.1 map units).

The IBRV gIII gene was chosen as the insertion site for FMDV sequences because it is expressed at high levels on the surface of virus particles and on the surface of virus-infected cells. Glycoprotein gIII is a 95,000 MW molecule containing both N-linked and O-linked oligosaccharides which assemble into homodimers and a macromolecular structure visible as a thin 20-25 nm spike protruding from the virion envelope. The flanking DNA sequences of the IBRV gIII gene with promoter, translational stop and polyadenylation signals are intact, so that the expression of an FMDV-gIII fusion protein is driven by the IBRV gIII promoter.

Two IBRV-FMDV recombinant viruses have at this writing been prepared. The first recombinant, designated IBRV-FMDV (monomer), has a DNA insert consisting of the bovine growth hormone (bGH) signal sequence linked to sequences encoding the FMDV (O1K) VP1 amino acids 200-213, a proline-proline-serine spacer, VP1 amino acids 141-158, and a second spacer (Fig. 1) (DiMarchi et al, 1986). Nucleotides encoding the first 38

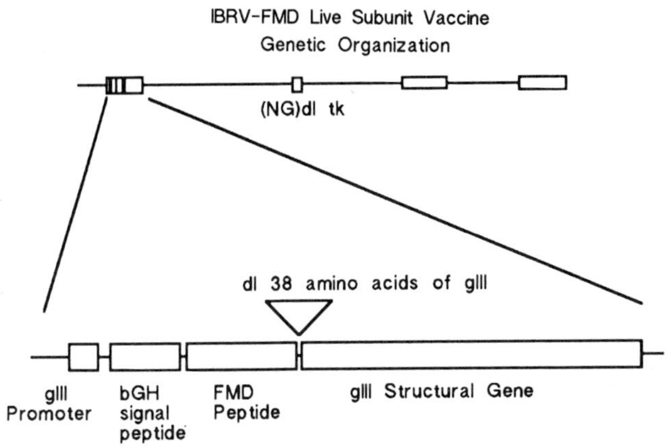

IBRV-FMD Live Subunit Vaccine
Genetic Organization

(NG)dl tk

dl 38 amino acids of gIII

gIII Promoter
bGH signal peptide
FMD Peptide
gIII Structural Gene

Fig. 1. Genetic organization of IBRV-FMDV(monomer) containing FMDV epitope sequences inserted in IBRV glycoprotein gIII and thymidine kinase (TK) genes at 0.11 to 0.12 map units and 0.405-0.32 map units, respectively. The unique short segment of the IBRV DNA is bounded by terminal repeat and inverted repeat sequences. NG is an oligonucleotide sequence inserted in place of deleted TK sequence. bGH is bovine growth hormone. Sequences encoding the first 38 amino acids of the IBRV gIII gene were deleted by unidirectional exonuclease digestion.

amino acids of the IBRV gIII have been deleted by unidirectional exonuclease digestion to avoid redundant signal sequences, but IBRV gIII epitope sequences recognized by monoclonal antibodies have been retained. In the case of the second recombinant, designated IBRV-FMDV (dimer), a head to tail dimer sequence of the VP1 O1K epitope was ligated to the bGH DNA signal sequence and inserted in the IBRV gIII gene, but sequences encoding only the first 21 amino acids of the IBRV gIII gene were deleted (Fig. 2).

The genotypes of the IBRV-FMDV recombinants were confirmed by DNA sequencing, restriction endonuclease, and Southern blotting experiments. The phenotypes were verified by _in situ_ immunostaining of recombinant virus plaques, immunocytochemical staining of nonpermeabilized- and permeabilized-infected cells, virus neutralization experiments, and immune electron microscopy.

The immunogold staining experiments demmonstrated that nonpermeabilized IBRV-FMDV-infected cells treated with guinea pig anti-FMDV sera exhibited circumferential labelling of the cell surface, but parental IBRV(NG)dltk-infected cells lacking the FMDV epitope sequences did not. However, as expected, both recombinant IBRV-FMDV- and IBRV(NG)dltk-infected cells were stained by anti-IBRV gIII monoclonal antibodies. Experiments on permeabilized cells gave similar specific staining results, except that the cytoplasm was stained. These experiments therefore demonstrated that the FMDV epitopes of FMDV-IBRV were expressed on the surface of infected cells.

```
CTGCAGCCGCGCGTGTGCTCAATCCCGGACCACGAAAGCACAAAACGGACGCCCTTAAAAATG

                                                              Oligo 315
TAGCCCGCGCGCGGTCGCGGCCATCTTGGATCCACCCGCGCGCACGACCGCCGAGAGACCGCC

                           bGH signal →
AGCCCGAGACCTCGCCGCGCGTCCGCC MET MET ALA ALA GLY PRO ARG THR SER
                            ATG ATG GCT GCA GGC CCC CGG ACC TCC

LEU LEU LEU ALA PHE ALA LEU LEU CYS LEU PRO TRP THR GLN VAL VAL
CTG CTC CTG GCT TTC GCC CTG CTC TGC CTG CCC TGG ACT CAG GTG GTG

        FMDV epitope (O1K amino acids 200-213) →
GLY ALA ARG HIS LYS GLN LYS ILE VAL ALA PRO VAL LYS GLN THR LEU
GGC GCC AGA CAC AAA CAG AAA ATT GTG GCA CCG GTG AAA CAG ACT CTG

__spacer___ FMDV epitope (O1K amino acids 141-158) →
PRO PRO SER VAL PRO ASN LEU ARG GLY ASP LEU GLN VAL LEU ALA GLN
CCC CCC TCC GTG CCC AAC TTG AGA GGT GAC CTT CAG GTG TTG GCT CAA

                                                FMDV epitope
                                                (O1K amino acids
                    ___spacer____               200-213) →
LYS VAL ALA ARG THR PRO ALA GLY GLY HIS ALA SER ARG HIS LYS GLN
AAG GTG GCA CGG ACG CCC GCC GGC GGC CAC GCG TCT AGA CAC AAA CAG

                                                    FMDV epitope
                                                    (O1K amino acids
                                        ___spacer__ 141-158) →
LYS ILE VAL ALA PRO VAL LYS GLN THR LEU PRO PRO SER VAL PRO ASN
AAA ATT GTG GCA CCG GTG AAA CAG ACT CTG CCC CCC TCC GTG CCC AAC

                                                            ___
LEU ARG GLY ASP LEU GLN VAL LEU ALA GLN LYS VAL ALA ARG THR PRO
TTG AGA GGT GAC CTT CAG GTG TTG GCT CAA AAG GTG GCA CGG ACG CCC

spacer                  amino acid 22 of gIII →
ALA GLY TYR ALA SER THR ARG ARG GLY LEU ALA GLU GLU ALA GLU ALA
GCC GGC TAT GCA TCT ACC CGG CGG GGG CTC GCC GAG GAG GCG GAA GCC

SER PRO SER PRO PRO PRO SER PRO SER PRO THR GLU THR GLU SER SER
TCG CCC TCG CCT CCG CCC TCC CCG TCC CCA ACC GAG ACG GAA AGC TCC
              Oligo 316 AGG GGT TGG CTC TGC CTT TCG AGG

ALA GLY THR THR GLY ALA THR PRO PRO THR PRO ASN SER PRO ASP ALA
GCT GGG ACC ACC GGC GCA ACG CCC CCC ACG CCC AAC AGC CCC GAC GCT
```

Fig. 2. Nucleotide sequence of recombinant IBRV-FMDV (dimer) gene showing bGH signal sequence, the FMDV(O1K) epitope sequences and spacer sequences inserted in the IBRV gIII gene. Sequences encoding the first 21 amino acids of the IBRV gIII gene were deleted by unidirectional exonuclease digestion. The complement of the transcribed strand and the predicted amino acid sequence of the bGH-FMDV-IBRV fusion protein are shown.

The virus neutralization experiments showed that plaque formation by IBRV-FMDV recombinant was inhibited by guinea pig anti-FMDV sera and by polyclonal anti-IBRV sera, but not by normal sera. In contrast, the parental IBRV(NG)dltk plaques were inhibited by anti-IBRV sera but not by guinea pig

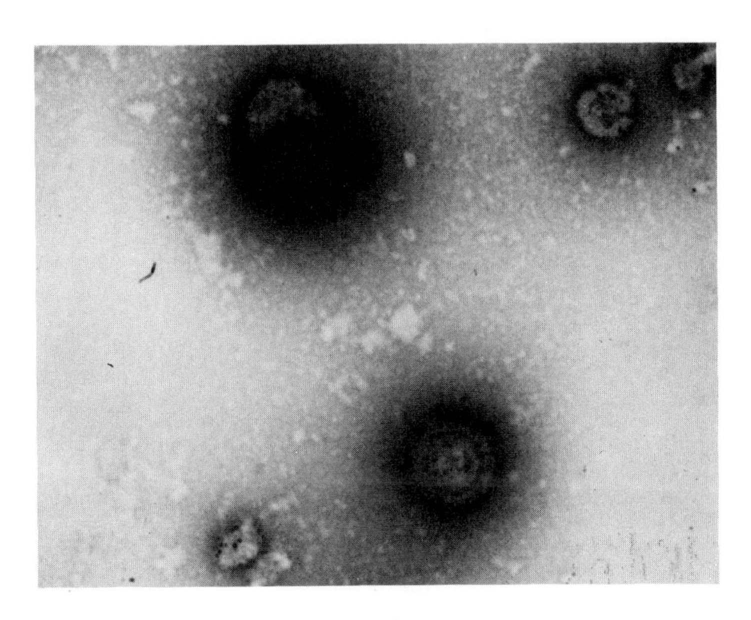

Fig. 3. Immunonegative staining and transmission electron-microscopy of recombinant IBRV-FMDV (dimer) particle (prepared by Advanced Biotechnologies, Inc., Columbia, MD). Partially purified virus particles were treated with guinea pig anti-peptide sera, biotin anti-guinea pig IgG and streptavidin gold, then fixed with glutaraldhyde and stained with uranyl acetate. Print magnification: about 120,000. IBRV-FMDV (dimer) particles were about 116-220 nm in size with electron dense cores. The primary and "bridge" antibodies formed a 65 nm halo around the labelled virus particles, many of which had over 45 individual gold particles attached. This halo was absent from the parental IBRV(NG)dltk particles used as the negative control.

antiFMDV sera. The virus neutralization experiments provided functional evidence that the FMDV epitopes were expressed on the surface of recombinant particles. Immunoelectron microscopy experiments provided conclusive evidence for this (Figs. 3 and 4). The first antibody was guinea pig anti-FMDV peptide sera. The bridge was biotin-anti-guinea pig sera and the indicator was streptavidin gold. Parental IBRV(NG)dltk particles were about 140-210 nm in size with electron dense cores, but did not show streptavidin gold staining (Fig. 4). The IBRV-FMD (dimer) (Fig. 3) as well as IBRV-FMDV (monomer) particles (not shown) were about the same size as parental particles, but in addition, exhibited a 65 nm halo around the virus particles due to the antibody binding. About 90-95% of all the recombinant IBRV-FMDV particles were labelled in this way, many with 45-50 or more individual gold particles attached. Thus, the FMDV epitopes were presented as repeated structures and were anchored in the hydrophobic herpesvirus envelope so as to enhance immunogenicity.

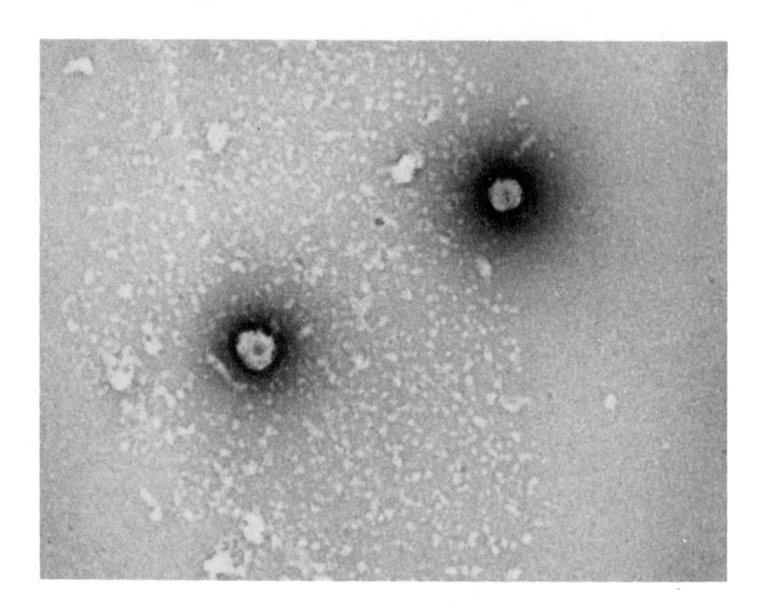

Fig. 4. Immunonegative staining and transmission electron-microscopy of parental IBRV(NG)dltk (control) particles. The particles measured between 140 and 210 nm in size and exhibited an electron dense core. The virus particles were not labelled with gold particles.

EXPERIMENTS IN CALVES

The safety and efficacy of the IBRV-FMDV recombinant vaccines have been tested by in vivo experiments performed in cattle. These experiments will be reported in detail elsewhere. Suffice it to say that intramuscular vaccination of calves with IBRV-FMDV recombinants induced IBRV neutralizing antibodies and protected the calves from challenge exposure to the virulent Cooper strain of IBRV furnished by the United States Department of Agriculture. Likewise, immunization of the calves with the monomer and dimer forms IBRV-FMDV induced protective levels of anti-FMDV antibodies as detected by ELISA assays. In addition, protective levels of FMDV neutralizing antibodies have been demonstrated in experiments carried out at the Institute for Animal Health, Pirbright, England, on the same sera obtained from calves immunized with the IBRV-FMDV (monomer) (Fig. 5). The anti-FMDV virus neutralization titers of sera from calves immunized with the IBRV-FMDV (dimer) are not as yet available. Finally, pilot studies recently completed have shown that cattle vaccinated intramuscularly with IBRV-FMDV (monomer) and challenged intradermolingually at 21 days post vaccination with infectious FMDV can be protected from the disease-causing virus.

In conclusion, the recombinants here described express the epitopes of the O1K strain of FMD. Recombinant IBRV vectors can be prepared with monovalent or trivalent VP1 epitope inserts of the A, O, C, or other serotypes. Such trivalent vaccines would be expected to have enhanced

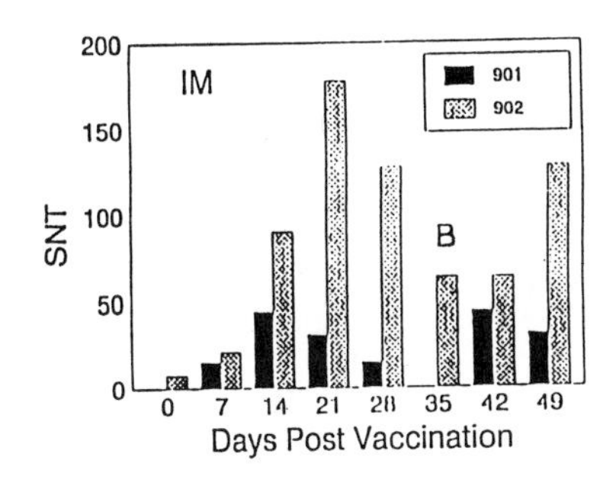

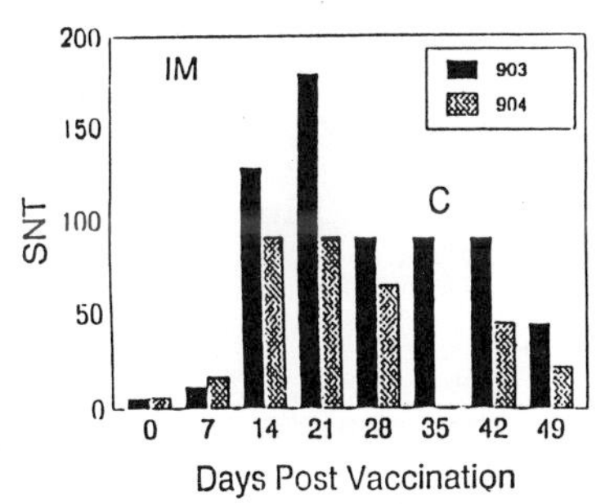

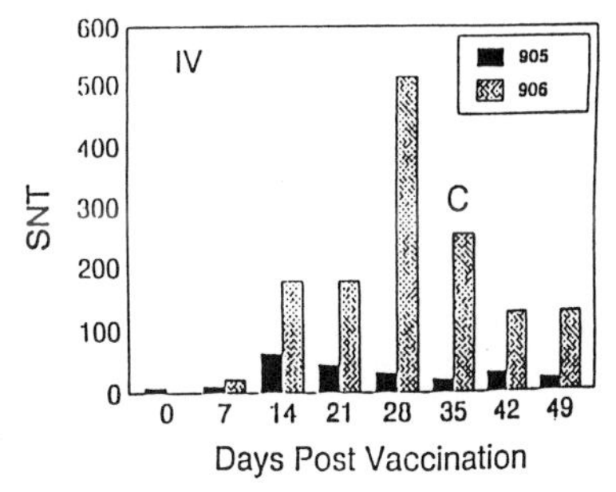

Fig. 5. Anti-FMDV serum neutralization titers (SNT) (T. Doel, Institute for Animal Health, Pirbright, England) after vaccinating cattle with IBRV-FMDV (monomer) intramuscularly (IM) or intravenously (IV). Animals were boosted (B) with the recombinant virus or challenged (C) with virulent IBRV (Cooper) at 35 days post vaccination.

efficacy. Additional studies have been planned to extend the <u>in vivo</u> challenge experiments to such trivalent vector vaccines.

REFERENCES

Bahnemann, H.G. 1990. Inactivation of viral antigen for vaccine preparation with particular reference to the application of binary ethylenimine. Vaccine 8:299.

Bittle, J.L., Houghten, R.A., Alexander, H., Shinnick, T.M., Sutcliffe, J.G., Lerner, R.A., Rowlands, D.J. and Brown, F. 1982. Protection against foot-and-mouth disease by immunization with a chemically synthesized peptide predicted from the viral nucleotide sequence. Nature 298:30.

Broekhuijsen, M.P., van Rijn, J.M.M., Blom, A.J.M., Pouwels, P.H., Enger-Valk, B.E., Brown F., and Francis, M.J. 1987. Fusion proteins with multiple copies of the major antigenic determinant of foot-and-mouth disease virus protect both the natural host and laboratory animals. J. Gen. Virol. 68:3137.

Carillo, C., Dopazo, J., Moya, A., Gonzalez, M., Martinez, M.A., Saiz, J.C., and Sobrino, F. 1990. Comparison of vaccine strains and the virus causing the 1986 foot-and-

mouth disease outbreak in Spain: epizootiological analyses. Virus Research, 15:45.
Clarke, B.E., Brown, A.L., Grace, K.G., Hastings, G.Z., Brown, F., Rowlands, D.J. and Francis, M.J. 1990. Presentation and immunogenicity of viral epitopes on the surface of hybrid hepatitis B virus core particles produced in bacteria. J. Gen. Virol., 71:1109.
Clarke, B.E., Newton, S.E., Carroll, A.R., Francis, M.J., Appleyard, G., Syred, A.D., Highfield, P.E., Rowlands, D.J. and Brown, F. 1987. Improved immunogenicity of a peptide epitope after fusion to hepatitis B core protein. Nature, 330, 381.
DiMarchi, R., Brooke, G., Gale, C., Cracknell, V., Doel, T., and Mowat, N. 1986. Protection of cattle against foot-and-mouth disease by a synthetic peptide. Science, 232, 639.
Francis, M.J. and Clarke, B.E. 1989. Peptide vaccines based on enhanced immunogenicity of peptide epitopes presented with T-cell determinants or Hepatitis B core protein. Methods in Enzymology, 178:659.
Kleid, D.G., Yansura, D., Small, B. Dowbenko, D., Moore, D.M., Grubman, M.J., McKercher, P.D., Morgan, D.O., Robertson, B.H. and Bachrach, H.L. (1981) Cloned viral protein vaccine for foot-and-mouth disease: responses in cattle and swine. Science, 214:1125.
Newton, S.E., Francis, M.J., Brown, F., Appleyard, G. and Mackett, M. 1986. Expression of a foot-and-mouth disease virus immunogenic site sequence in vaccinia virus. Vaccines 86, Cold Spring Harbor Laboratory, Cold Spring Harbor, NY. p 303.
Roosien, J., Belsham, G.J., Ryan, M.D., King, A.M.Q. and Vlak, J.M. 1990. Synthesis of foot-and-mouth disease virus capsid proteins in insect cells using baculovirus expression vectors. J. Gen. Virol. 71:1703.
Rowlands, D.J. 1989. Enhancement of the immunogenicity of peptide vaccines. Biochem. Society Transactions, 17:945.
Strohmaier, K., Franze, R., and Adam, K.H. 1982. Location and characterization of the antigenic portion of the FMDV mmunizing protein. J. Gen. Virol., 59:295.
Wiesmuller, K-H., Jung, G. and Hess, G. 1989. Novel low-molecular-weight synthetic vaccine against foot-and-mouth disease containing a potent B-cell and macrophage activator. Vaccine, 7:29.

DEVELOPMENT OF NON-TOXIGENIC VACCINE STRAINS OF *BORDETELLA PERTUSSIS* BY GENE REPLACEMENT

S. Cockle, G. Zealey, S. Loosmore, R. Yacoob, R. Fahim, Y.-P. Yang, G. Jackson, H. Boux, L. Boux and M. Klein

Connaught Centre for Biotechnology Research
Willowdale, Ontario, Canada M2R 3T4

INTRODUCTION

During the next decade, it is likely that current whole-cell whooping cough vaccines prepared from inactivated *Bordetella pertussis* will be replaced by component vaccines of defined purity and composition that offer higher efficacy and reduced reactogenicity. The principal *B. pertussis* antigens under consideration for such vaccines are pertussis toxin (PT), filamentous hemagglutinin (FHA), pertactin (69 kDa protein) and fimbrial agglutinogens. Although PT is a potent immunogen, it is also a major virulence factor of *B. pertussis*. However, the chemical treatments now used to inactivate PT can reduce its immunogenicity, and depending on the method used, may be susceptible to reversion. An ideal approach to PT detoxification is the genetic replacement or removal of a few critical functional amino acid residues so that three-dimensional structure and immunogenicity are only minimally impaired.

This communication describes the construction of isogenic derivatives of wild-type *B. pertussis* that secrete inactive forms of PT, yet can also be used as the source of all other antigens necessary for an effective acellular vaccine. A four-component recombinant vaccine has been produced in this way and shown to be strongly immunogenic and protective in laboratory animals.

PERTUSSIS TOXIN MUTATIONS

PT is a non-covalent hexameric protein of molecular weight 106 kDa with two functional moieties. The catalytic A or S1 subunit is an ADP-ribosyltransferase that inactivates host regulatory G_i proteins, while the B oligomer containing subunits S2-S5 (with two copies of S4) is responsible for binding to target cells. Since many of the biological properties of PT are mediated by ADP-ribosylation, the effect of point mutations on the enzymic activity of subunit S1 has been extensively studied by ourselves and others (Barbieri and Cortina, 1988; Burnette et al., 1988; Pizza et al., 1988; Cockle et al., 1989; Locht et al., 1989; Loosmore et al., 1990). Specifically, we have shown that replacement of Arg9, Glu58 or Glu129 can give rise to stable holotoxin analogues of low activity that provide a promising basis for vaccine development. All three residues occur in conserved sequences common to S1 and the catalytic subunits of cholera toxin and *Escherichia coli* heat labile toxin. Glu129 was indeed proven to be part of the active site of S1 by photocrosslinking to radiolabelled NAD (Cockle, 1989).

CONSTRUCTION OF RECOMBINANT *B. PERTUSSIS* STRAINS

The *Tox* operon of *B. pertussis* is a polycistron comprising a single promoter and tandem structural genes encoding the five subunits in the order S1, S2, S4, S5, S3. The S1 gene was excised, subcloned into M13 and subjected to site-directed mutagenesis by the phosphorothioate method. Reconstructed Tox^* operons were then incorporated into a broad-host-range plasmid for conjugation into *Bordetella parapertussis*. This provided a convenient method of screening recombinant products for assembly and activity, since *B. parapertussis* does not express its own PT.

Suitable Tox^* operons were reintegrated into the *B. pertussis* chromosome as depicted in Figure 1 (Zealey et al., 1990). Cells were transformed sequentially by electroporation with two different linearized plasmids and screened for double homologous recombination events. In the first step, the *Tox* operon of a spontaneous Str^r mutant of the Connaught *B. pertussis* vaccine strain was replaced by a Tc^r-*S12* gene cartridge sandwiched between extensive *Tox* 5'- and 3'-non-coding sequences. The resulting Tc^r, Str^s cells were Tox^- but continued to secrete other *B. pertussis* antigens. The Tc^r-*S12* cartridge was in turn displaced by a mutated Tox^* operon to reestablish the Tc^s, Str^r phenotype in a strain secreting an inactive PT analogue. To select against cells that became Tc^s, Str^r as a result of spontaneous excision of the labile Tc^r-*S12* cartridge, the transforming DNA carried an Ap^r gene that rendered transformed cells transiently Ap^r before chromosomal integration. In this way, about 10-50% of the putative transformants were indeed Tox^+ as a result of allelic exchange at the *Tox* locus.

Southern hybridization analysis verified that the endogenous *Tox* operon had been replaced in a site-specific manner. Mutated regions of the S1 gene were also amplified by the polymerase chain reaction and sequenced to confirm the mutations. The Tox^* operon remained stable through at least 70 generations on serial transfer in liquid culture. Moreover, when recombinant strains were grown in 10, 35 or 300 litre bioreactors, rates of bacterial growth and production of PT analogue, FHA, pertactin and agglutinogens were equivalent to those of the original wild-type strain, with toxin concentrations reaching 10-20 mg/l. Recombinant strains were thus identical to the parent except for the mutations deliberately introduced into the S1 gene.

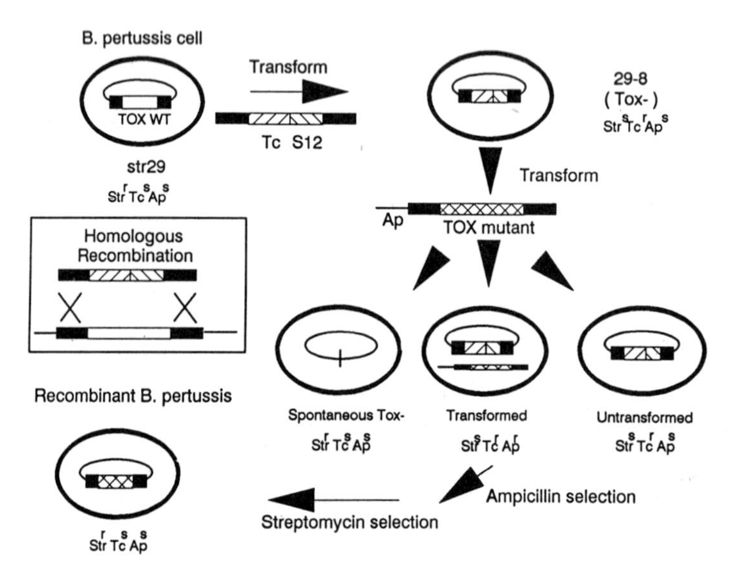

Figure 1. Allelic exchange in *B. pertussis* by homologous recombination.

Table 1. Biological Properties of Pertussis Toxin Mutants

S1 Mutation	S1 Content %	CHO Cell Activity %	ADPR Activity %	LP[a] Activity ED μg	HS[b] Activity LD_{50} μg	S1 Epitope %	Mouse Protection ED_{50} μg
Wild-Type	100	100	100	0.1	0.1	100	2
R9→H	100	0.8	0.2	1	5	90	4
R9→K	100	0.8	0.2	-	6	80	2
R58→E	100	0.5	1.0	-	2	90	4
E129→G	100	0.2	0.2	4	7	90	3
K9G129	100	≤0.0005	≤0.0002	>50	>50	80	3
E58G129	75	≤0.0005	0.0002	>50	>40	80	2
E58S129	75	≤0.0005	0.0004	>40	>40	80	2
K9E58G129	75	<0.005	0.0002	>50	>50	50	3

[a] Leukocytosis stimulation. ED is the dose giving rise to double the number of circulating leukocytes.
[b] Histamine sensitization.

CHARACTERIZATION OF PERTUSSIS TOXIN MUTANTS

PT analogues were purified from culture supernatants under mild conditions by chromatography on perlite followed by hydroxyapatite (Tan et al., 1991). Properties of a series of toxin analogues with mutations at S1 positions 9, 58 and 129 are summarized in Table 1. Single mutants and the double mutant Lys9Gly129 contained the five dissimilar PT protomers in stoichiometric proportions as judged by SDS-PAGE and reverse-phase HPLC (Figure 2). From this and ELISA evidence, these analogues were concluded to be fully assembled. Other multiple mutants were partially depleted in S1, probably as a result of incomplete assembly in the periplasmic space.

ADP-ribosyltransferase activity was determined directly using trans-

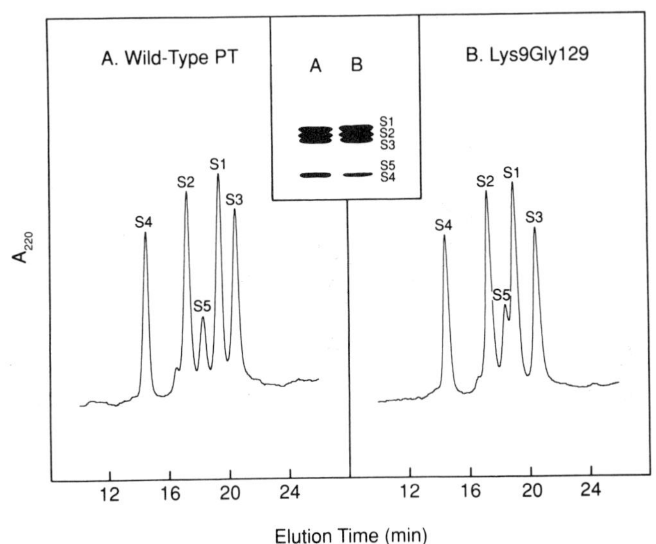

Figure 2. Subunit analysis of PT analogue Lys9Gly129 by reverse-phase HPLC and SDS-PAGE. **A**, wild-type PT; **B**, Lys9Gly129 mutant.

ducin and [*adenylate*-^{32}P]NAD as substrates, and indirectly from the ADPR-dependent clustering of Chinese hamster ovary (CHO) cells. Whereas single mutants retained 0.1-1% of the activity of wild-type PT, the residual activity of double mutants was less than 0.001%, indicating a synergistic effect of the second mutation. Most noticeably, the double mutant Lys9Gly129 displayed no detectable enzymic activity. In accordance with these results, double mutants displayed little or no toxicity in mice at doses up to 40-50 μg, as measured by leukocytosis promotion and histamine sensitization assays.

DEVELOPMENT OF A RECOMBINANT PERTUSSIS VACCINE

PT analogues were screened by ELISA for the presence of an immunodominant protective S1 epitope, using a murine anti-PT monoclonal antibody that inhibits PT-induced CHO cell clustering and affords passive protection in mice against intracerebral challenge with virulent *B. pertussis*. This epitope was slightly affected in all cases but was not severely compromised unless positions 9, 58 and 129 were mutated simultaneously (Table 1).

Mice and guinea pigs immunized with PT analogues in the range 0.1-3 μg showed strong dose-dependent primary and secondary anti-PT, anti-S1 and anti-B oligomer antibody responses. The optimum dose in guinea pigs for the Lys9Gly129 mutant was about 3 μg for induction of neutralizing antibodies that inhibited PT-induced CHO cell clustering, compared with 10 μg for glutaraldehyde-detoxified PT. All PT analogues also efficiently protected mice against intracerebral challenge at doses below 5 μg (Table 1), whereas the ED_{50} for glutaraldehyde-detoxified PT is about 10 μg.

Since the characteristics of the Lys9Gly129 mutant were particularly suitable for vaccine purposes, the *B. pertussis* strain producing this analogue was used as the source of all antigens for a prototype recombinant alum-adsorbed vaccine comprising PT analogue (3 μg per dose), FHA (5 μg), pertactin (5 μg) and agglutinogens (3 μg). Antibody responses in guinea pigs to all four antigens compared favourably with those of the current Canadian whole cell vaccine and a non-recombinant four-component vaccine containing 10 μg of glutaraldehyde-treated PT per dose (Figure 3).

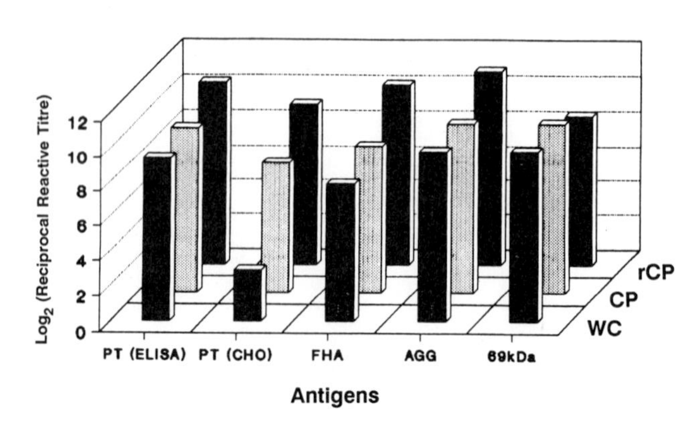

Figure 3. Comparative immunogenicity of recombinant (rCP) and non-recombinant (CP) four-component and whole-cell (WC) pertussis vaccines in guinea pigs. Primary antibody responses after 28 days as measured by ELISA and CHO cell clustering inhibition for PT, and by ELISA for FHA, agglutinogens and pertactin (69 kDa).

DISCUSSION

An efficient method has been developed for making isogenic derivatives of a licensed *B. pertussis* vaccine strain in which the endogenous *Tox* operon has been replaced by *in vitro*-mutated alleles coding for inactive forms of PT. Since these strains are otherwise genetically identical to the parent, they can be grown under standard vaccine conditions to produce all antigens necessary for an acellular vaccine requiring no chemical detoxification. An excellent candidate for inclusion in such a vaccine is a PT analogue with the S1 mutations Arg9→Lys and Glu129→Gly, which is totally inactive in several toxicity assays, yet remains strongly immunogenic and protective in laboratory animals. Pizza et al. (1989) have also obtained isogenic strains of *B. pertussis* by homologous recombination and described the advantageous properties of the Lys9Gly129 mutant (Nencioni et al., 1990). We have now formulated a vaccine containing this analogue and three other *B. pertussis* antigens isolated from the same strain, which in guinea pigs shows superior immunogenicity to non-recombinant chemically inactivated products. We therefore believe that this approach represents a major advance towards the development of a new generation of safe, effective whooping cough vaccines.

REFERENCES

Barbieri, J.T. and Cortina, G., 1988, ADP-ribosyltransferase mutations in the catalytic S-1 subunit of pertussis toxin, *Infect. Immun.*, 56:1934-1941.

Burnette, W.N., Cieplak, W., Mar, V.L., Kaljot, K.Y., Sato, H. and Keith, J.M., 1988, Pertussis toxin S1 mutant with reduced enzyme activity and a conserved protective epitope, *Science*, 242,:72-74.

Cockle, S., 1989, Identification of an active-site residue in subunit S1 of pertussis toxin by photocrosslinking to NAD, *FEBS Lett.*, 249:329-332.

Cockle, S., Loosmore, S., Radika, K., Zealey, G., Boux, H., Phillips, K. and Klein, M., 1989, Detoxification of pertussis toxin by site-directed mutagenesis, *Adv. Exptl. Biol.*, 251:209-214.

Locht, C., Capiau, C. and Feron, C., 1989, Identification of amino acid residues essential for the enzymatic activities of pertussis toxin, *Proc. Natl. Acad. Sci. USA*, 86:3075-3079.

Loosmore, S.M., Zealey, G.R., Boux, H.A., Cockle, S.A., Radika, K., Fahim, R.E.F., Zobrist, G.J., Yacoob, R.K., Chong, P.C.-S., Yao, F.-L. and Klein, M.H., 1990, Engineering of genetically detoxified pertussis toxin analogues for development of a recombinant whooping cough vaccine, *Infect. Immun.*, 58:3653-3662.

Nencioni, L., Pizza, M., Bugnoli, M., De Magistris, T., Di Tommaso, A., Giovannoni, F., Manetti, R., Marsili, I., Matteucci, G., Nucci, D., Olivieri, R., Pileri, P., Presentini, R., Villa, L., Kreeftenberg, J.G., Silvestri, S., Tagliabue, A. and Rappuoli, R., 1990, Characterization of genetically inactivated pertussis toxin mutants: candidates for a new vaccine against whooping cough, *Infect. Immun.*, 58:1308-1315.

Pizza, M., Bartoloni, A., Prugnola, A., Silvestri, S. and Rappuoli, R., 1988, Subunit S1 of pertussis toxin: mapping of the regions essential for ADP-ribosyltransferase activity, *Proc. Natl. Acad. Sci. USA*, 85:7521-7525.

Pizza, M., Covacci, A., Bartoloni, A., Perugini, M., Nencioni, L., De Magistris, M.T., Villa, L., Nucci, D., Manetti, R., Bugnoli, M., Giovannoni, F., Olivieri, R., Barbieri, J.T., Sato, H. and Rappuoli, R., 1989, Mutants of pertussis toxin suitable for vaccine development, *Science*, 246:497-500.

Tan, U.L., Fahim, R.E.F., Jackson, G., Phillips, K., Wah, P., Alkema, D., Zobrist, G., Herbert, A., Boux, L., Chong, P., Harjee, N., Klein, M. and Vose, J., 1991, A novel process for preparing an acellular pertussis vaccine composed of non-pyrogenic toxoids of pertussis toxin and filamentous hemagglutinin, *Molec. Immunol.*, in press.

Zealey, G.R., Loosmore, S.M., Yacoob, R.K., Cockle, S.A., Boux, L.J., Miller, L.D. and Klein, M.H., 1990, Gene Replacement in *Bordetella pertussis* by transformation with linear DNA, *Bio/Technology*, 8:1025-1029.

IMMUNOGENICITY OF LIPOPOLYSACCHARIDE DERIVED FROM BRUCELLA ABORTUS: POTENTIAL AS A CARRIER IN DEVELOPMENT OF VACCINES FOR AIDS

J. Goldstein[1], D. Hernandez[1], C. Frasch[2], P.R. Beining[1], M. Betts[1], T. Hoffman[1] and B. Golding[1]

Divisions of Hematology[1] and Bacterial Products[2], Center for Biologics Evaluation and Research, USFDA, Bethesda Md 20892

ABSTRACT

In view of its unique ability to stimulate human B cells, we have considered using Brucella abortus (BA) as a carrier for human vaccines. Recently we showed that HIV-1 coupled to BA, but not unconjugated HIV-1, was able to stimulate murine responses even in the relative absence of CD4+ T cells. This result suggested that HIV-BA may be useful in boosting the immunity of individuals infected with HIV-1 and who have impaired CD4+ T cell function. In order to refine this carrier we purified lipopolysaccharide (LPS) from BA and examined its effects on immune responses. Similar to LPS from E. coli (LPS-EC), LPS-BA was capable of stimulating mouse B cells to proliferate. In addition, LPS-BA could activate mouse spleen cells to secrete antibodies in vitro. Isotype analysis revealed that IgM and all the IgG subclasses were elicited. When comparing these responses to those of LPS-EC, LPS-BA induced a greater percentage of IgG2a and LPS-EC evoked more IgG3. IgG2a is probably important in protection against murine viral infection. LPS-BA was haptenated with trinitrophenol TNP-LPS (BA) and tested for carrier effect. Similar to TNP-BA and TNP-LPS (EC), TNP-LPS (BA) triggered anti-TNP antibody of the IgM and all IgG subclasses. In contrast, TNP-ficoll induced mainly IgM and only small amounts of IgG3. These results suggest that LPS-BA, like intact BA, behaves as a T-independent type 1 carrier, and as such may be advantageous as a carrier for human vaccines development.

INTRODUCTION

Brucella abortus (BA) has several attributes which may be useful in the development of vaccines in certain immunodeficiency states. BA behaves as a T-independent type 1 (TI-1) in both mouse (Mond et al, 1978) and human (Golding et al, 1981, 1984) responses. Unlike T-dependent antigens (TD) such as soluble proteins BA does not require cognate interaction between CD4+

Immunobiology of Proteins and Peptides VI
Edited by M.Z. Atassi, Plenum Press, New York, 1991

helper T cells and antigen presenting cells. However, T cell factors are probably required for BA responses (Mond et al, 1983), especially for isotype switching (Snapper and Paul, 1987). Unlike T-independent type 2 (TI-2) antigens, such as polysaccharides, BA can activate neonatal B cells as well as B cells from mice and humans with X-linked immunodeficiency (CBA/N and Wiskott-Aldrich Syndrome, respectively).

Immunization of mice with BA was also shown to induce high IgG2a titers (Finkelman et al, 1988). This isotype has the highest affinity for Fc receptors, compared to other murine isotypes (Unkeless, 1977), and is potent in complement activation (Johnson et al, 1985). These findings suggested that BA may be particularly well-suited as a carrier in promoting anti-HIV immunity in asymptomatic individuals infected with HIV-1. These individuals have impaired CD4+ T cell responses to soluble protein antigens (Lane and Fauci, 1985). We reasoned that BA would bypass the requirement for CD4+ T cells and stimulate antibody responses. In order to test this hypothesis, we immunized mice, some of which had been depleted of CD4+ T cells, with HIV-1 and with HIV-1 conjugated to BA (HIV-BA) (Golding et al, 1990). Both forms of the HIV-1 antigen evoked antibody responses in intact mice. In contrast to HIV-1, HIV-BA was able to elicit responses in the CD4+ T cell depleted mice. However, the HIV-BA response in CD4+ T cell depleted mice was reduced, compared to that of intact mice, and consisted mainly of IgG2a. Importantly, syncytia formation, which followed addition of gp160-vaccinia to human T cells, was inhibited by sera from CD4+ T cell depleted mice immunized with HIV-BA. Thus, HIV-BA immunized mice could generate anti-HIV neutralizing antibodies even in the relative absence of CD4+ T cells (Golding et al, 1990).

Human IgG3 appears functionally similar to mouse IgG2a in that it has high affinity for Fc receptors and complement and is more potent than other isotypes in viral neutralization assays (Beck, 1981). Individuals infected with HIV-1 were found to express all IgG anti-HIV-1 isotypes during the asymptomatic stage, but when they developed AIDS, IgG3 levels decreased, whereas other IgG isotypes remained unchanged (Mcdougal et al, 1987).

In this study we examine whether lipopolysaccharide derived from BA displays the same immunogenic properties as the intact bacillus. This would help considerably in refining the carrier for conjugation purposes and would probably limit potential side effects.

RESULTS

Purification and chemical analysis of LPS from Brucella abortus

Live BA organisms were subjected to butanol extraction and the LPS was lyophilized from the water phase following treatment with proteinase K. The LPS contained <2% protein by the Lowry method, 1.0% 2-ketone deoxyoctanate (KDO) by the Osborn method, and <1% nucleic acids by spectrophotometry at 260 nm. KDO is unique to LPS and the 1.0% level is consistent with a highly purified LPS preparation (Phillips et al 1989). SDS-PAGE (16%) was performed using a silver stain, modified according to Tsai and Frasch, for detecting LPS. Four different LPS preparations, with similar migration patterns are shown in Figure 1. All preparations showed faster migrating bands, seen at the bottom of the gel in Figure 1, which represent lipid A and core oligosaccharide, and a slower migrating smear which represents the complete LPS including lipid A, core and O-chain polysaccharide.

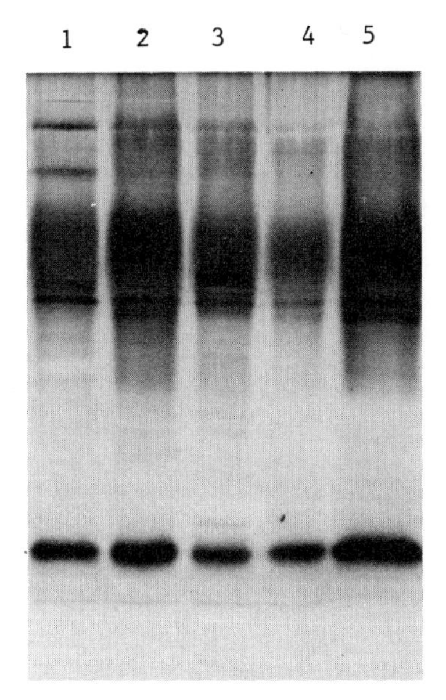

Figure 1. SDS-PAGE of different LPS preparations (modified silver stain). LPS-BA, 10 ug, was added to lanes 1-4. Lane 5 was loaded with the same LPS-BA as lane 4, but 20 ug was added.

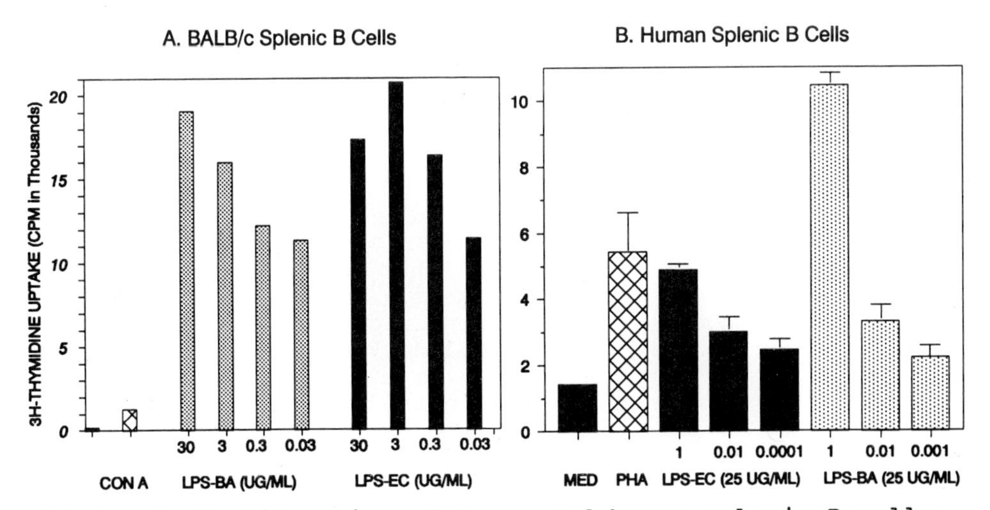

Figure 2. Proliferation of mouse and human splenic B cells. Mouse T cells were removed by treatment with anti-Thy 1.2 and complement, whereas human T cells were removed by E-rosetting. Cells were cultured at 10^6 cells per well, pulsed for 12 hr with ^{3}H-thymidine and harvested at 72 hr.

The reason for the smear is that the O-chain unit, which is added on to BA LPS in varying numbers, consists of only one sugar molecule. LPS from other bacteria, e.g. meningococcus exhibit multiple bands because the O-chain unit contains four sugar molecules (data not shown).

LPS-BA induces mouse and human splenic B cells to proliferate

LPS-BA was added to mouse and human splenic B cell cultures and compared with LPS derived from E.coli (LPS-BA) for its ability to induce B cell proliferation. As can be seen from Figure 2, LPS-BA, stimulated murine and human B cells to proliferate and the dose response curves for LPS-BA and LPS-EC were similar in each case. The Con A and PHA responses were low suggesting that the T cell depletion methods were effective both for the mouse and human cells. This was confirmed by flow cytometry which showed that the splenic B cells contained fewer than 2% CD3+ T cells (data not shown).

LPS-BA induces mouse spleen cells to secrete antibody

LPS-BA was also assessed for its ability to induce antibody responses in mouse spleen cell cultures and was compared to LPS-EC. Figure 3 shows the percentages of IgM and IgG isotypes induced by each preparation of LPS at 14 days. Interestingly, the isotype patterns were different, in that LPS-BA elicited relatively more IgG2a, but less IgG3, than LPS-EC. This may be due to the fact that these LPS preparations stimulate B cells via different cellular pathways. LPS-BA may stimulate T cells to secrete more IFNg, which would favor IgG2a production.

LPS-BA functions as a carrier for anti-TNP antibody responses

TNP was conjugated to LPS-BA, TNP-LPS (BA), which was used to immunize BALB/c mice. The anti-TNP antibody response to TNP-LPS (BA) was compared with those of DNP-KLH, TNP-LPS (EC), DNP-ficoll and saline. TNP-LPS (BA) elicited both IgM and IgG responses, similar to the other antigens, except that TNP-ficoll only induced IgM responses (Figure 4). Specific anti-TNP IgG subclass antibody was also determined and the patterns observed were similar to those seen in Figure 3, that is, TNP-LPS (BA) induced more IgG2a than TNP-LPS (EC).

DISCUSSION

Previously we had shown that BA behaves as a T-independent type 1 carrier in humans (Golding el al., 1982, 1984). We have shown, in a murine model, that HIV-BA can elicit neutralizing antibodies in CD4+ T cell depleted mice (Golding et al., 1990), suggesting that BA may be useful in boosting the antibody responses in persons infected with HIV-1 and with impaired CD4+ T cell responses. In considering BA for development of human vaccines, we decided to purify the LPS from BA, to determine if it exhibited the same characteristics as the bacillus on immune responses. This would facilitate conjugation procedures and potentially reduce the toxicity of administring the whole organism. In this report we demonstate that LPS purified from intact BA stimulates murine B cells to proliferate and secrete antibody. The isotype pattern of the secreted

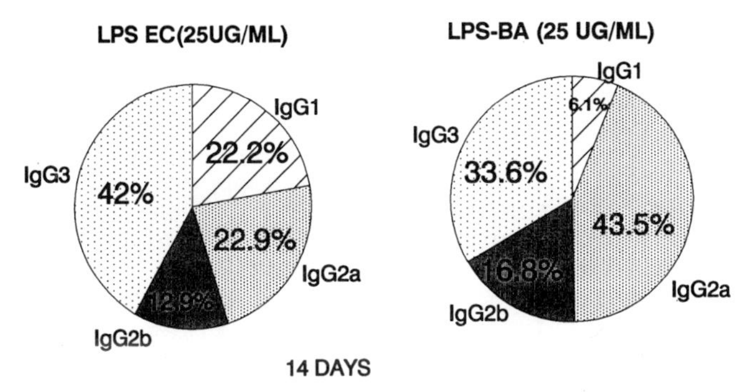

Figure 3. IgG isotype responses of BALB/c spleen cells to LPS-BA and LPS-EC. The spleen cells were cultured at 10^6 cells per well and supernatants assessed by using an IgG subclass-specific ELISA. Isotype standards were used to calculate concentrations of each subclass, so that total IgG and the percentage of each isotype could be calculated.

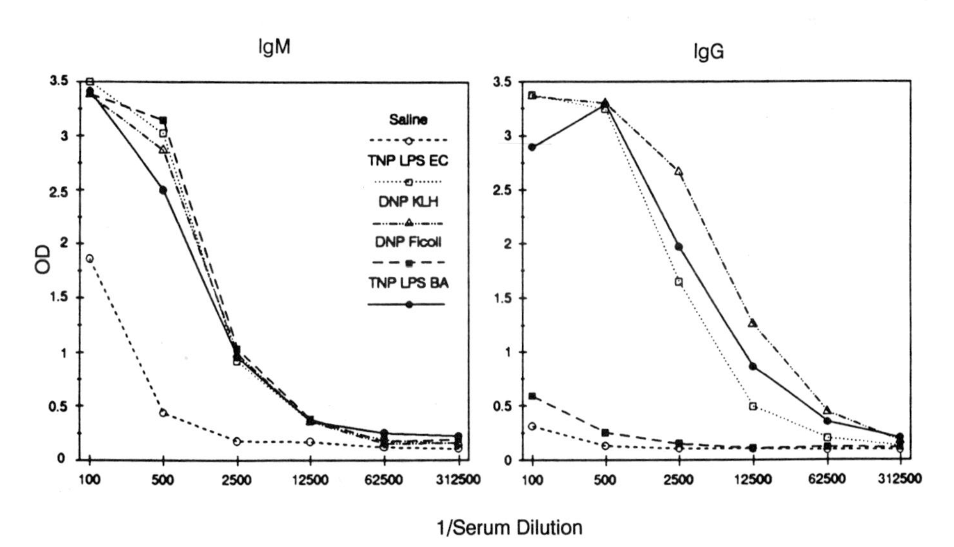

Figure 4. LPS-BA functioned as a carrier in anti-TNP responses. BALB/c mice (5 per group) were immunized i.p. with each antigen at 25 ug/ml. Sera were collected 5 days after the third bleed. Anti-TNP antibody in the sera were assessed by ELISA using TNP-coated plates. Mu and gamma chain specific reagents were used to determine IgM and IgG levels.

antibody was also similar to that previously observed with BA. In addition when a hapten, TNP, was coupled to LPS-BA the resulting conjugate elicited anti-TNP responses in mice which included IgM and all the IgG subclasses. These results suggest that LPS is the immunodominant component of BA, similar to LPS from enterobacteria such as E. coli.

The effects of LPS-BA on murine B cells was the focus of this work. However, the B cell responses probably reflect effects of LPS-BA on other cells such as macrophages and T cells. It has been shown that switching from IgM to IgG2a is dependent on interferon-gamma (IFNg) (Snapper and Paul, 1987). This would imply that the switching to IgG2a we observed in the whole spleen cell cultures, was a consequence of LPS-BA triggering T cells to release this cytokine. Isotype analysis of LPS-BA and LPS-EC responses revealed that LPS-BA induced a higher percentage of IgG2a, whereas LPS-EC evoked more IgG3. This finding may be a consequence of LPS-BA being more active than LPS-EC in inducing IFNg release from T cells. Indeed, we recently compared the effect of LPS-BA and LPS-EC on IFNg induction and found that LPS-BA evoked release of this cytokine to a greater extent than LPS-EC (unpublished data).

LPS from enterobacteria have been implicated in the pathogenesis of endotoxic shock because of their effects on inducing IL-1 and TNF release from monocytes. On comparing the effects of LPS-BA with LPS-EC on human monocytes, we found that LPS-EC elicited 1400 fold more IL-1 beta and 490 fold more TNF alpha than LPS-BA (manuscript in preparation). These results suggest that LPS-BA would be less likely to cause side effects than LPS-EC if given to humans.

The different effects of LPS-BA and LPS-EC on monocytes and T cells may have a structural basis. LPS-BA has the same basic structure as LPS from E. coli; that is lipid A, core oligosaccharide and O-chain polysaccharide (Westphal et al., 1983). However, there are differences in fatty acid composition of the lipid A and in the sugar molecules comprising the O-chain polysaccharides.

To apply these results for vaccine development it will be necessary to determine the effects of LPS-BA on human B cells. These studies have been initiated and LPS-BA can stimulate human B cells to proliferate (Figure 2B). Effects on antibody responses will also be determined. In addition, various antigens which are important in human infections, such as HIV-1 or components of the virus will be conjugated to LPS-BA and tested for immunogenicity. It will be important to determine whether such conjugates elicit responses in the relative absence of CD4+ T cells, as was the case for HIV-BA.

References

Beck, O.E. 1981. Distribution of virus antibody activity among human IgG subclasses. Clin. Exp. Immunol. 43:626.

Finkelman, F.D., I.D. Katona, T.R. Mosmann and R.L. Coffman. 1988. IFN-gamma regulates the isotypes of Ig secreted during in vivo humoral immune responses. J. Immunol. 140:1022.

Golding, B., S.P. Chang, H. Golding, R.E., Jones, K.L. Pratt, D.R. Burger and M.B. Rittenberg, 1981. Human lymphocytes can generate thymus-independent as well as thymus-dependent anti-hapten plaque-forming cell responses in vitro, J. Immunol. 127:220.

Golding, B., A.V. Muchmore, and R.M. Blaese. 1984. Newborn and Wiskott-Aldrich patient B cells can be activated by TNP-Brucella abortus: evidence that TNP-Brucella abortus behaves as a T-independent type 1 antigen in humans. J. Immunol. 133:2966.

Golding, B., H. Golding, S. Preston, D. Hernandez, P.R. Beining, J. Manischewitz, E.F. Lizzio, and T. Hoffman. 1990. Mice immunized with HIV-1 conjugated to Brucella abortus, develop anti-HIV antibodies that are predominantly IgG2a and inhibit gp160-vaccinia-induced syncytia. In F. Brown, R.M. Chanock., H.S. Ginsberg, and R.A. Lerner (Ed). Modern Approaches to New Vaccines Including Prevention of AIDS. Cold Spring Harbor Press. Vaccines 90:249.

Johnson, W.J., Z. Steplewski, H. Koprowski, and D.O. Adams. 1985. Destructive interactions between murine macrophages tumor cells and antibodies of the IgG2a isotype, p. 75. In P. Henkart and E. Martz (eds.), Mechanisms of cell-mediated cytotoxicity II. Plenum, New York.

Lane, H.C., and A.S. Fauci. 1985. Immunologic abnormalities in the acquired immunodeficiency syndrome. Ann. Rev. Immunol. 3: 477.

McDougal, J.S., M.S. Kennedy, J.K.A. Nicholson, T.J. Spira, H.W. Jaffe, J.E. Kaplan, D.B. Fishbein, P. O'Malley, C.H. Aloisio, C.M. Black, M. Hubbard, and C.B. Reimer. 1987. Antibody responses to human immunodeficiency virus in homosexual men. Relation of antibody specificity, titer and isotype to clinical status, severity of immunodeficiency and disease progression. J. Clin. Invest. 80:316.

Mond, J.J., I. Scher, D.E. Mosier, R.M. Blaese and W.E. Paul. 1978. T-independent responses in B cell-defective CBA/N mice to Brucella abortus and to trinitrophenyl (TNP) conjugates of Brucella abortus, Eur. J. Immunol. 8:459.

Mond, J.J., J. Farrar, W.E. Paul, J. Fuller-Farrar, M. Schaefer and M. Howard. 1983. T cell dependence and factor reconstitution of in vitro antibody responses to TNP-B. abortus and TNP-ficoll: Restoration of depleted responses with chromatographed fractions of a T cell-derived factor. J. Immunol. 131:633.

Phillips, M., G.W. Pugh, and B.L. Deyoe. 1989. Chemical and protective properties of Brucella lipopolysaccharide obtained by butanol extraction. Am. J. Vet. Res. 50:311

Snapper M.C., and W.E. Paul. 1987. Interferon-gamma and B cell stimulatory factor-1 reciprocally regulate Ig isotype production. Science 236:944.

Tsai, C. and C. Frasch. A sensitive silver stain for detecting lipopolysaccharides in polyacrylamide gels. Anal. Biochem. 1982. 19:115.

Unkeless, J.C. 1977. The presence of two Fc receptors on mouse macrophages: evidence from a variant cell line and differential trypsin sensitivity. J. Exp. Med. 145:931.

Westphal., O., K. Jann, and K. Himmelspach. 1983. Chemistry and immunochemistry of bacterial lipopolysaccharides as cell wall antigens and endotoxins. Prog. Allergy 33:9.

ANTIGENIC MAPPING OF LIGHT CHAINS AND T CELL RECEPTOR β CHAINS

J.J.Marchalonis, F.Dedeoglu, V.S.Hohman, K.McGee, S.F. Schluter, and A.B.Edmundson*

Microbiology and Immunology, University of Arizona, Tucson AZ 85724 and * Harrington Cancer Ctr., Amarillo, TX 79106

INTRODUCTION

Rearranging immunoglobulins form a closely homologous family of recognition and defense proteins specified by variable (V), joining (J) and constant (C) gene segments that encode peptide segments showing >50%, >70% and approximately 40% identity in phylogenetic comparisons from shark to man (1,2). This family also includes the rearranging T cell receptors that are formed from the same type of elements and show comparable degrees of identity to the corresponding segments of immunoglobulins (1,3,4). There has been considerable interest in antigenic determinants associated with the combining sites of antibodies (idiotypes) that are determined by the complementarity determining regions (CDR). However, overall antigenic mapping of immunoglobulin molecules has not been performed as it has for other proteins including lysozyme (5), myoglobulin (6) and albumin (7).

Here, we chose to focus on the human λ myeloma protein Mcg, a molecule of known sequence and three-dimensional structure (8), and to compare its peptide antigenic determinants with those of a human TCR β chain (YT35) predicted from gene sequence (3) and λ light chains of the sandbar shark (*Carcharhinus plumbeus*) for which we have peptide sequence (9) and complete sequence derived from cDNA clones (2,10). Our approach was to construct a series of overlapping 16-mer synthetic peptides that completely modelled the Mcg and β chain sequences. We then assessed the capacity of polyclonal antibodies to react with individual peptides by ELISA and Western blot analyses. We provide evidence that widely recognized antigenic determinants correspond to segments of the V region that are exposed on the outer face of Mcg in its 3-dimensional conformation (FR1, FR3 and FR4), with the exposed C-terminal stretch of the constant domain and with internal peptides within the S-S loop of the constant region. Similar V region determinants occurred in the shark λ chain and the predicted human TCR β chain. The J-region defined FR4 peptide segment generated antigenic determinants shared among the two λ chains and TCR β chains.

METHODS AND MATERIALS

Synthetic Peptides

The λ light chain Mcg was modeled by a series of 20 synthetic peptides that overlapped by 5 residues. All except the most C-terminal peptide consisted of 16 residues. The gene sequence of human TCR β YT35 (11) was used to construct a series of 22 synthetic 16-mer peptides that overlapped by 5 residues. The most C-terminal peptide consisted of 12 residues. These were synthesized by the University of Arizona Biotechnology Center using an Applied Biosystems Peptide Synthesizer. Purity was determined by amino acid composition and sequence analysis. When necessary, peptides were purified by high performance liquid chromatography under reverse phase conditions.

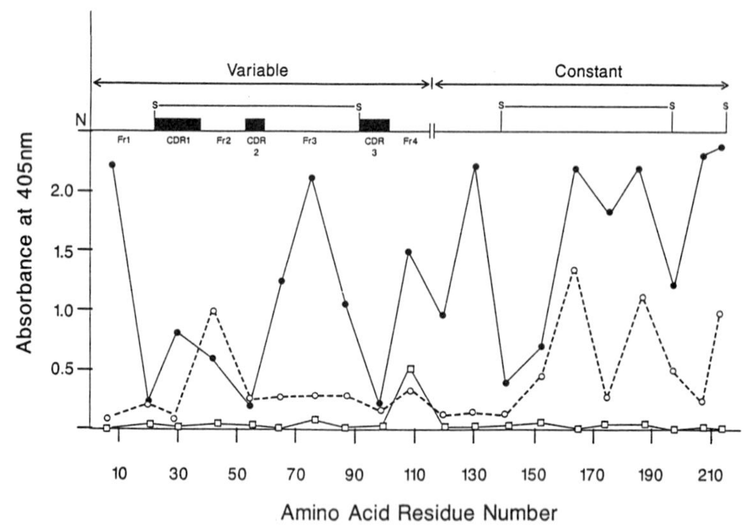

Figure 1. Reaction of synthetic, overlapping Mcg peptides with rabbit anti-serum specific for human λ light chains (—●—), goat anti-serum specific for human λ chain, (---o---), and affinity purified rabbit anti-serum against synthetic Jβ peptide (—□—). Anti-sera were tested in ELISA at a dilution 1:1000. Normal rabbit serum gave no reaction with any of the peptides under these conditions. The data points shown represent the midpoints of the peptides tested. The ELISA data are mapped onto a schematic diagram illustrating the individual regions of the λchain Mcg.

Antibodies

Commercial rabbit and goat antisera specific for human λ chains were purchased from DAKOPATTS (Carpinteria, CA). These were reported to react with both free λ chains and λ chains covalently associated with heavy chains. Rabbit and goat antisera directed against an intact human λμ Waldenstrom macroglobulin and the $(Fab')_1$ produced from this molecule were prepared as previously described (12). Affinity-purified rabbit antibody to the synthesit $J\beta$ peptide ANYGYTFGSGTRLTVV were prepared as previously described (13).

Immunological Assays

ELISA was carried out under previously described conditions (13). Western blot analysis was performed using procedures previously described in detail (13,14).

RESULTS

A panel of polyclonal rabbit and goat antibodies and commercial murine monoclonal antibodies were tested for their capacity to react with Mcg and YT35 peptides under solid-phase binding conditions. No reactivity was observed with the MAbs, but a number of peptides were reactive with the goat and rabbit antibodies to human λ chains. Figure 1 illustrates the binding of rabbit anti-human λ, goat anti-human λ and rabbit anti-$J\beta$ peptide with the set of overlapping Mcg peptides. The antisera were tested here at a dilution of 1/1000 and the plates were coated with 10 μg of peptide per well. We confirmed the individual reactions by titration of the antisera versus the peptides and by competitive inhibitions using both peptides and intact Mcg. The titers of rabbit anti-λ, for example, were 64,000 against peptide Mcg #1, 16,000 against peptide #7 and 32,000 against peptide #20. As shown in Figure 2, peptides and intact λ chain inhibited the binding to the

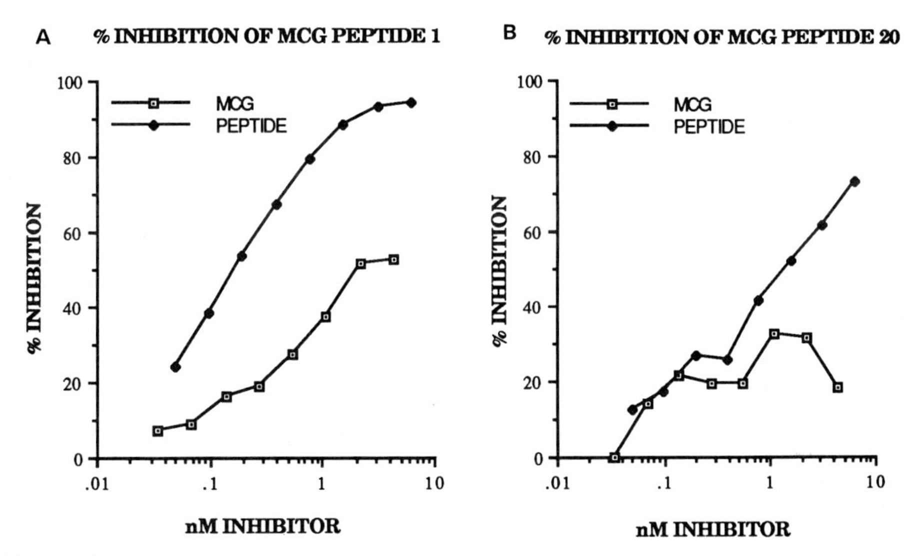

Figure 2. Inhibition of binding of rabbit anti-serum directed against human λ chain to individual Mcg peptides 1(A) and 20 (B) with either the respective peptide or intact Mcg λ chain.

peptides. Fifty percent inhibition by the peptides and maximal inhibition by the proteins occurred at approximately 1 nM under these conditions. Normal rabbit serum did not react with any of the Mcg peptides under these conditions.

Mapping of the points corresponding to the mid-points of the peptides to the schematic Mcg region showed that peptides corresponding to FR1 and FR3 from the V region and the beginning and end of the C region bore stronger antigenic determinants. The segment of the C-domain ranging from residues 155 to 192 likewise contained strong antigenic markers for both the rabbit and goat anti-λ. The FR4 region showed moderate antigenic reactivity and was recognized by both the goat and rabbit anti-λ sera. Rabbit anti-Jβ peptide reacted with only one peptide, the expected sequence of peptide #10.

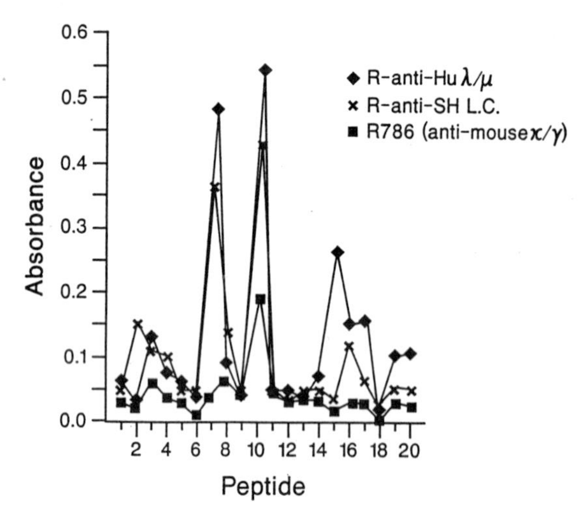

Figure 3. Reaction of synthetic, overlapping Mcg peptides with rabbit antiserum to human λ chains, antiserum to shark light chains, and antiserum to normal mouse IgG (κ, λ).

Rabbit antiserum to isolated light chains of the galapagos shark (*Carcharhinus galapagenesis*) reacted in ELISA and Western blots with human λ light chains (Wang, H. and McGee, K., personal communication). Figure 3 compares the reactions of rabbit anti-human λμ, anti-shark λ and anti-mouse κγ with the Mcg peptides. The anti-human λμ reacted strongly with V region peptides 7 and 10 and less strongly with C-region peptides 15-17. The anti-shark λ reacted with the V region peptides as strongly as did the anti-human. It reacted extremely weakly with one C-region peptide. The anti-mouse κγ reacted only with peptide 10, and this weakly.

The rabbit anti-Jβ reacted strongly with β peptides #10 and #11 as would be expected from their sequences which overlap that of the immunizing peptide. The rabbit anti-human λ reacted with β peptides 8, 11 and 17 (Figure 4). By contrast with the results with the Mcg peptides, normal rabbit serum gave reactions with β peptides 3 and 21 as did the specific antisera. Therefore, these peptides reactivities were considered nonspecific. This conclusion was verified by Western blot analysis because these questionable peptides were unreactive in this assay.

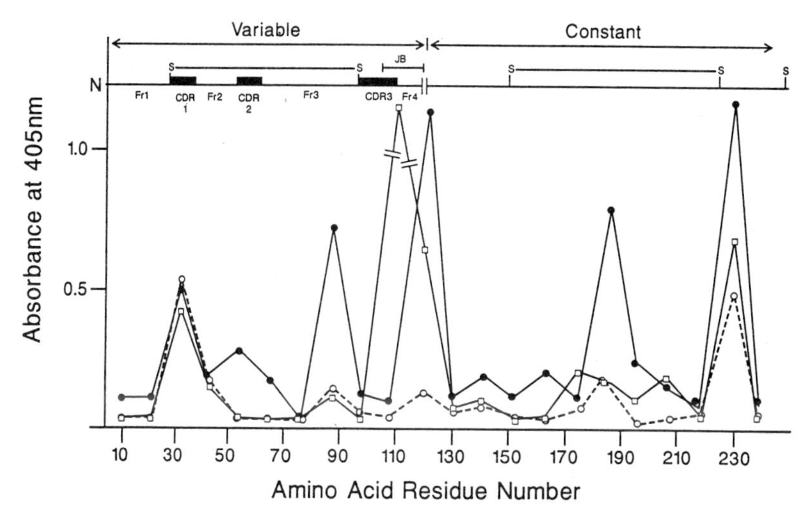

Figure 4. Reaction in ELISA of rabbit antiserum against synthetic Jβ peptide (—□—), rabbit antiserum against human λ light chain (—●—), and normal rabbit serum (---o---) with synthetic overlapping peptides predicted from the gene sequence of human TCR β chain. The data points represent the midpoints of the peptides. Sera were tested at a dilution of 1:1000. The peptide data are mapped upon a schematic representation of the TCR B chain. The position of the synthetic peptide is indicated by JB.

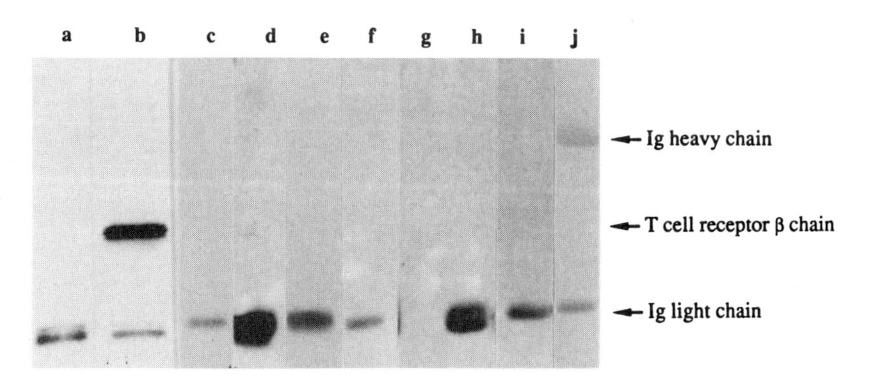

Figure 5. Western blot showing reactivity of affinity purified rabbit antibody to the synthetic Jβ peptide with various immunoglobulin and T cell receptor preparations. Lanes: A, mouse serum; B, CHAPS extract of mouse spleen; C, trout serum; D, trout IgM; E, shark serum; F, shark IgM; G, turkey IgM; H, toad IgM; I, murine myeloma protein TEPC15; J, human serum. The assay was carried out as previously described (14,15). Under the conditions used, no reactivity was obtained when the samples were reacted with an equal mass of normal rabbit IgG immunoglobulin.

The reaction in Western blots of affinity-purified anti-Jβ antibody with TCR β chain and certain light chains is shown in Figure 5. The only components detected in mouse serum (lane a) are κ and λ light chains. κ and λ light chains and TCR β chain are detected in a CHAPS extract of murine spleen lymphocytes (lane b). These components were positively identified by the use of purified light chains (13-15) and cross-absorptions of spleen extracts with antibody to murine β chain and CD3 (complexed with α/β TCR). Anti J_β reacts with some light chains, e.g. those of mouse (lanes a,b), shark (lanes e, f) but not those of others such as turkey (lane g). This result was shown elsewhere (14) to correlate with amino acid sequence, notably the presence or absence of positively charged lysyl or arginyl residues at position 12 of the 16-mer Jβ peptide.

DISCUSSION

We carried out an overall mapping of λ chain peptide determinants using polyclonal antisera directed against λ chain markers. The major antiserum used (the commercial anti λ serum) is a specific diagnostic reagent for λ light chains. It reacts with both V and C region peptides, but does not detect peptides from the CDR segments. It and rabbit antiserum to shark light chains both react with peptides from the FR3 and FR4 regions. These regions should be exposed on the outside of the folded Mcg and shark light chain structure (8) and show considerable sequence homology in comparison between the corresponding stretches of shark and human λ sequence; viz.

(Peptide #7, 67-82)

Mcg	S	K	S	G	N	T	A	S	L	T	V	S	G	L	Q	A
Shark	S	S	S	S	N	K	M	H	L	T	I	T	N	V	Q	S

(Peptide #10, 100-115)

Mcg	V	F	G	T	G	T	K	V	T	V	L	G	Q	P	K	A
Shark	I	F	G	S	G	T	K	L	N	V	L	G	N	P	R	S

Although it is commonly regarded that isotypic specificities are defined by C-region determinants, these results show that V regions can specify λ chain specific antigens. Consistent with this, commercial rabbit anti-κ chain did not react with Mcg or these peptides under these conditions.

Constant region peptides were also detected. Two of these were associated with regions expected to be exposed (#12, and #19, 20) in the folded 3-dimensional structure of the Mcg dimer. Three others (15, 16, and 17) are localized to a β-band portion of the folded structure that would be involved in forming the contact surface between the two dimers. The fact that these are detected by antisera raised against intact immunoglobulin chains suggests that there is more flexibility in the 3-dimensional structure than is apparent in the crystallographic analysis. Alternatively the immunization result giving rise to antibodies to these C-region peptides might reflect the manner in which the λ chain is degraded into peptides and presentated to helper T cells.

Cross-reactions with certain TCR β peptides suggest that these molecules have FR3 and FR4 structures comparable to those of λ chains. The Jβ segment showed strong homology to Jλ and Jκ sequences, particularly in the portion following the FG residues indicating the beginning of FR4. This is illustrated as follows (with FR4 in Uppercase)

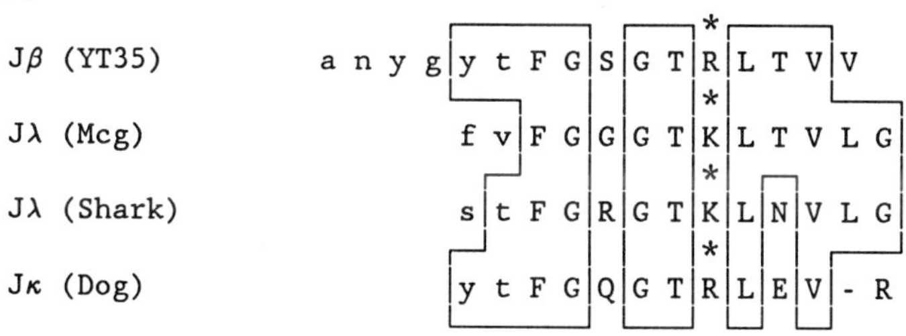

This peptide segment proved to be a useful antigen for the production of antibodies widely reactive with rearranging immunoglobulin light chains and TCR chains (14, 15). The initial portion of the segment is part of CDR3 and would be associated with the combining site for antigen, whereas the FR4 portion would be on the outer face (16) forming a hydrogen bonded β-sheet with exposed residues of FR1.

We found that antibodies against synthetic peptides predicted from TCR gene sequence could react with proteins. Moreover, we have also found that the affinity-purified anti-Jβ reacted in immunocytofluorescence with surface immunoglobulins of some human B cells and TCR of some human T cells (17). These results confirm the prediction that the C-terminal portion of the joining segment is exposed in the native conformation of immunoglobulins and T cell receptors.

Supported in part by USPHS grant CA-42049 from the National Cancer Institute.

REFERENCES

1. Marchalonis, J.J. and Schluter, S.F. Evolution of variable and constant domains and joining segments of rearranging immunoglobulins. The FASEB J. 3:2469-2479, (1989).
2. Schluter, S.F., Edmundson, A.B. and Marchalonis, J.J. Evolution of immunoglobulin light chains: complementary DNA clones specifying sandbar shark constant regions. Proc. Natl.Acad.Sci.USA 86:9961-9965, (1990).
3. Toyonaga, B. and Mak, T.W. Genes of the T-cell antigen receptor in normal and malignant T cells. Ann. Rev. Immunol. 5:585, 1987.
4. Kronenberg, M., Siu, G., Hood, L.E. and Shastri, N. The molecular genetics of the T-cell antigen receptor and T-cell antigen recognition. Ann. Rev. Immunol. 4:529 (1986).
5. Atassi, M.Z., Determination of the entire antigenic structure of lysozyme by surface-simulation synthesis: a novel concept in molecular recognition. CRC Crit.Rev. Biochem. 6:371, (1979).

6. Atassi, M.Z. Antigenic structure of myoglobin. THe complete immunochemical anatomy of a protein and conclusion relating to antigenic structures of proteins. Immunochem. 14:423 (1975).
7. Sakata, S., and Atassi, M.Z. Immunochemistry of serum albumin. X. Five major antigenic sites of human serum albumin are extrapolated from bovine serum albumin and confirmed by synthetic peptides. Mol. Immunol. 17:139 (1980).
8. Edmundson, A.B., Ely, K.R., Abola, E.E., Schiffer, M. and Panagotopoulos, N. Rotational allomerism and divergent evolution of domains in immunoglobulin light chains. Biochem. 14:3953 (1975).
9. Schluter, S.F., Beischel, C.J., Martin, S.A. and Marchalonis, J.J. Sequence analysis of homogeneous peptides of shark immunoglobulin light chains by tandem mass spectrometry: Correlation with gene sequence and homologies among variable and constant region peptides of sharks and mammals. Mol. Immunol. 26:611-624, 1990.
10. Marchalonis, J.J., Hohman, V.S. and Schluter, S.F. The usefulness of heterologous antibody and molecular probes in studies of the evolution of antibodies. in: "Evolution of the Immune System", Charlemagne, J. and Tournefier, eds., Telford Press, Caldwell, N.J., in press.
11. Yanagi, Y., Yoshikai, Y., Leggett, K., Clark, S.P., Aleksander, I., Mak, T.W. A human T cell specific cDNA clone encodes a protein having extensive homology to immunoglobulin chains. Nature 308:145-149 (1984).
12. Marchalonis, J.J., Warr, G.W., Wang, A.-C., Burns, W.H. and Burton, R.C. A Fab-related surface component of normal and neoplastic human and marmoset T cells: Demonstration, functional analysis and partial characterization. Mol. Immunol. 17:877-891 (1980).
13. Schluter, S.F. and Marchalonis, J.J. Antibodies to synthetic joining segment peptide of the T-cell receptor β-chain: serological cross-reaction between products of T-cell receptor genes, antigen binding T-cell receptors and immunoglobulins. Proc. Natl. Acad. Sci. USA 83:1872-1876, (1986).
14. Marchalonis, J.J., Schluter, S.F., Hubbard, R.A., McCabe, C., and Allen, R.C. Immunoglobulin epitopes defined by synthetic peptides corresponding to joining-region sequence: Conservation of determinants and dependence upon the presence of an arginyl or lysyl residue for cross-reaction between light chains and T cell receptor chains. Mol. Immunol. 25:771-784 (1988).
15. Schluter, S.F., Rosenshein, I.L., Hubbard, R.A. and Marchalonis, J.J. Conservation among vertebrate immunoglobulin chains detected by antibodies to a synthetic joining segment peptide. Biochem. Biophys. Res.Comm. 145:699-705 (1987)
16. Beale, D., and Coadwell, J. Unusual features of the T-cell receptor C domains are revealed by structural comparisons with other members of the immunoglobulin superfamily. Comp. Biochem. Physiol. 85B:205 (1986).
17. Shankey, T.V., Schluter, S.F. and Marchalonis, J.J. Flow cytometric analysis of human lymphocytes using affinity purified antibody to T cell receptors synthetic J region. Cell Immunol 118:526-531 (1989).

EPITOPE MAPPING STUDIES OF SNAKE VENOM PHOSPHOLIPASE A_2 USING MONOCLONAL ANTIBODIES

Bradley G. Stiles* and John L. Middlebrook

Department of Toxicology
Pathophysiology Division
U.S. Army Medical Research Institute of Infectious Diseases
Frederick, Maryland 21702-50112

ABSTRACT

Fifteen different monoclonal antibodies developed against pseudexin, a snake venom phospholipase A_2 with presynaptic neurotoxicity, were screened for linear epitope recognition. Peptides (9-mers) spanning pseudexin were synthesized by using alanine-derivatized polyethylene pins and subsequently probed with antibody. Four antibodies bound to toxin peptides and were detected with an enzyme-linked immunosorbent assay. Three of the bound antibodies recognized a site important in calcium binding and the interlocking of dimeric forms of snake venom phospholipase A_2. Analogous regions from other phospholipases were synthesized and probed with the four reactive antibodies. A good correlation was found between the reactivity of whole molecule phospholipases and peptide regions with the antibodies. Monoclonal antibodies neutralizing the lethal or enzymatic effects of pseudexin did not recognize any linear epitopes.

INTRODUCTION

Snake venom phospholipase A_2 (PLA_2) are single or multichained enzymes dependent on Ca^{++} for activity. They cause many different physiological effects, including presynaptic neurotoxicity, myonecrosis, and anticoagulant activity (1). Mouse LD_{50} values range from 1 µg/kg for textilotoxin (a pentameric complex) to the nontoxic, dimeric PLA_2 from _Crotalus atrox_. Previous antibody studies show cross-reactivity and protective immunity among different PLA_2s, suggesting that epitope mapping may yield useful information about neutralizing epitopes (2). This study provides epitope mapping information for pseudexin, a neurotoxic PLA_2 from venom of the Australian Redbellied Black Snake (_Pseudechis porphyriacus_).

MATERIALS AND METHODS

Monoclonal Antibodies

Antibodies were produced against pseudexin isoforms A, B, and C (peak V) as previously described (3). Monoclonal antibodies were purified from ascites fluid by using Protein A-agarose.

Peptide Synthesis

Peptides (9-mers) were synthesized with a commercially available pin procedure (Cambridge Research Biochemicals, Valley Stream, NY) originally described by Geysen (4) using published PLA_2 sequences (5-9). Purified peptides for competition studies were purchased (Peptide Technologies Inc., Washington D.C.).

Enzyme Linked Immuno Sorbent Assay (ELISA) Procedures

The procedure for ELISA with peptides on pins was done as described in the manual from Cambridge Research Biochemicals. Purified monoclonal antibody (750 ng/ml) was used for epitope mapping studies. Antibodies not reacting with peptide sequences were assayed again at 7.5 μg/ml.

The competitive ELISA consisted of coating Immulon II plates overnight at $4^{o}C$ with pseudexin peak V using 100 ng/ml or 250 ng/ml in carbonate buffer, pH 9.6. Antibody 2 (0.31 μg/ml) or 14 (1.89 μg/ml) was preincubated 1 h at $37^{o}C$ with varying concentrations of peptide diluted in phosphate buffered saline (PBS) containing 0.1% Tween 20 and 0.1% gelatin. The antibody-peptide mixture was added to pseudexin-coated plates for 1 h at $37^{o}C$. Plates were washed with PBS containing 0.1% Tween (PBST) and anti-mouse, alkaline phosphatase conjugate added for 1 h at $37^{o}C$. After washing with PBST, substrate was added and absorbance read (405 nm) in 30 minutes. Percent competition was calculated by comparing absorbance means (samples done in triplicate) of antibody/peptide mixtures to those with antibody alone.

RESULTS AND DISCUSSION

Epitope Mapping of Monoclonal Antibodies

The amino acid sequence of pseudexin B, the most lethal and antibody-reactive pseudexin isoenzyme (3,9), was synthesized and screened with 15 different pseudexin monoclonal antibodies. Four antibodies recognized linear sites. Antibody 2 recognized one linear site, VDELD (38-42); while antibodies 5, 13, and 14 each recognized the same two sites, YGCYCGPGG (25-33) and FPKLTLYSW (61-69) (Table 1). The YGCYCGXGG is a highly conserved region in all PLA_2s important in the binding of Ca^{++}(10).

Heterologous Sequence Recognition

Since pseudexin monoclonal antibodies cross-react with other

TABLE 1. REACTIVE PSEUDEXIN B SEQUENCES WITH ANTIBODIES 2, 5, 13, AND 14*

Antibody	Sequence	OD_{405}
2	GHGTPVDEL	0.055
	HGTPVDELD	1.975
	GTPVDELDR	1.915
	TPVDELDRC	1.949
	PVDELDRCC	1.591
	VDELDRCCK	1.744
	DELDRCCKI	0.707
	ELDRCCKIH	0.064
5 / 13 / 14	YADYGCYCG	0.047/0.145/0.017
	ADYGCYCGP	0.186/0.310/0.130
	DYGCYCGPG	0.385/0.350/0.180
	YGCYCGPGG	0.510/0.514/0.450
	GCYCGPGGH	0.147/0.202/0.080
	CYCGPGGHG	0.188/0.246/0.047
	YCGPGGHGT	0.060/0.125/0.016
	CFPKLTLYS	0.117/0.290/0.045
	FPKLTLYSW	0.364/0.525/0.510
	PKLTLYSWK	0.090/0.155/0.036

* Antibodies 2 and 14 were tested at 750 ng/ml and 5 and 13 at 7.5 μg/ml.

TABLE 2. PLA_2 CROSS-REACTING SEQUENCES WITH ANTIBODIES 2 AND 14

Antibody	Sequence	OD_{405}
2	VDELD (Pseudexin B)	2.800
	VDDLD	0.825
	IDDLD	0.136
	RDATD	0.031
	IDALD	0.025
	KDATD	0.021
	QDASD	0.020
	VTLDL (-)*	0.034
	DKQSA (-)	0.019
	IDATD (-)	0.019
14	YGCYCGPGG(Pseudexin B)	2.065
	YGCYCGVGG	2.311
	YGCYCGLGG	2.301
	YGCYCGRGG	2.059
	YGCYCGAGG	2.030
	YGCYCGWGG	1.874
	YGCYCGKGG	0.739
14	FPKLTLYSW(Pseudexin B)	0.865
	TPYTSLYTW	2.418
	NTKWDIYGY	2.268
	APYWTLYSW	1.712
	NTKWDIYRY	1.166
	FPKMSAYDY	1.063
	SPYTDRYKF	0.740
	EPNNDTYSY	0.587
	NPYTESYSY	0.535
	NPKTSQYSY	0.531
	WPYFKTYSY	0.307
	FPKEKICIP	0.097
	GIKKNPVGH (-)	0.109
	CDDGHIKAD (-)	0.068
	EEHIKNAEG (-)	0.067
	IMKGNESIP (-)	0.048

*(-) = Negative Control Peptides

PLA_2s, the best possible sequence alignment was made and suspected reactive sites synthesized. Antibodies 2 and 14 were used for this study because both were more reactive in epitope mapping experiments with homologous antigen. Antibody 2 was more discriminating in recognizing sequences than antibody 14 (Table 2). Antibody 14 reacted more strongly with some heterologous sequences than homologous YGCYCGPGG or FPKLTLYSW. Interestingly, some dimeric forms of PLA_2, like crotoxin and mojave toxin, reacted strongly with antibody 14 in the pin system and weakly with whole molecule. The YGCYCGWGG sequence, found at the interface of the dimeric C. atrox PLA_2, is buried within the molecule and not readily accessible (11).

Competitive Studies with Soluble Peptides

Soluble VDELD (0.125-2 mM), representing the antibody 2 recognition site, competitively inhibited antibody binding (9-40%) to pseudexin compared to an equal concentration of control peptide IDALD (2-15%). Studies with antibody 14 and the YGCYCGPGG and FPKLTLYSW peptides showed good competitive inhibition with the former (0.125-2 mM = 38-62% competiton) but not latter peptide (1-17% competition). This was unexpected since the FPKLTLYSW peptide gave high readings in the epitope mapping studies. Peptide-mixing experiments showed an additive effect with 2 mM each of YGCYCGPGG/FPKLTLYSW (81% competition) compared to a control YGCYCGPGG/FPKEKICIP (63% competition) mixture. This data suggests that the two peptide regions are probably contact sites in a larger conformational epitope.

Computer Modeling of Porcine PLA_2

In order to determine if both sites recognized by antibodies 5, 13, and 14 were in proximity to each other, molecular modeling of closely related porcine pancreatic PLA_2 was kindly done by Dr. Dallas Hack from this Institute. Both analogous sites (25-33 and 67-75) lie near each other (closest distance is 6Å between Leu 31 and Tyr 69) and encompass an area of 610$Å^2$. This is less than the 690$Å^2$ and 774$Å^2$ antibody binding region for two different discontinuous epitopes found in lysozyme (12,13).

REFERENCES

1. Kini, R.M. and Evans, H. (1989) Toxicon **27**, 613-635.
2. Middlebrook, J. L. and Kaiser I. I. (1989) Toxicon **27**, 965-977.
3. Middlebrook, J. L. (1990) Toxicon, In Press.
4. Geysen, H.M., Meloen, R.H., and Barteling, S.J. (1984) Proc. Natl. Acad. Sci. USA **81**, 3998-4002.
5. Aird, S.D., Kaiser, I.I., Lewis, R.V. and Kruggel, W.G.(1985) Biochemistry **24**, 7054-7058.

6. Bieber, A.L., Becker, R.R., McParland, R., Hunt, D.F., Shabanowitz,J., Yates, J.R., Martino, P.A., and Johnson, G.R. (1990) Bioch. Biophys. Acta **1037**, 413-421.
7. Dufton, M.J. and Hider, R.C. (1983) Eur. J. Biochem. **137**, 545-551.
8. Mebs, D. (1985) List of Biologically Active Components from Snake Venoms. Pergamon Press, Oxford, England.
9. Schmidt, J.J., and Middlebrook, J.L. (1989) Toxicon **27**, 805-818.
10. Keith, C., Feldman, D.S., Deganello, S., Glick, J., Ward, K.B., Jones, E.O., and Sigler, P.B. (1981) J. Biol. Chem. **256**, 8602-8607.
11. Renetseder, R., Brunie, S., Dijkstra, B.W., Drenth, J., and Sigler, P.B. (1985) J. Biol. Chem. **260**, 11627-11634.
12. Amit, A.G., Mariuzza, R.A., Philips, S.E.V., and Poljak, R.J. (1986) Science **233**, 747-753.
13. Padlan, E.A., Silverton, E.W., Sheriff, S., Cohen, G.H., Smith-Gill, S.J., and Davies, D.R. (1989) Proc. Natl. Acad. Sci. USA **86**, 5938-5942.

IDENTIFICATION OF EPITOPES OF THE RECEPTOR BINDING SUBUNIT OF CHOLERA TOXIN BY SYNTHETIC PEPTIDE AND CBIB APPROACHES

Mohammad Kazemi* and Richard A. Finkelstein

Department of Molecular Microbiology and Immunology
School of Medicine, University of Missouri-Columbia
Columbia, MO 65212

INTRODUCTION

Cholera toxin (CT) and related diarrheagenic heat-labile enterotoxins (LTs) are produced, among others, by strains of *Vibrio cholerae* as well as strains of *Escherichia coli* of human (H) and porcine (P) origin (reviewed in Finkelstein, 1988). These enterotoxins are structurally, immunologically and functionally related. All CT-related toxins consist of a homopentamer receptor-binding B-subunit protein and an enzymatically active A-subunit. The B-subunit proteins, designated as CT-B-1 and CT-B-2 (from classical and El Tor biotype strains of *V. cholerae* 569B and 3083, respectively) as well as H-LT-B and P-LT-B (from *E. coli* strains) bind to G_{M1} ganglioside on the surface of target cells. Receptor-binding is essential for the internalization and the intracellular activity of the A-subunit protein. Identification of the epitopes of the immunodominant CT-related B-subunit proteins, particularly those that are involved in the receptor-binding process, is of great interest for vaccine development. In this report we describe two different approaches that we have taken to analyze the epitopic regions of CT-B-1.

EXPERIMENTAL APPROACHES

Checkerboard immunoblotting (CBIB)

This technique (Kazemi and Finkelstein, 1990a) provides an efficient means of analyzing the immunological cross-reactivities of multiple antigen/antibody combinations. In the CBIB assay, antigens are immobilized on nitrocellulose membranes in the form of parallel lanes which are then probed with antibodies in similarly applied parallel lanes perpendicular to the antigen lanes. Following development with appropriate secondary antibody and substrate, positive reactions appear as small colored squares at the intersections of reactive antibodies and antigens. CBIB is rapid, requires small amounts of reagents and, most importantly, there are no quantitative variations between samples within each antigen or antibody lane. Additionally the developed blots provide permanent records of the patterns of cross-reactivities between antigens and antibodies. Simultaneous comparisons of such reactions can be effectively utilized to examine the effect of structural differences between antigens on their immunological behavior.

Synthetic overlapping peptides-on-pins

This is a technique developed by Geysen et al. (1987) which allows one to scan the

*Present address: Baxter Healthcare Corp., Hyland Division, 1720 Flower Ave., Duarte, CA 91010

Immunobiology of Proteins and Peptides VI
Edited by M.Z. Atassi, Plenum Press, New York, 1991

primary structure of protein for identification of potential antigenic determinants. Overlapping peptides spanning the entire amino acid sequence of an antigen are synthesized on the tips of derivatized plastic pins (Cambridge Research Biochemicals, Wilmington, DE) and are then used to examine their reactivities with antibodies of interest. Peptides covalently-bound on pins can be reused many times by disrupting the antibody-peptide bond after each reaction using SDS and sonication. The patterns of reactivity of the peptides with antibodies are indicative of the structure of the reactive epitopes. This technique is particularly suitable for studying the continuous (sequence-related) epitopes, but may also provide information about segments of the antigen that comprise the interactive parts of discontinuous (conformational) epitopes.

We synthesized overlapping hexapeptides and octapeptides spanning the entire amino acid sequence of the CT-B-1 molecule and from their patterns of reactions with antibodies to CT-B-related peptides and proteins various epitopic regions were identified. A detailed structural analysis of one epitope in the conserved region of the CT-B-1 molecule was carried out by amino acid substitutions. The strategy involved in such study was to synthesize all possible peptide analogues of the epitope in which every single amino acid residue was replaced with all 20 different naturally occurring amino acids. Reactivities of the parent epitope peptide and its analogues with antibodies of interest were compared to assess the essentiality of each amino acid residue in the epitope.

RESULTS AND DISCUSSION

CBIB assay was performed on various combinations of CT-related antigens and antibodies. In addition to various purified B-subunit proteins, four chimeric P-LT-Bs (pDL-2, 3, 5 and 7), with single or two amino acids substitution(s) of the corresponding H-LT-B residues, as described by Finkelstein et al. (1987), were also used for reaction. Antigens were immobilized on nitrocellulose membrane in their native state as well as in variously denatured forms. Numerous CT-related mAbs, human sera, and polyclonal antibodies, including antibodies that were raised against synthetic peptides (CTPs) representing different portions of CT-B-1 (Jacob et al., 1983; Finkelstein et al., 1987) were used to react with the antigens.

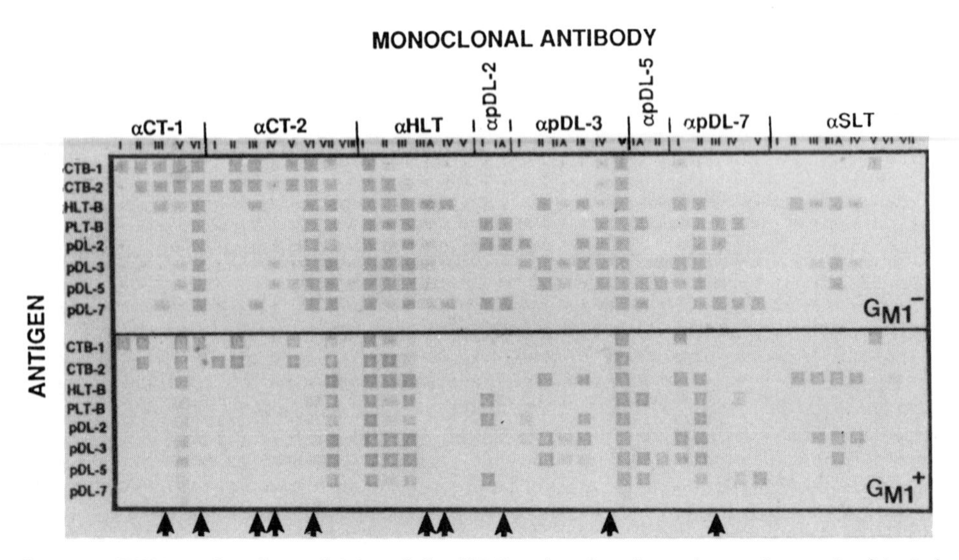

Figure 1. CBIB results of reactivities of the CT-B-related antigens (top to bottom) with their corresponding mAbs (left to right). Antigens were applied at 100 ng/spot in their native state and reacted with 1:10 dilution of culture supernatants of mAb producer hybridoma, either directly (G_{M1}^- panel) or after saturating with receptor (G_{M1}^+ panel). Reactions that are affected by G_{M1} treatment are marked with an arrow.

CBIB results of various antigen/antibody cross-reactivities are reported elsewhere (Kazemi and Finkelstein, 1990b), and only selected results will be presented here. The top panel (G_{M1}^-) of Figure 1 represents the reactivities of various mAbs raised against CT-related proteins (Finkelstein et al., 1987; Kazemi and Finkelstein, 1990b) with corresponding antigens in their native state. None of these mAbs reacted with antigens that were denatured or fragmented with CNBr (not shown). Comparison of the amino acid sequence variations among various CT-related antigens, as shown by Kazemi and Finkelstein (1990b), and their pattern of reactivities with mAbs, reveals rather detailed information about the involvement of some amino acid residues in the formation of epitopes. For example, class III αCT-1 and independently derived class III αCT-2 mAbs (Fig. 1) reacted identically only with CT-B-1, CT-B-2, H-LT-B-1 and pDL-7, all of which share a His residue at position 13. Unreactive antigens, H-LT-B-2 (not shown), P-LT-B, pDLs 2, 3 and 5, contain an Arg in that position. In contrast to the reaction of these two mAbs, αpDL-3 class IV mAb recognized only proteins with Arg^{13}, i.e., P-LT-B, and pDL-2, 3, and 5. It is therefore likely that residue 13 plays a dominant role in the formation of these epitopes. The bottom panel of Figure 1 (G_{M1}^+) shows the CBIB assay of same antibody/antigen reactions except that the antigens immobilized on the membrane were incubated with a saturating level of receptor (G_{M1} ganglioside) prior to reaction with antibodies in order to identify epitopes that are blocked by the receptor binding process. As a result of this treatment, reactivity of several mAbs (among them class III of αCT-1 and αCT-2) was abolished indicating that their epitopes are likely to be located at or near a G_{M1} binding site. It is also conceivable that the receptor binding process affects the epitope by altering the native conformation of the protein. The patterns of reactivity of these G_{M1} affected reactions (G_{M1}^+ panel) identify certain amino acid residues to be involved in epitopes which in turn implicate the region around such residues as potential receptor binding site(s). Such regions occurred around residues 4, 13, 18, 46, 54 and possibly 102. Whether any of these residues are directly involved in receptor binding requires further study.

The human convalescent sera examined by CBIB reacted primarily with the native form of the toxin proteins (Kazemi and Finkelstein, 1990b). Sera from individuals challenged with a virulent El Tor biotype strain of *V. cholerae* were generally more reactive than those who had received strain CVD 103 - a genetically engineered *ctx* A^-B^+ candidate vaccine strain derived from classical wild-type strain 569B Inaba (Levine et al., 1988). Figure 2 shows the reactivities of unadsorbed as well as variously adsorbed sera from two individuals challenged with El Tor biotype *V. cholerae*, with native and denatured forms of CT-B-1 and CT-B-2. Results of this CBIB assay further confirmed the broader antigenicity of the toxin from the El Tor strain of *V. cholerae* relative to CT-1 --native CT-B-2 adsorbed the reactivity of antibodies to both toxins while CT-B-1 only adsorbed the reactivity of CT-B-1 . This type of CT-B-2 superiority over CT-B-1 in adsorption of antibody reactivities was also observed with hyperimmune sera.

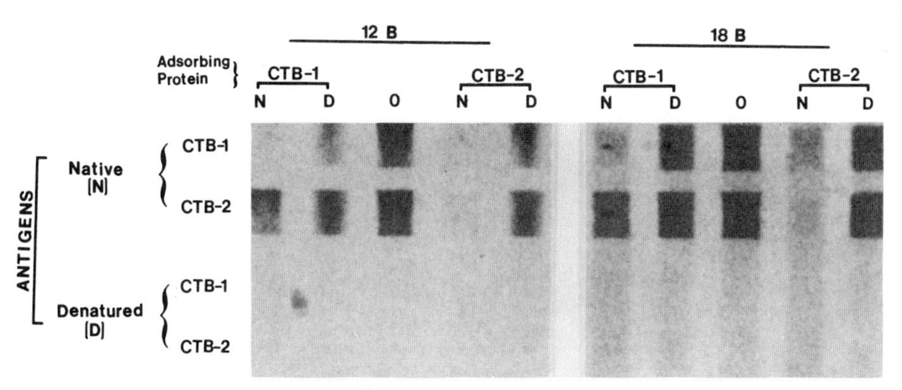

Figure 2. CBIB results of reactivities of CT-B-1 and CT-B-2 in their native (N) and denatured (D) forms with two human sera convalescent from challenge with El Tor strain of *V. cholerae*. Sera were used at 1:50 dilution as unadsorbed (O) or adsorbed with the native or the denatured forms of antigens.

At a much higher resolution, potential epitopic regions of CT-B-1 were identified by the synthetic peptide-on-pin approach of Geysen et al., 1987, as reported in detail by Kazemi and Finkelstein (1991). Reliability of the assay was assessed by the reactions of several antisera (from Dr. C.O. Jacob, Stanford University) that were raised against synthetic peptides (CTPs) representing different portions of CT-B-1 (Fig. 3 bottom 3 panels). Such antibodies reacted only with hexapeptides that were within the region of the immunizing CTP. An apparent immunological cross-reactivity between two of these peptides (CTP-1 and CTP-7) is attributed to the presence of three identical (but in different order) amino acids, Gln^{16}-Ile^{17}-His^{18} (of CTP-1) and Gln^{56}-His^{57}-Ile^{58} (of CTP-7), in these two peptides. Interestingly, antibodies to the peptides representing these two regions of CT-B-1 were shown by Jacob et al. (1984) to partially neutralize the biological activity of native toxin. This is specially true for the antibodies to its most conserved region (the CTP-7 region). It was also shown by Jacob et al. (1986a and b) that immunization of rabbits with a single dose of a synthetic peptide of this region (CTP-3, residues 50-64 of CT-B-1) primed for a vigorous immune response to subsequent administration of a subimmunizing dose of any of the three CT-related enterotoxins tested. Analysis of the structure of antigenic determinants of this region is, therefore, highly desirable for the development of peptide vaccines.

Anti-CTP-7 antiserum recognized multiple epitopes in the conserved region of the CT-B-1 sequence, one of which was the tetramer Ser-Gln-His-Ile (residues 55-58) represented by the reactivity of hexapeptides 53-55 (peptide numbers correspond to the number of their N-terminus residues in the CT-B-1 sequence). This epitope which contains the CTP-7/CTP-1 cross-reactive Gln, His, and Ile residues was also recognized, although at different degrees of reactivities, by most antisera raised against CT-B-related peptides and proteins including convalescent human sera. The top three panels in Figure 3 show the reactivity of hyperimmune

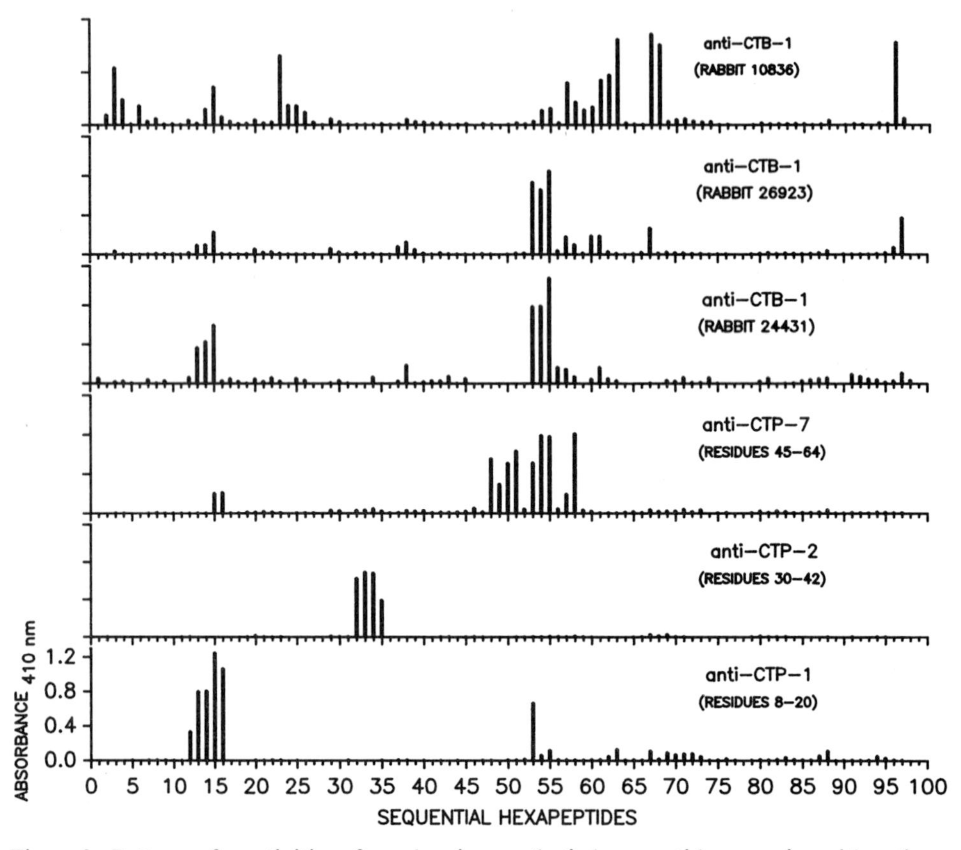

Figure 3. Patterns of reactivities of overlapping synthetic hexapeptides-on-pins with antisera raised against synthetic peptides (CTPs) representing different portions of CT-B-1 (1:50 dilution), and against purified CT-B-1 (1:1000 dilution).

sera raised against purified CT-B-1 in rabbits. While the serum from one rabbit (#10836) recognized multiple epitopes along the CT-B-1 sequence, sera from the other two rabbits (#24431 and #26923) were exclusively reactive with the tetrameric epitopes Ser-Gln-His-Ile (residues 55-58) and Thr-Gln-Ile-His (residues 15-18). These two epitopes correspond to the CTP-1 and CTP-7 epitopes containing Gln, His and Ile residues in different orders. The fact that antibodies to these regions were generated with either the whole CT-B-1 protein itself or the synthetic peptides representing these regions (CTP-1 and CTP-7) indicates that the two regions contain distinct epitopes that are independently recognized by antibody-producing lymphocytes. Such recognition does not necessarily require the native conformation of protein.

Further studies were conducted to evaluate the role of each amino acid residue in the reactivity of the Ser-Gln-His-Ile (SQHI) epitope with antibodies. All possible analogues of the pentapeptide GSQHI (the SQHI epitope containing the extra N-terminus Gly^{54} of the CT-B-1 sequence) in which every residue was replaced with all 20 different amino acids were synthesized and reacted with the antibodies. Figure 4 depicts the reaction pattern of the pentapeptide analogues with αCT-B-1 antiserum from rabbit 24431. The results indicated that residues Gln and His are essential for reactivity and are not replaceable with any other amino acid. It is likely that these two residues are the principal, and specific, binding sites with antibody. The role of His^{57} in binding to antibody has also been previously shown by Anglister et al. (1988) through their NMR evaluation of the antibody/antigen complex. Results of the amino acid replacement studies also indicated that Ser^{55}, and to a lesser degree Ile^{58}, can be replaced with a few other amino acids with minimal effect on reactivity of the epitope. The extra Gly^{54}, on the other hand, can be replaced with many other amino acids with little or no effects on the reactivity of the epitope with the exception of the negatively charged Glu and Asp: these substitutions abolished the reaction. It is also interesting to note that the tetramer SQHI by itself gave the strongest reaction (Figure 4, bar shown with arrow) suggesting that the tetramer most closely approximates the epitope.

In many instances, the specificity of the peptide reactions was assessed by adsorption studies. In such cases, the antibodies were adsorbed, in the fluid phase, with saturating levels of adsorbing antigens (CT-related proteins in native or fragmented forms), and reacted with peptides-on-pins. Such adsorption studies demonstrated that reactivity with the SQHI epitope was significantly adsorbed with both the native and fragmented forms of proteins (not shown). This indicates that SQHI is a continuous epitope that is exposed on the surface of the native protein. The CT (holotoxin) was a more effective adsorbent than the purified B-subunit protein. This suggests that the conformation of the B-subunit protein is different in its purified form from its conformation in the holotoxin.

Antibodies that on CBIB assay were primarily reactive with the native form of proteins (conformational epitopes) failed to identify distinctive, strongly reactive, peptides when reacted with overlapping peptides-on-pins. Among these were many classes of mAbs, as well as human sera. They generally gave weak and diffuse patterns of reactions. Human sera appeared to identify regions of reactive peptides that might indeed represent portions of conformational epitopes that give partial (weak) reactions (Kazemi and Finkelstein, 1991). Further analyses

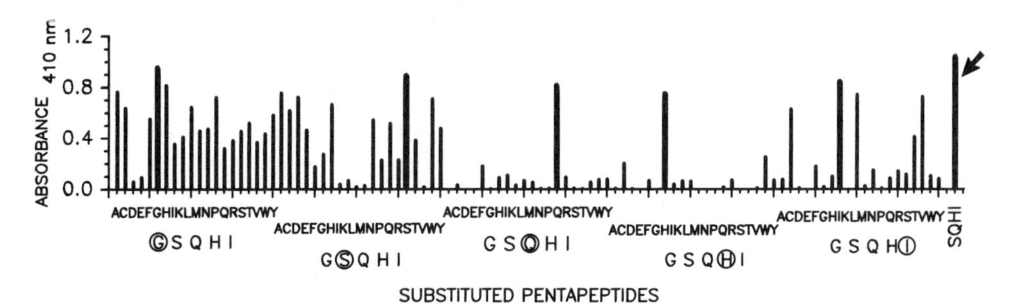

Figure 4. Reactivities of αCT-B-1 antiserum from rabbit #24431 with pentapeptide GSQHI and with all its amino acid substituted analogues (see text). The replaced residue is circled and bars representing the reaction of parent peptide are highlighted. The reaction of the full length original epitope SQHI by itself is also marked by an arrow.

of CT-B-1 epitopes were carried out by synthesizing octapeptides spanning the CT-B-1 sequence. Results were, in general, in agreement with the results of hexapeptide reactions. It was, however, noticed that octapeptides appeared to identify other larger size epitopes that were not detectable with hexapeptides. Some octapeptides containing the entire structure of certain epitopes failed to react with antibodies to those epitopes. Presumably the extra amino acid residues in those octapeptides had negative effects on the reactivity of the epitopes. It is therefore important to validate the structure of epitopes by scanning the protein with different window sizes (different sizes of overlapping peptides). Anti-CT-2 class III mAb exhibited weak reactivity with a number of different octapeptides from different regions of CT-B-1, among them octapeptides #12 and 13 containing His^{13} in their sequence. The same mAb in CBIB assay (Fig. 1) recognized only the CT-B-related antigens containing His (and not Arg) at position 13. Therefore, this result suggests, but not prove, that at least some of these peptides (such as peptides #12 and 13) may represent portions of the conformational epitope that is recognized by this antibody.

In conclusion, the synthetic peptide-on-pin approach coupled with CBIB assays can be effectively utilized to study the epitopic sites on proteins of known amino acid sequences. We learned that a suitably designed series of reactions with unadsorbed as well as selectively adsorbed antibodies can give detailed information about the structure of epitopes.

ACKNOWLEDGMENT

This work was supported in part by Public Health Service grants AI 16776 and AI 17312 from the National Institute of Allergy and Infectious Diseases.

REFERENCES

Anglister, J., Jacob, C.O., Assulin, O., Ast, G., Pinker, R., and Arnon, R., 1988, NMR study of the complex between a synthetic peptide derived from the B subunit of cholera toxin and three monoclonal antibodies, Biochem., 27:717.

Finkelstein, R.A., Burks, M.F., Zupan, A., Dallas, W.S., Jacob, C.O., and Ludwig, D.S., 1987, Epitopes of the cholera family of enterotoxins, Rev. Infect. Dis., 9:544.

Finkelstein, R.A., 1988, Cholera, the cholera enterotoxins, and the cholera enterotoxin-related enterotoxin family, In: Immunochemical and molecular genetic analysis of bacterial pathogens, P. Owen, and T.J. Foster ed., Elsevier Science Publishers, Amsterdam, The Netherlands.

Geysen, H.M., Rodda, S.J., Mason, T.J., Tribbick, G., and Schoofs, P.G., 1987, Strategies for epitope analysis using peptide synthesis, J. Immunol. Methods, 102:259.

Jacob, C.O., Arnon, R., and Finkelstein, R.A., 1986a, Immunity to heat-labile enterotoxins of porcine and human *Escherichia coli* strains achieved with synthetic cholera toxin peptides, Infect. Immun., 52:562.

Jacob, C.O., Grossfeld, S., Sela, M., and Arnon, R., 1986b, Priming immune response to cholera toxin induced by synthetic peptides, Eur. J. Immunol., 16:1057.

Jacob, C.O., Sela, M., and Arnon, R., 1983, Antibodies against synthetic peptides of the B-subunit of cholera toxin: cross-reaction and neutralization of the toxin, Proc. Natl. Acad. Sci. USA, 80:7611.

Jacob, C.O., Sela, M., Pines, M., Hurwitz, S., and Arnon, R., 1984, Both cholera toxin-induced adenylate cyclase activation and cholera toxin biological activity are inhibited by antibodies against related synthetic peptides, Proc. Natl. Acad. Sci. USA, 81:7893.

Kazemi, M., and Finkelstein, R.A., 1990a, Checkerboard immunoblotting (CBIB): an efficient, rapid, and sensitive method of assaying multiple antigen/antibody cross-reactivities, J. Immunol. Methods, 128:143.

Kazemi, M., and Finkelstein, R.A., 1990b, Study of epitopes of cholera enterotoxin-related enterotoxins using checkerboard immunoblotting, Infect. Immun., 58:2352.

Kazemi, M., and Finkelstein, R.A., 1991, Mapping epitopic regions of cholera toxin B-subunit protein, Molec. Immunol., (In press).

Levine, M.M., Herrington, M.D., Losonsky, G., Tall, B., Kaper, J.B., Ketley, J., Tacket, C.O., and Cryz, S., 1988, Safety, immunogenicity of recombinant live oral cholera vaccines, CVD 103 and CVD 103-HgR, Lancet, ii:467.

AUTOIMMUNE RECOGNITION PROFILE OF THE ALPHA CHAIN OF HUMAN ACETYLCHOLINE RECEPTOR IN MYASTHENIA GRAVIS

Tetsuo Ashizawa*, Minako Oshima+,
Ke-He Ruan+, M. Zouhair Atassi+

Department of Biochemistry (+) and
Neurology (*) Baylor College of Medicine
Houston, Texas

INTRODUCTION

Myasthenia gravis (MG) is an autoimmune disease with skeletal muscle weakness and easy fatigability due to autoantibodies directed against acetylcholine receptor (AChR) molecules at the motor endplates. The AChR consists of 2 alpha (α) subunits and one each of beta, gamma and delta subunit in a circular arrangement. Studies have shown that major populations of the autoantibodies are directed against extracellular domains of the α subunit which bears binding sites for the ligands and α-toxin (reviewed in Ashizawa and Appel, 1985; Atassi *et al.*, 1987).

The entire amino acid sequence of the human α subunit has been deduced from cDNA nucleotide sequence (Noda *et al.*, 1983). Based on the sequence, we synthesized (Mulac-Jericevic *et al.*, 1988) 18 overlapping peptides representing the main extracellular domain (residues α1-210) of the α chain (Figure 1) to delineate the peptide recognition profile of autoantibodies and T-cells from MG patients within this domain.

	Peptide	Structure		Peptide	Structure
1	α1-16	S-E-H-E-T-R-L-V-A-K-L-F-K-D-Y-S	10	α100-115	F-A-I-V-K-F-T-K-V-L-L-Q-Y-T-G-H
2	α12-27	F-K-D-Y-S-S-V-V-R-P-V-E-D-H-R-Q	11	α111-126	Q-Y-T-G-H-I-T-W-T-P-P-A-I-F-K-S
3	α23-38	E-D-H-R-Q-V-V-E-V-T-V-G-L-Q-L-I	12	α122-138	A-I-F-K-S-Y-G-E-I-I-V-T-H-F-P-F-D
4	α34-49	G-L-Q-L-I-Q-L-I-N-V-D-E-V-N-Q-I	13	α134-150	H-F-P-F-D-E-Q-N-G-S-M-K-L-G-T-W-T
5	α45-60	E-V-N-Q-I-V-T-T-N-V-R-L-K-Q-Q-W	14	α146-162	L-G-T-W-T-Y-D-G-S-V-V-A-I-N-P-E-S
6	α56-71	L-K-Q-Q-W-V-D-Y-N-L-K-W-N-P-D-D	15	α158-174	I-N-P-E-S-D-Q-P-D-L-S-N-F-M-E-S-G
7	α67-82	W-N-P-D-D-Y-G-G-V-K-K-I-H-I-P-S	16	α170-186	F-M-E-S-G-E-W-V-I-K-E-S-R-G-W-K-H
8	α78-93	I-H-I-P-S-E-K-I-W-R-P-D-L-V-L-Y	17	α182-198	R-G-W-K-H-S-V-T-Y-S-C-C-P-D-T-P-Y
9	α89-104	D-L-V-L-Y-N-N-A-D-G-D-F-A-I-V-K	18	α194-210	P-D-T-P-Y-L-D-I-T-Y-H-F-V-M-Q-R-L

Fig. 1. Covalent structures of the synthetic overlapping peptides encompassing the extracellular part of the α chain of human AChR. The regions of overlaps between consecutive peptides are underlined.

Immunobiology of Proteins and Peptides VI
Edited by M.Z. Atassi, Plenum Press, New York, 1991

This comprehensive synthetic peptide approach (Kazim and Atassi, 1980, 1982) proved to be very useful for the localization of *continuous* regions of the AChR involved in autoantibody binding and T-cell recognition in MG. Indeed, T-cell lines (Pachner *et al.*, 1989) and antibodies (Mulac-Jericevic *et al.*, 1987) from EAMG have been shown to recognize the peptides of the extracellular α subunit of *Torpedo* AChR. Difficulties in obtaining pure human AChR in amounts sufficient for passage *in vitro* of autoreactive T-cells from MG patients and limited (80%) homology of *Torpedo* AChR with human AChR (Noda *et al.*, 1982, 1983) also give advantages to the synthetic peptide approach. Passage of protein-primed lymphocytes by peptides *in vitro* is an effective approach for the preparation of site specific T-cell lines (Yoshioka *et al.*, 1983). Any *discontinuous* antigenic sites of AChR that might be formed by residues distant in the sequence but coming into close spatial proximity (Atassi and Smith, 1978) in the three-dimensional structure cannot be investigated by the comprehensive synthetic peptide approach (Kazim and Atassi, 1980, 1982). However, the current lack of knowledge of the three-dimensional structure of the AChR leaves studies of *discontinuous* antigenic sites, should they indeed exist in AChR, for the future. In this paper, we present data concerning autoantibody and T-cell recognition profiles of *continuous* regions of the α subunit.

METHODS

Human AChR-specific T-cell lines were prepared as described by Oshima *et al.* (1990). PBL from MG patients were cultured with an equimolar mixture of the 18 peptides for 4-5 days followed by isolation of lymphoblasts by density gradient centrifugation. Isolated lymphoblasts underwent one or two more passages with the optimum dose of peptide mixture, 10-13 days apart. X-irradiated autologous B-cells transformed with Epstein-Barr viruses (EBV) were used as antigen presenting cells (Neitzel, 1986). T-cell lines were cultured in the presence of individual antigen(s) for 5 days, pulsed with ^{3}H-thymidine and then harvested to assay radioactivities.

Determination of autoantibodies against synthetic peptides was performed by attaching the peptides to 96-well-Co-bind plates. The wells were blocked with 0.1-0.5% casein, washed with PBS, incubated with plasma or serum at 37°C for 4 hours, washed again, and then incubated with the second antibodies at 4°C overnight. After washing, ^{125}I-protein A was added at room temperature for 3 hours. The plates were washed and the contents were assayed for the radioactivities.

RESULTS

The Profile of Peptides Recognized by the T-Cell Lines

Proliferative responses of PBL from 11 MG patients and 4 controls to varying doses of an equimolar mixture of the 18 synthetic peptides determined the optimum stimulating dose to be 40 μg/ml. The viability of the cells and the specificity of the responses were confirmed using phytohemagglutinin, nonsense peptides and ovalbumin. Except for one MG sample out of eight examined, the response of PBL was greater for the peptide mixture than *Torpedo* AChR in all MG patients studied.

Using 40 μg/ml of the equimolar mixture, T-cell lines were prepared from PBL of 5 MG patients by repeated passage *in vitro* of a population of T-cells with an equimolar mixture of the 18 peptides in the presence of the X-irradiated autologous B-cell line (Oshima *et al.*, 1990).

The profile of peptides recognized by the T-cell lines were variable; the five T-cell lines did not recognize any common immunodominant region while peptide α146-162 elicited the most frequent response in 3 out of the 5 T-cell lines (Figure 2). The HLA phenotype was variable among these T-cell lines while DR w52 was the common MHC class II in all 5. In LBA, 82% of T-cells were $CD4^+$, whereas BB showed that 90% of T-cells were $CD8^+$, suggesting both helper and cytotoxic T-cells can play a role in autoimmune recognition of a limited number of regions on AChR.

The Profile of Peptides Recognized by Human Autoantibodies in MG

Antibody recognition sites were studied in 15 plasma samples from MG patients using the 18 synthetic peptides and one of the inter-transmembrane regions (peptide α262-276). The binding profiles of autoantibodies from 9 MG patients and 9 normal controls are summarized in Figure 3. MG autoantibodies predominantly recognized 4 regions broadly localized within, but not necessarily including all of, residues α10-30,

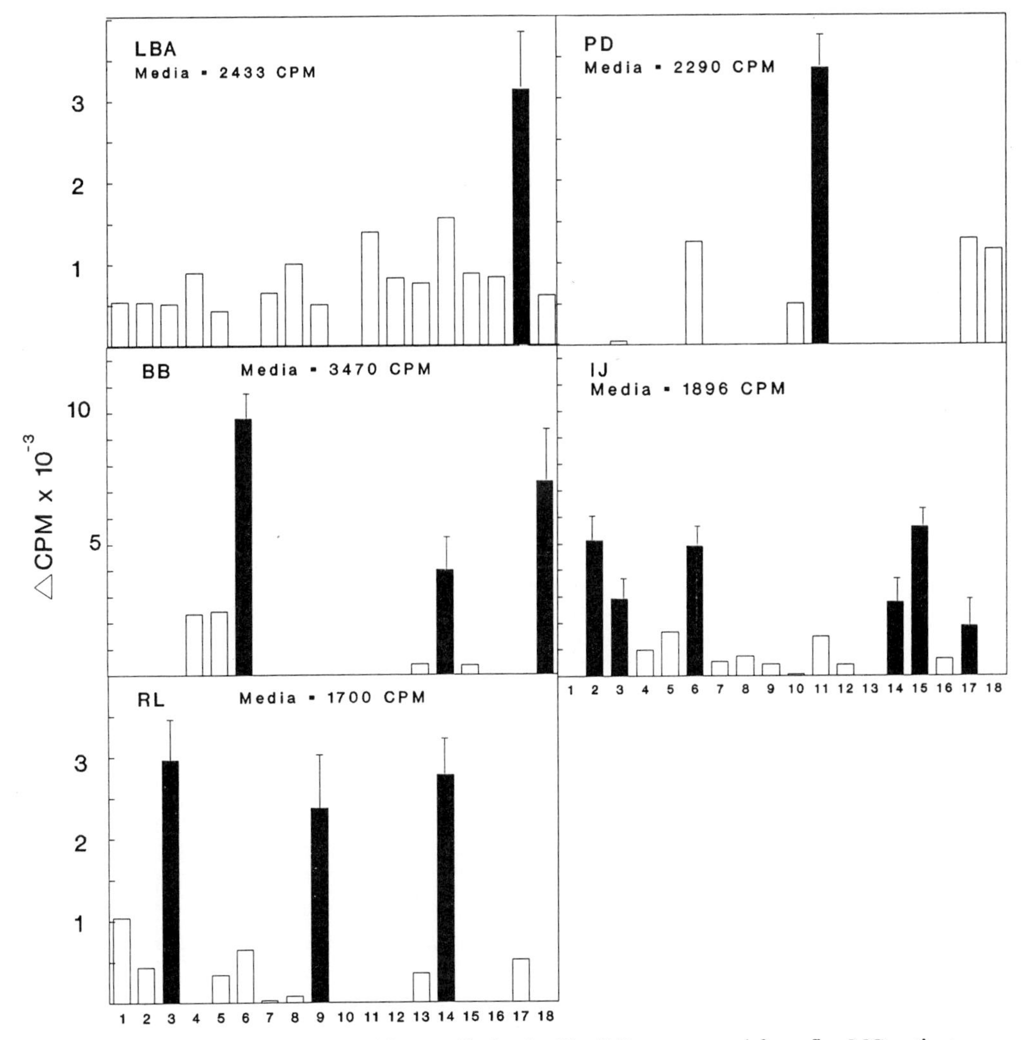

Fig. 2. Autoimmune peptide recognition profile by the T-cell lines prepared from five MG patients using an equimolar mixture of the 18 synthetic peptides of human AChR.

α111-145, α175-198 and, less frequently, residues α45-77. The pattern of recognition was highly variable from one MG patient to another. In general, the region residing within α175-198 was recognized by all MG autoantibodies studied and was immunodominant in most cases. Binding of autoantibodies in 9 MG to each of the peptides α12-27, α111-126, α122-138 and α182-198 showed, when compared to 9 control plasma samples, that the mean binding activity for each peptide was significantly higher in MG plasma than normals. However, some overlaps were observed between MG and control groups. When an equimolar mixture of these four peptides was used, the mean autoantibody binding activities were significantly higher in MG than normal and disease controls with no overlap ($p<0.001$, Figure 4).

Comparison of the Profiles of Peptide Recognition by T-Cells and Autoantibody

The autoantibody and T-cell recognition profiles were compared in 3 MG patients (Figure 5). Autoantibodies of donor BB recognized essentially only one major autodeterminant within region α182-198, while autoimmune T cells of the same donor

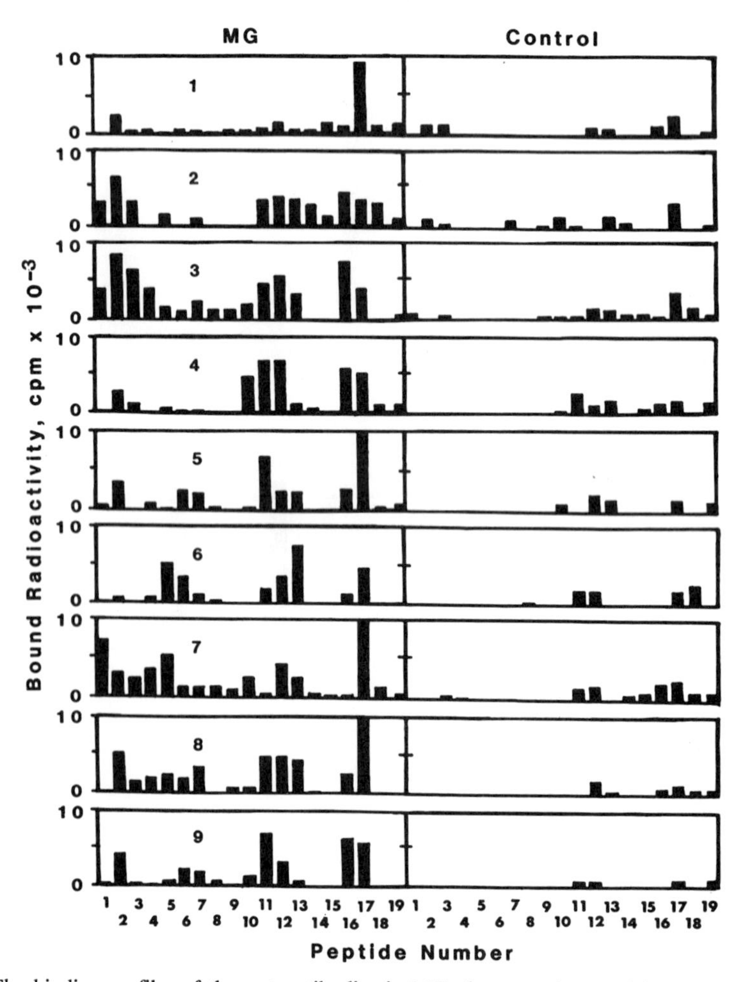

Fig. 3. (left) The binding profiles of the autoantibodies in MG plasma and normal human plasma to the synthetic peptides.

responded strongly *in vitro* to three regions of which only one (α194-210) overlapped with the region recognized by the autoantibodies. The other two regions within α56-71 and α146-162 were uniquely T-cell autodeterminants with little or no detectable antibodies directed to them. The autoimmune T-cell line from the patient RL recognized 4 peptides, only one of which (peptide α23-38) was also recognized by autoantibodies. Two T-cell recognition regions, α1-17 and α89-104, shared an overlap with antibody recognition regions within residues α12-27 and α100-115, respectively. One T-cell auto-determinant within region α146-162 was recognized exclusively by the T-cells in this patient. The T-cell line from LBA recognized three auto-determinants, two of which (α111-126 and α182-198) coincided completely with regions recognized by autoantibodies while the third (α146-162) shared an overlap with the autoantibody binding region within residues α134-150.

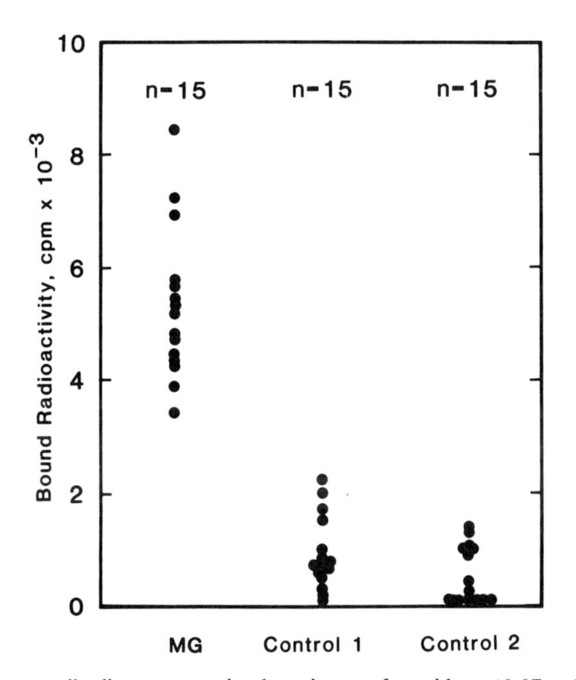

Fig. 4. Binding of autoantibodies to an equimolar mixture of peptides α12-27, α111-126, α122-138 and α182-198 in MG patients, normal controls and disease controls.

DISCUSSION

Our results showed that human AChR-specific autoimmune T-cell lines (both $CD4^+$ and $CD8^+$ cells) can be prepared from MG peripheral blood lymphocytes using a mixture of overlapping synthetic peptides corresponding to the extracellular region of the human AChR α subunit. The peptide recognition profiles by the T-cell lines are highly variable from one MG patient to another. The profiles of the autoantibody binding to individual peptides were also variable, but four regions (residues α10-30, α111-145, α175-198, and less frequently, residues α45-77) were particularly reactive with the autoantibodies. The variability of these recognition sites is consistent with genetic control operating at the antigenic site level as suggested by the variability of HLA phenotype/genotype of the each T-cell lines. In a given individual, autoantibodies and autoimmune T-cells tend to recognize similar sites although there are sites recognized

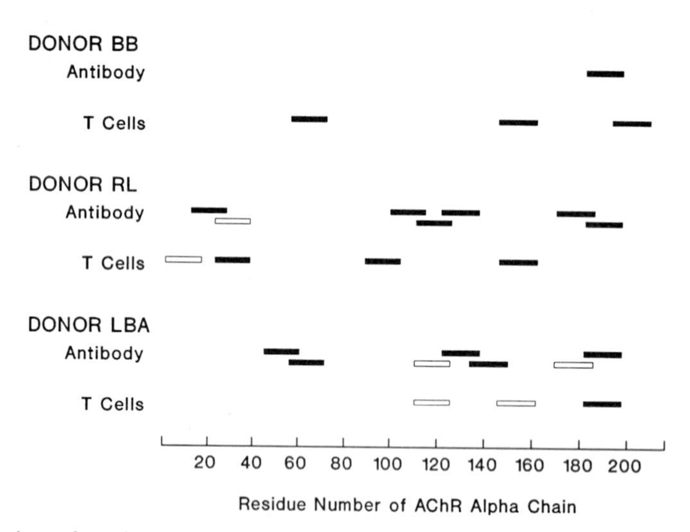

Fig. 5. Comparison of autoimmune antibody and T-cell peptide recognition profiles of the extra-cellular part of human AChR α-chain in 3 MG patients. The bars indicate the locations of the stimulating peptides in the sequence. Solid bars represent a strong response while open bars denote an intermediate response.

exclusively either by autoantibodies or by T-cells. One of the peptides recognized by MG autoantibodies corresponded to α122-138, a major region involved in α-neurotoxin binding. An equimolar mixture of four autoantibody-binding peptides provides a highly specific diagnostic radioimmunoassay for detection of autoantibodies in MG. Thus, this equimolar mixture of the 4 peptides may provide reliable clinical diagnosis of MG.

REFERENCES

Ashizawa, T., and Appel, S. H., 1985, Immunopathological events at the end-plate in myasthenia gravis, *Springer Sem. Immunopathol., 8*:177.

Atassi, M.Z. and Smith, J.A., 1978, A proposal for the nomenclature of antigen sites in peptides and proteins, *Immunochem., 15*:609.

Atassi, M.Z., Mulac-Jericevic, B., Yokoi, T. and Manshouri, T., 1987, Localization of the functional sites on the α chain of acetylcholine receptor, *Fed. Proc., 46*:2538.

Kazim, A.L. and Atassi, M.Z., 1980, A novel and comprehensive synthetic approach for the elucidation of protein antigenic structures. Determination of the full antigenic profile of the α-chain of human hemoglobin, *Biochem. J., 191*:673.

Kazim, A.L. and Atassi, M.Z., 1982, Structurally inherent antigenic sites: Localization of the antigenic sites of the α-chain of human haemoglobin in three host species by a comprehensive synthetic approach, *Biochem. J., 203*:201.

Mulac-Jericevic, B., Kurisaki, J., and Atassi, M. Z., 1987, Profile of the continuous antigenic regions on the extracellular part of the α chain of an acetylcholine receptor, *Proc. Natl. Acad. Sci. USA, 84*:3633.

Mulac-Jericevic, B., Manshouri, T., Yokoi, T., and Atassi, M. Z., 1988, The regions of α-neurotoxin binding on the extracellular part of the α-subunit of human acetylcholine receptor, *J. Prot. Chem., 7*:173.

Neitzel, H., 1986, A routine method for the establishment of permanent growing lymphoblastoid cell lines, *Hum. Genet., 73*:320.

Noda, M., Takahashi, H., Tanabe, T., Toyosato, M., Furutani, Y., Hirose, T., Asai, M., Inayama, S., Miyata, T., and Numa, S., 1982, Primary structure of α-subunit precursor of *Torpedo californica* acetylcholine receptor deduced from cDNA sequences, *Nature, 299*:793.

Noda, M., Furutani, Y., Takahashi, H., Toyosato, M., Tanabe, T., Shimizu, S., Kikyotani, S., Kayano, T., Hirose, T., Inayama, S., and Numa, S., 1983, Cloning and sequence analysis of calf cDNA and human genomic DNA encoding α-subunit precursor of muscle acetylcholine receptor, *Nature, 305*:818.

Oshima, M., Ashizawa, T., Pollack, M.S., and Atassi, M.Z., 1990, Autoimmune T cell recognition of human acetylcholine receptor: The sites of T cell recognition in myasthenia gravis on the extracellular part of the α subunit, *Eur. J. Immunol. 20*:2563.

Pachner, A. R., Kantor, F. S., Mulac-Jericevic, B., and Atassi, M. Z., 1989, An immunodominant site of acetylcholine receptor in experimental myasthenia mapped with T lymphocyte clones and synthetic peptides, *Immunol. Lett., 20*:199.

Yoshioka, M., Bixler, G.S., and Atassi, M.Z., 1983, Preparation of T-lymphocyte lines and clones with specificities to preselected protein sites by *in vitro* passage with free synthetic peptides: demonstration with myoglobin sites, *Mol. Immunol. 20*:1133.

PAUCITY OF HUMORAL RESPONSE IN PATIENTS TO GLIOMA-ASSOCIATED ANTIGEN(S): ANTIGEN LOCALIZATION BY IMMUNOFLUORESCENCE

Duncan K. Fischer[1,2], Masafumi Matsuda,[1,2]
Fatma Shaban[2], Raj K. Narayan[1], and
M. Zouhair Atassi[2]

Departments of Neurosurgery[1]
and Biochemistry[2], Baylor College
of Medicine, Houston, Texas - 77030, USA

ABSTRACT

Xenogeneic immunization of freshly-prepared human glioma extracts into goats has yielded a polyclonal antiserum, which after multiple absorptions specifically identifies antigenic entities only in glioma extracts, and not in appropriate controls, both by radioimmunoassays (RIAs) and Western immunoblots. The results from the absorbed polyclonal antiserum have been confirmed by the successful generation of six stable murine monoclonal antibodies (MAbs) which recognize a subset of the same antigens with high specificity on immunoblots and with no apparent cross-reactivities by RIA to normal brain, serum, liver, muscle, kidney, spleen, or melanoma tissues. Moreover, the tested murine MAbs (B12C4) reveal a striking and abundant glial filament protein, possibly related to glial fibrillary acidic protein (GFAP) or other intermediate filament proteins, by frozen-section immunofluorescence. This is seen only in gliomas and is absent, or dramatically reduced, in normal human cortex. Use of potent immortalizing strain (FF41) of Epstein-Barr virus (EBV) to establish antibody-secreting human lymphoblastoid lines, and the generation of mouse-human chimeric fusions, have yielded lines possessing variable supernatant human antibody secretion. Radioimmunoassays using culture supernatants, and sera from glioma patients and an normal individual, have demonstrated surprisingly similar reactivity profiles, even after a sensitive sandwich RIA employing the B6C6 murine MAb. These results suggest that, although human glioma-associated antigens, including possibly the up-regulation of GFAP expression, clearly exist, there seems to be a muted humoral response as evidenced by a paucity of tumor-specific B-cells. This may be due to antigenic shielding by the blood-brain barrier, or due to a form of immunological compromise in patients harboring these malignancies.

Immunobiology of Proteins and Peptides VI
Edited by M.Z. Atassi, Plenum Press, New York, 1991

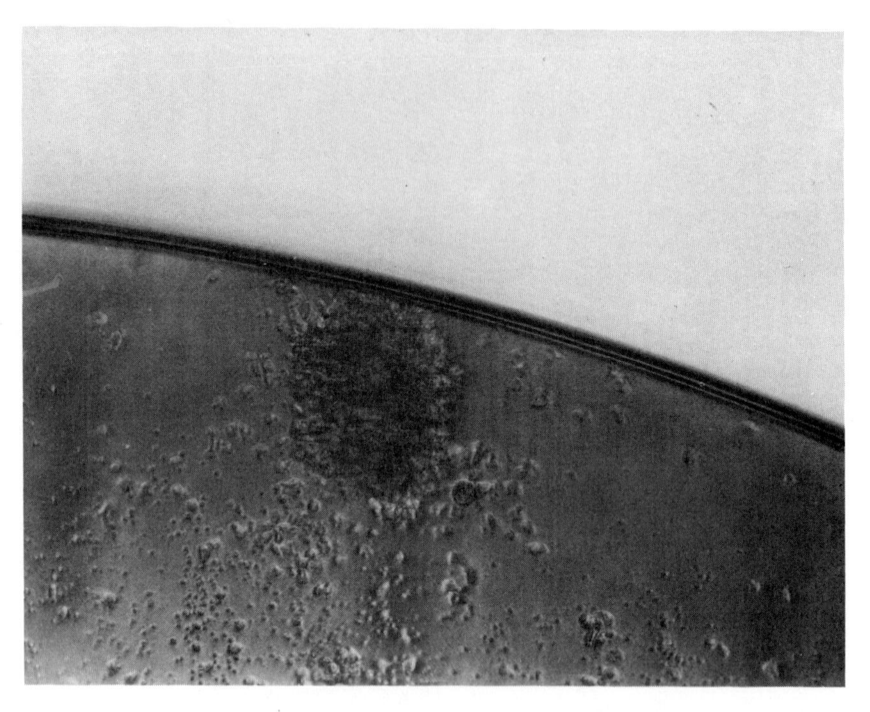

Fig. 1A. Inverted phase contrast photomicrograph of limiting dilutions of EBV strain FF41-immortalized lymphoblastoid lines from glioma patients often reveal only one colony per microtiter plate well.

INTRODUCTION

The annual incidence of primary central nervous system neoplasms in the United States approaches 15,000, yet, despite all currently available therapies, the 2-year survival rate for high grade gliomas is less than 15-20% (1). Molecular and immunological approaches are needed to understand the gene regulation and antigen expression patterns in these tumors, as well as the patients' apparently often delayed humoral response to their growth.

Our laboratory has been successful in preparing a polyclonal goat antiserum against glioma extracts which was highly reactive with astrocytoma-associated antigens in radioimmunoassays (RIAs) and Western immunoblots, and following extensive absorptions did not react with normal human brain, human serum, or human serum albumin. Recently, ongoing laboratory efforts have generated a panel of six murine monoclonal antibodies (MAbs) (B6C6, B6G7, B12B4, B12C3, B12C4, & B12F10) strongly reactive with, and entirely specific for, glioma extracts on radioimmunoassay. In contrast to previously reported monoclonal reagents (1), these murine MAbs exhibited little or no reactivity on RIA with normal human cerebral cortex, serum, liver, spleen, kidney, muscle, or melanoma tissue (2). The above specificity detected by both polyclonal and monoclonal reagents derived from xenogeneic immunization strongly supports the existence of human glioma-associated antigens, including possibly the up-regulation of GFAP expression.

MATERIALS AND METHODS

To determine whether patients mounted a specific immunological response against their tumors *in vivo*, sera samples and peripheral blood lymphocytes isolated on sterile Ficoll-Hypaque gradients were obtained from glioma patients 4 to 6 weeks after craniotomy for initial tumor resection. The B-lymphocytes were then immortalized (3, 4) in culture with the FF41 strain (5) of Epstein-Barr virus (EBV), or fused to the nonsecreting mouse myeloma line X63/Ag8.653 employing polyethylene glycol (PEG) and HAT (hypoxanthine, aminopterin, thymidine) selection in the hope of establishing

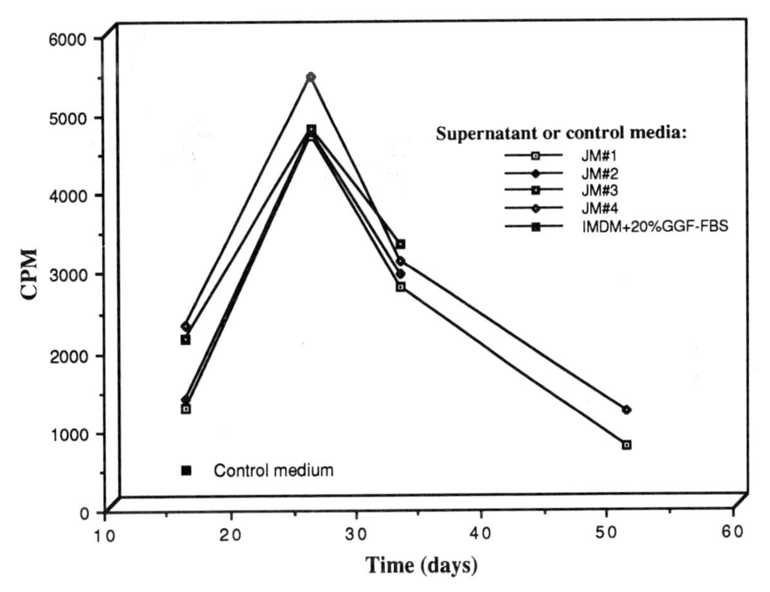

Fig. 1B. Time course radioimmunoassays document variable supernatant human antibody production by these lymphoblastoid cell lines, often maximal at 3.5 weeks. The above represents four lymphoblastoid line (JM #1, #2, #3, #4) supernatants, as well as immunoglobulin-free fetal bovine control medium.

human antibody-producing lymphoblastoid or mouse-human chimeric hybridoma clones. Tumor and tissue extracts were always freshly prepared in cold phosphate-buffered saline (PBS, pH 7.2) at the time of surgery in the presence of phenylmethylsulfonyl fluoride protease inhibitor. Radioimmunoassays (RIAs) were performed on 2.5 to 10 μg/well of antigen extract, often using 0.2% casein in PBS to block nonspecific reactivity, and diluted rabbit or goat antihuman or antimouse IgG and IgM as an amplifying antibody bridge. These steps were followed by ^{125}I-labeled staphylococcal protein A (200,000 cpm/well in 50 μl PBS/0.02% casein) for detection. For immunofluorescence, frozen-section slides of glioma and control cortex tissue were rapidly fixed in cold 100% ethanol, air-dried, and then incubated for 30 min. at 37° C with both diluted primary antibody and fluorescein-tagged antihuman or antimouse Ig. Intervening washes of PBS were performed between these steps.

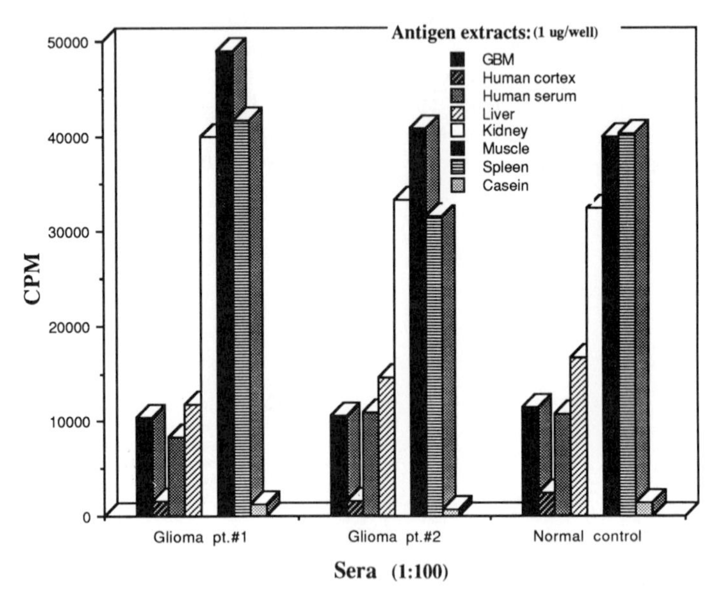

Fig. 2. Radioimmunoassays demonstrating the concordance of strikingly similar reactivity profiles of sera from two representative glioma patients, and a normal individual as a control.

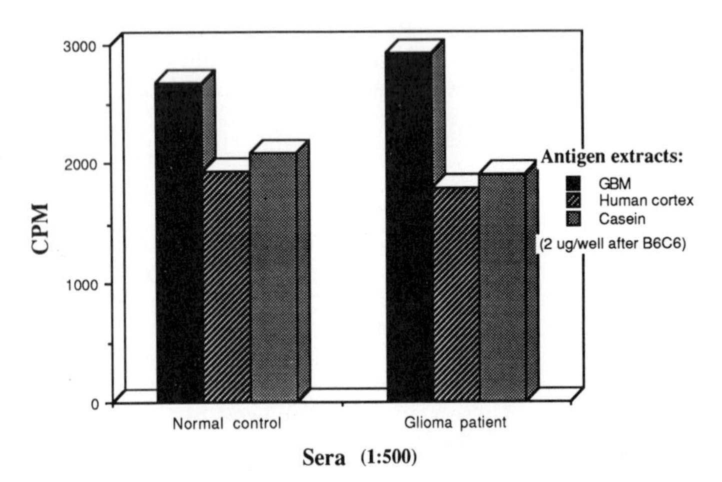

Fig. 3. The reactivity profiles of a glioma patient and a normal individual control retain their concordance, even after a sandwich radioimmunoassay using the B6C6 murine MAb.

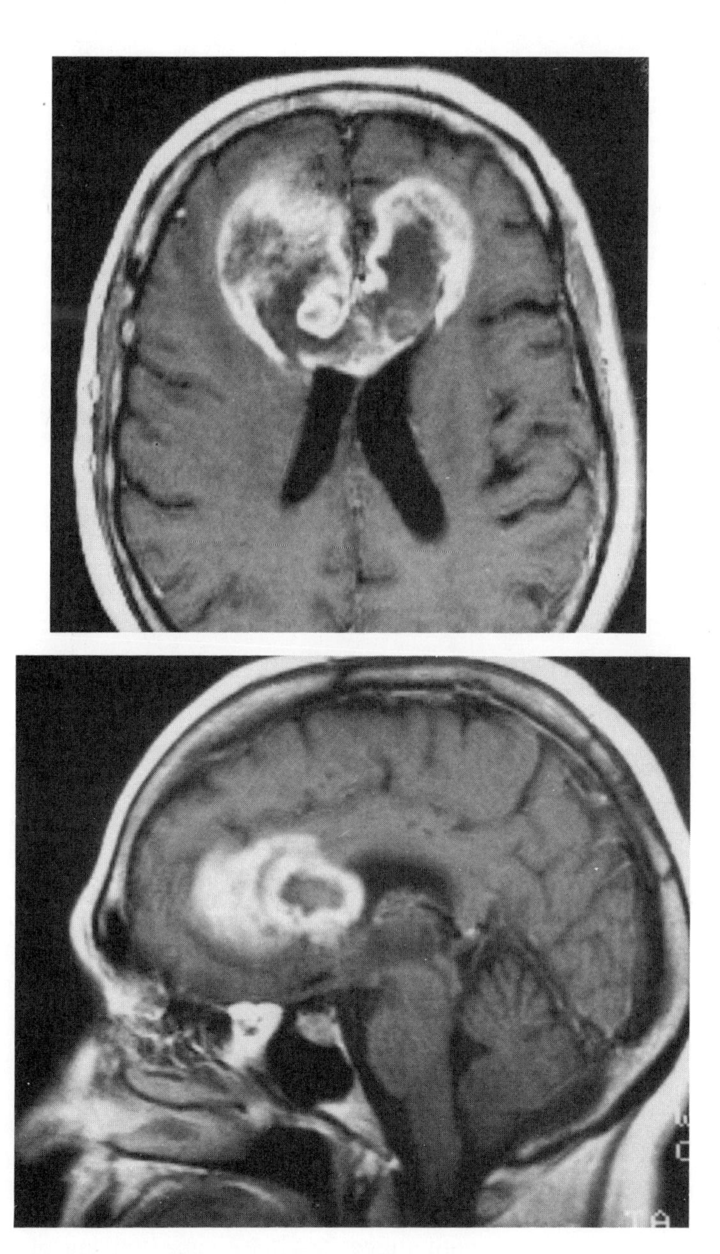

Fig. 4. Representative axial (top) and sagittal (bottom) brain magnetic resonance imaging (MRI) studies with gadolinium enhancement reveal a high grade "butterfly" glioma involving the corpus callosum and both hemispheres. This actual fresh surgical tumor tissue was stained by immunofluorescence in Figure 5.

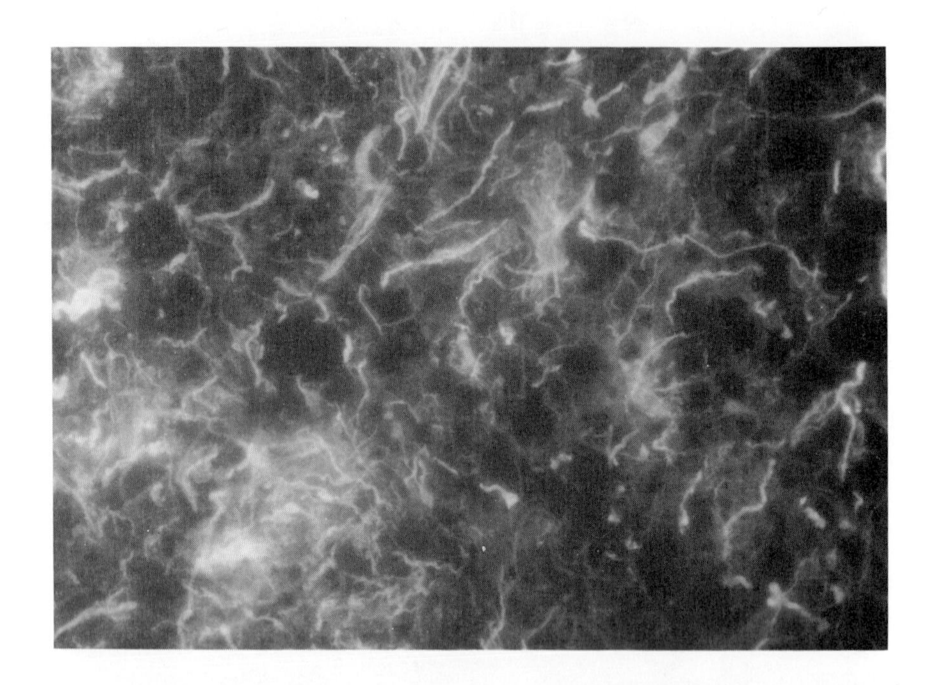

Fig. 5. Frozen-section immunofluorescence of the glioma (top) in Figure 4, employing the B12C4 murine MAb, detects a striking glial filament protein, possibly human GFAP, which is absent, or in dramatically reduced abundance (bottom), in normal human cortex or when a control murine MAb to myoglobin is employed. The B12C4 MAb was used at a 1/10 dilution, and the fluorescein-tagged secondary antibody employed at 1/30. The top exposure was for less than 10 sec., whereas the bottom control photomicrograph needed a 50 sec. exposure to adequately visualize the field.

RESULTS

After immortalization and fusion, individual colonies were clearly visible within 14 days (6). The spectrum and time course of supernatant antibody reactivity were assayed by RIA on normal human cerebral cortex and glioblastoma multiforme (GBM) extracts, as well as on human liver, spleen, kidney, and muscle as tissue controls. High reactivity was found by RIA to GBM antigen extracts in growing culture (Figure 1A) supernatants (6), even after limiting dilution. Yet, supernatant antibody production was variable over time and maximal at 3.5 weeks (Figure 1B). Although strong reactivities were noted to tumor extracts and far less to normal brain, the same concordant reactivity profiles (Figure 2), including to normal tissue antigens, were noted in the sera of glioma patients and a normal individual, despite multiple manipulations of the assay and blocking conditions to minimize cross-reactivities and nonspecific binding (7). These results were confirmed with a glioma and a normal human serum by a sensitive sandwich radioimmunoassay (Figure 3) using the B6C6 murine monoclonal.

Frozen-section immunofluorescent data suggest that the murine MAbs identify a glial filament protein in abundant quantity in tissue from gliomas (Figure 4; Figure 5 top), and absent, or present in dramatically reduced amounts, in normal brain controls (Figure 5 bottom). The relationship of this filament protein to glial fibrillary acidic protein (GFAP), or other intermediate filament proteins, and its characterization are currently in progress.

DISCUSSION

Radioimmunoassays using culture supernatants and sera from glioma patients and a normal individual have demonstrated surprisingly similar, concordant reactivity profiles, including to normal tissue antigens, even after a sandwich RIA employing the B6C6 murine MAb. These results suggest that, although glioma-associated antigens, including a probable augmented expression of GFAP, are clearly known to exist by both polyclonal and monoclonal assay systems and now by frozen-section immunofluorescence (Figure 5), a deficient humoral response represented by a paucity of tumor-specific B-cells is present in the potentially immunocompromised patient harboring an intracranial malignancy. The patients' tolerance or blocked recognition of a new or an up-regulated antigen complex known to exist by the above assays provides additional evidence in support of their immunocompromised status and/or a sequestering of glioma antigens. The latter may, in part, be due to leukocyte-induced protective pericellular halos composed of glycosaminoglycans, as reported surrounding glioma cells in culture (8).

Combined immunological and biomolecular techniques represent some of the most promising strategies to dissect the humoral repertoire and complexities of tumor antigen expression in glioma patients. Our conclusions have implications about the restricted immunological repertoire of glioma patients, either due to B-cell or antigen presentation deficits, and suggest that future directions in therapy should address this deficiency. Nevertheless, the panel of murine MAbs, highly reactive with a glial filament protein, possibly GFAP, on a rapid frozen-section immunofluorescent assay, may provide a rapid diagnostic test for intracranial tumors of uncertain glial origin, or sensitively to detect "spilled" antigenic components across a disrupted blood-brain barrier within the sera or cerebrospinal fluid of patients. In contrast to the numerous cross-reactivities of the previously reported monoclonals (1), these reagents display excellent specificity by both immunofluorescence and radioimmunoassay (2) for human glioma tissue. However, since the relative concentration and amounts of tumor-associated antigen(s) vary with extracts, probably depending on the stage of the tumor, it would be presently difficult to provide, for general use, sufficient quantities of standardized antigen reagent for serologic tests.

ACKNOWLEDGEMENTS

This research was supported by the Herbert M. Karol Foundation, the Welch Foundation, the Robert A. Welch Chair of Chemistry (to M.Z.A.), and by an American College of Surgeons scholarship (to D.K.F.). We gratefully appreciate the advice and encouragement of Professor Robert G. Grossman.

REFERENCES

1. Fischer, D.K., Chen, T.L., and Narayan, R.K., Immunological and biochemical strategies for the identification of brain tumor-associated antigens, *J. Neurosurg* 68:165 (1988).
2. Matsuda, M., Fischer, D.K., Narayan, R.K. and Atassi, M.Z., Preparation and characterization of antisera and of murine monoclonal antibodies to human glioma-associated antigen(s), *Adv. in Exp. Med. and Biol.*, 1991, in press.
3. Miller, G., Robinson, J., Heston, L. and Lipman, M., Differences between laboratory strains of Epstein-Barr virus based on immortalization, abortive infection, and interference, *Proc. Natl. Acad. Sci. USA* 71:4006 (1974).
4. Steinitz, M., Klein, G., Koshimies, S. and Makela, O., EB virus-induced B lymphocyte cell lines producing specific antibody, *Nature* 269:420 (1977).
5. Fischer, D.K., Miller, G., Gradoville, L., Heston, L., Westrate, M.W., Wright, J., Brandsma, J. and Summers, W.C., Genome of a mononucleosis Epstein-Barr virus contains DNA fragments previously regarded to be unique to Burkitt's lymphoma isolates, *Cell* 24:543 (1981).
6. Fischer, D.K., Narayan, R., Abaza, M.-S., Matsuda, M. and Atassi, M.Z., Epstein-Barr virus-immortalized peripheral blood lymphocytes from astrocytoma patients contain supernatant antibodies highly reactive with glioma antigens, *Surgical Forum* 40:501 (1989).
7. Fischer, D.K., Shaban, F., Abaza, M.-S., Matsuda, M., Narayan, R.K., and Atassi, M.Z., Dissection of human glioma humoral response using direct serological, EBV-proliferation, and chimeric mouse-human fusion assays, *Surgical Forum 41*:513 1990.
8. Gately, C.L., Muul, L.M., Greenwood, M.A., Papazoglou, S., Dick, S.J., Kornblith, P.L., Smith, B.H., and Gately, M.K., *In vitro* studies on the cell-mediated immune response to human brain tumors. II. Leukocyte-induced coats of glycosaminoglycan increase the resistance of glioma cells to cellular immune attack, *J. Immunol.* 133:3387 (1984).

PREPARATION AND CHARACTERIZATION OF ANTISERA AND OF MURINE MONOCLONAL ANTIBODIES TO HUMAN GLIOMA-ASSOCIATED ANTIGEN(S)

Masafumi Matsuda[1,2], Duncan K. Fischer[1,2]
Raj K. Narayan[2] and M. Zouhair Atassi[1]

Departments of Biochemistry[1]
and Neurosurgery[2], Baylor College
of Medicine, Houston, Texas-77030, USA

ABSTRACT

Human glioma-associated markers can be exploited for the development of new diagnostic strategies and treatment modalities for these malignancies. A goat antiserum was first raised against human anaplastic astrocytoma (AC or AA) and glioblastoma multiforme (GB or GBM) extracts. Extensive sequential absorptions with normal brain tissue, normal serum, and human serum albumin (HSA) gave an antibody fraction specific for glioma. Balb/c mice were subsequently immunized with these glioma extracts. B-cell hybridomas from these mice were then cloned and subcloned by limiting dilution, yielding six monoclonal antibodies (MAbs) that were entirely specific for tumor tissues, and did not react with normal human serum or with normal human brain, liver, kidney, spleen, or muscle. Moreover, the murine MAbs did not cross-react with certain other human tumors, including melanoma. The fully absorbed antiserum and the murine MAbs both identify a polypeptide pattern possibly related to human glial fibrillary acidic protein (GFAP) or other intermediate filament proteins on immunoblots. These immunological reagents could serve as powerful tools for the diagnosis and possibly therapy of these uniformly fatal tumors.

INTRODUCTION

The identification of malignant brain tumor markers is important for the design of molecular diagnostic methods and possibly new therapeutic approaches. An immunological approach presents the advantages of specificity and sensitivity. Mouse monoclonal antibodies (MAbs) have been made against astrocytoma, medulloblastoma and glioma antigens, but these reported MAbs exhibited broad cross-reactivities (23). Rabbit polyclonal antisera have been prepared against astrocytomas and glioblastoma and, after absorption with normal brain tissue, were reported to be specific for tumor tissue (1-3). However, these studies often did not consider the problem that tumor tissues contain large amounts of serum proteins, and particularly human serum albumin (HSA), since the blood-brain barrier is compromised at the sites of these tumors.

Immunobiology of Proteins and Peptides VI
Edited by M.Z. Atassi, Plenum Press, New York, 1991

Therefore, the recorded specificities of the absorbed sera for reaction with presumed tumor-associated antigens could have been due to reaction with HSA and other serum proteins.

In the present work, we wish to report results with goat polyclonal antisera subjected to exhaustive absorption strategies. Also, mouse MAbs have been prepared, screened, and selected so that they exhibit exquisite specificity for human glioma tissue. These goat polyclonal and murine monoclonal reagents help clarify the existence of a tumor-associated antigen system, possibly related to GFAP, in human gliomas.

MATERIALS AND METHODS

Preparation of tissue extracts

Wet brain tissue samples (2 gm) were homogenized by hand using a glass homogenizer in 10 ml of 10 mM $NaPO_4$ buffer pH 8.0 at 4°C with 100 mM phenylmethylsulfonyl fluoride (PMSF) as protease inhibitor. After centrifugation (30,000 rpm, 90 min. at 4°C) of the samples, protein concentrations in the supernatants were determined by the method of Lowry (4) using bovine serum albumin (Sigma) as the standard.

Preparation of goat antisera

Polyclonal antisera were raised in goats. One goat was immunized with a Freund's adjuvant extract of normal human brain tissue and another with a mixture of brain tumor extracts. For injections of normal brain tissue, rapidly prepared autopsy samples from two patients dying of cardiovascular causes were employed. The tumor extract used for immunization was a mixture of surgical tissue samples from a patient with glioblastoma multiforme (GB or GBM) and a second patient with anaplastic astrocytoma (AC or AA). The final protein concentration of the extracts was 5 mg/ml. Each goat was injected intramuscularly with 5 mg of the appropriate extract in PBS and complete Freund's adjuvant. They were boosted in a similar manner at 3 weeks and thereafter monthly. Each animal was bled 7 times, with a three week interval between bleeds.

Preparation of monoclonal antibodies

Female Balb/c mice, 7 to 8 weeks old, were obtained from Jackson Laboratories (Bar Harbor, Maine, USA). Each mouse was immunized in the footpad with 50 μl of an emulsion (containing 50 μg protein) of equal volumes of anaplastic astrocytoma extract and complete Freund's adjuvant. Two weeks later, the mice were boosted intraperitoneally with a similar antigen emulsion. After three weeks, the mice received a similar dose of booster composed of an emulsion in complete Freund's adjuvant of two anaplastic astrocytomas. Test bleeds a week later showed that the mice had very high titers of antibodies against glioma antigens by radioimmunoassay.

Three days before fusion, the mice were injected intraperitoneally with a booster of 200 μg of AA extracts in complete Freund's adjuvant. For preparation of hybridomas, the spleen cells were fused with P3 x 63 - Ag8.653 mouse myeloma. The procedures for fusion and limiting dilution have been described (5). The murine monoclonal antibodies described here were prepared by two limiting dilutions. Attempts to generate human monoclonal antibodies using the immortalizing FF41 strain of Epstein-Barr virus (14) will be reported elsewhere.

Radiolabeling of antibodies and protein A

Antibodies and purified protein A (Pharmacia) were radiolabeled with ^{125}I (Amersham Corp., Arlington Heights, IL) using the chloramine-T method (6). Unbound ^{125}I was separated from the radiolabeled protein A by gel filtration on Sephadex G-25 (Pharmacia). At least 95% of the protein A-associated ^{125}I was precipitable with 10% (w/v) trichloracetic acid.

Gel electrophoresis

Gel electrophoresis (7) was done in 10% acrylamide sodium dodecyl sulfate gels (SDS-PAGE) for 3 to 5 hrs at a constant current of 25 mA. All samples were reduced using beta-mercaptoethanol. Samples were loaded in the gel at 30 - 40 μg of protein per lane. Gels (1.02 mm. thick) were strained for 1 hour in 0.25% Coomassie Brilliant Blue R - 250 (Sigma) in ethanol-acetic acid-water (5:0.5:4.5 v/v). Certain gels were silver-stained using Merril's technique with a minor modification (8).

Western immunoblots

SDS-PAGE gels were transferred to nitrocellulose (0.45 μm.) in transfer buffer (192 mM glycine, 25 mM Tris base, 0.1% SDS, 25% methanol) for 4 hours in the cold room at 200 mA (Bio-Rad apparatus), using only minor modifications of current electro-blotting techniques (9,10). Nitrocellulose filters were blocked overnight at 4°C with 0.2% casein in PBS, and then washed 5 times (10 min. each) with PBS. Antinormal brain and antitumor sera were used (at dilutions of 1/100 in 0.02% casein/PBS) for 3 hours at room temperature, followed by incubation (45 min.) at room temperature with ^{125}I-labeled rabbit anti-goat IgG (200,000 - 300,000 cpm/ml in 0.02% casein/PBS). In experiments with fully absorbed antitumor sera, incubation with labeled IgG was done using 800,000 - 1,000,000 cpm/ml and the antisera diluted only 1/10 to 1/20 for increased sensitivity. Immunoblots were also developed using the murine MAbs at 1/2,000 dilution, followed by horseradish peroxidase (HRP)-conjugated rabbit anti-mouse IgG at 1/5,000.

Preparation of immunoglobulin fraction

The immunoglobulin fraction of the antisera was prepared by precipitation with 40% $(NH_4)_2SO_4$, followed by washing on the centrifuge with the same solvent. The precipitates were dissolved in and dialyzed against several changes of PBS, pH 7.2.

Preparation of adsorbents and absorption experiments

Coupling of proteins to cyanogen bromide-activated Sepharose CL-4B (Pharmacia) was done as previously described (11). The adsorbents were suspended in PBS and poured into syringes to make 25 ml (packed volume) columns. Sequential absorptions on different immunoadsorbents were done following the same protocol. Binding to adsorbent was done for 24 hours at 4°C, after which it was run out of the adsorbent and an aliquot removed for analysis. The effluent was then applied onto the next immunoadsorbent and so on. At each step an aliquot was saved for analysis. Whenever necessary, more than one cycle on a given adsorbent was done until all undesired reactivities were depleted. The specificity of the products was determined by radioimmunoassays, SDS-PAGE electrophoresis, or Western immunoblots. In order to obtain the desired fraction from goat antitumor antiserum, the latter had to be passed three times each through immunoadsorbents of human serum albumin, normal human serum, and normal brain tissue in that order.

Solid-phase radioimmune assay of antisera and monoclonal antibodies

Solid-phase radioimmunoassay (RIA) of hybridoma supernatants was done as previously described (12,13). A mixture of tumor extracts from one glioblastoma (GB or GBM) and three anaplastic astrocytomas (AC or AA) or of normal human brain tissue (50 μg protein in 50 μl per well) were used as the plate antigens (3 hrs, 37°C). After washing with PBS, the wells were blocked (1 hr, 37°C) with 100 μl of 0.5% casein in PBS, followed by washing five times with PBS. Hybridoma supernatant, or appropriately diluted antisera (1:500 to 1:128,000, v/v, in PBS/0.02% casein), were added (50 μl per well) and the plates incubated at 4°C overnight and then washed 5 times with PBS. Prediluted (1:1500, v/v in PBS/0.02% casein) rabbit anti-mouse IgG + IgM (50 μl) was added and the plates incubated at 37°C for 2 hrs. The plates were then washed (5 times with PBS) and ^{125}I-labeled protein A was added to each well (2 x 10^5 cpm in 50 μl PBS/0.02% casein) and allowed to bind at 37°C for 1 hr. The plates were then washed, dried, and the wells cut out and counted on a gamma counter.

Determination of antibody class

Solid-phase RIA was used to determine the class of each antibody preparation obtained by this procedure. In these assays, the aforementioned tumor extract was used as the plate antigen for the test antibody and the latter was followed by a panel of rabbit anti-mouse immunoglobulin class-specific antisera (Litton Bionetics), which were used to amplify the murine anti-tumor antibody binding, or the tumor extract was followed by normal rabbit serum or PBS-casein in control assays. The antisera were diluted 1:1,000 - 1:3,000 in PBS-0.02% casein. The specificities of the anti-class antisera were first confirmed by RIA and then in Ouchterlony immunodiffusion plates in which class-specific myeloma proteins (Zymed Laboratories, San Francisco, CA) were used as antigens.

RESULTS

Goat antisera against normal brain tissue and tumor extracts

Antisera were raised in one goat against normal human brain tissue and in another goat against a mixture of glioma extracts. In solid-phase RIA, antisera against normal brain tissue showed approximately the same reactivity with normal brain tissue or tumor extracts. Antisera against glial tumor tissue showed considerably higher reactivity with tumor extract than with normal brain tissue (Figure 1). Anti-tumor antisera, however, also reacted strongly with human serum albumin (HSA) (Figure 1). The tumor extract was subsequently absorbed three times on an affinity adsorbent of anti-normal brain tissue antibodies. After three absorptions, the effluent of the tumor extract did not react with anti-normal brain antibodies, yet retained almost full reactivity with anti-tumor antisera (Figure 1). It should be noted that the presence of abundant contaminating 67 Kd HSA in the tumor extract was confirmed by PAGE (Figure 5B), and explains why albumin-absorbed anti-glioma serum often reacts less strongly with necrotic glioblastoma extracts containing serum proteins which permeate through the compromised blood-brain barrier (Figure 2).

Absorption of anti-tumor antisera

To remove the population of antibodies that react with normal brain tissue, HSA, and normal human serum proteins from anti-glioma antisera, the latter were absorbed on affinity adsorbents of HSA, normal human sera and normal brain tissue extract.

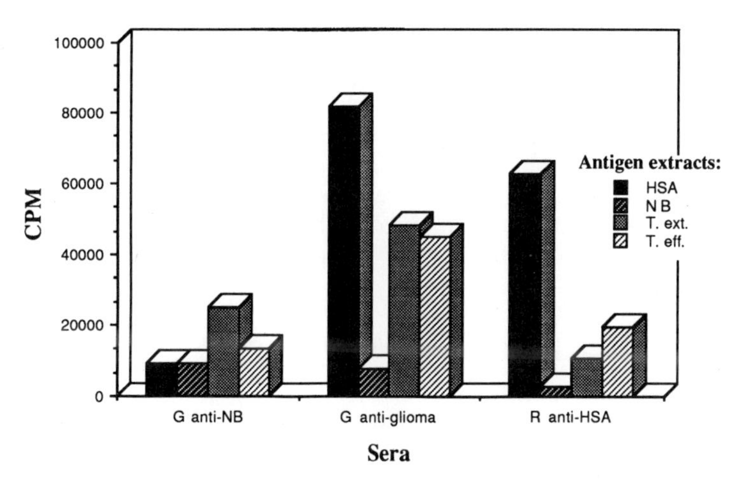

Fig. 1 Binding of the goat anti-normal brain antiserum (G anti-NB), goat anti-glioma antiserum (G anti-glioma), and rabbit (R) anti-HSA with human serum albumin (HSA), extract of normal brain (NB), a mixture of extracts of GBM and AA tumors (T. Ext.), and the tumor extract mixture after absorption on an affinity column of anti-normal brain tissue antibodies (T. Eff.). Note the high reactivity to serum albumin of unabsorbed anti-tumor serum.

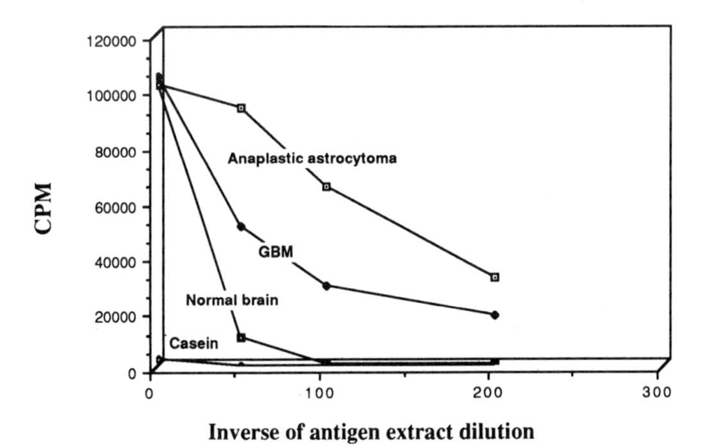

Fig. 2 RIA's employing HSA-absorbed goat anti-glioma serum often display less reactivity to the necrotic glioblastoma multiforme (GBM) extracts because of abundant contaminating serum proteins, including serum albumin, in these extracts.

The effluent from a given adsorbent (e.g. HSA) was re-absorbed on another adsorbent of the same antigen, and the process was then repeated until the antiserum was depleted of the population of antibodies which reacted with that antigen. Three absorption cycles on each type of adsorbent were needed to achieve specificity for tumor-associated antigens. Figures 3 and 4 demonstrate the specificity of this fully absorbed antiserum. The results show that this absorbed antiserum reacted only with tumor extracts [of both AA or GBM (Figure 4)], or with tumor extracts that had been absorbed on anti-normal tissue antibodies (Figure 3), but did not react with normal brain tissue, normal human serum, or with HSA (Figures 3 and 4). Western immunoblots with the absorbed antisera revealed at least six potential tumor-associated markers in glioblastoma and astrocytoma extracts, and only very small quantities, if any, in the non-glial tumor medulloblastoma or in control tissue and human serum (Figure 5A).

Tumor-specific mouse monoclonal antibodies

Culture supernatants of hybridomas of spleen B-cells from mice that had been immunized with tumor extracts were screened by solid-phase RIA both with normal brain tissue and with glioma extracts, thus permitting the selection of tumor-specific antibodies. After two limiting dilutions, over 30 tumor-specific MAbs were obtained. Six of these were studied in detail.

The MAbs reacted strongly with glioma extracts and showed little or no reactivity with HSA, normal human serum, or with extracts of normal human brain, liver, spleen, kidney, and muscle tissues (Table 1 and Figure 6). On Western immunoblots the murine MAbs identified a similar 43 - 52 Kd polypeptide pattern, possibly related to

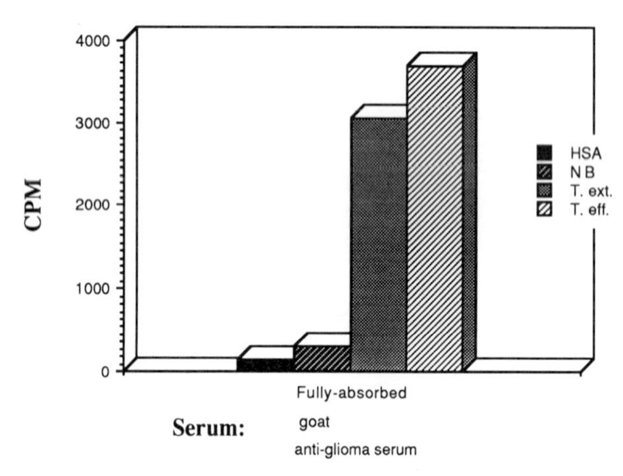

Fig. 3 Binding of goat anti-glioma antiserum (which had been absorbed with HSA, normal human serum and normal brain tissue) to HSA, normal brain tissue extract (NB), a mixture of extracts of GBM and AA tumors (T. Ext.), and the tumor extract after absorption on an affinity column of anti-normal brain tissue antibodies (T. Eff.). Note that after exhaustive absorptions to reduce cross-reactivities, persistent reactivity remains to the tumor extract and the column tumor effluent.

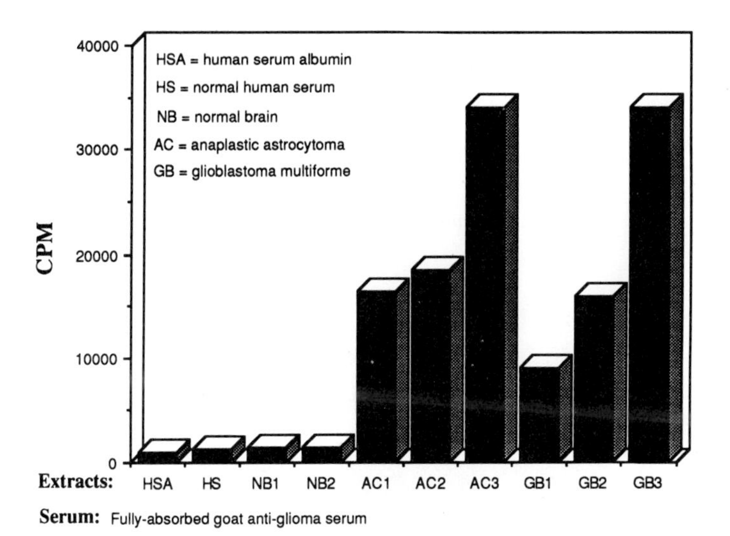

Fig. 4 Reactivity of the fully absorbed goat anti-glioma antiserum with HSA, normal human serum (HS), two extracts of normal human brain tissue (NB1, NB2), extracts of three anaplastic astrocytomas (AA1 or AC1, AC2, AC3), and extracts of three glioblastomas multiforme (GBM1 or GB1, GB2, GB3).

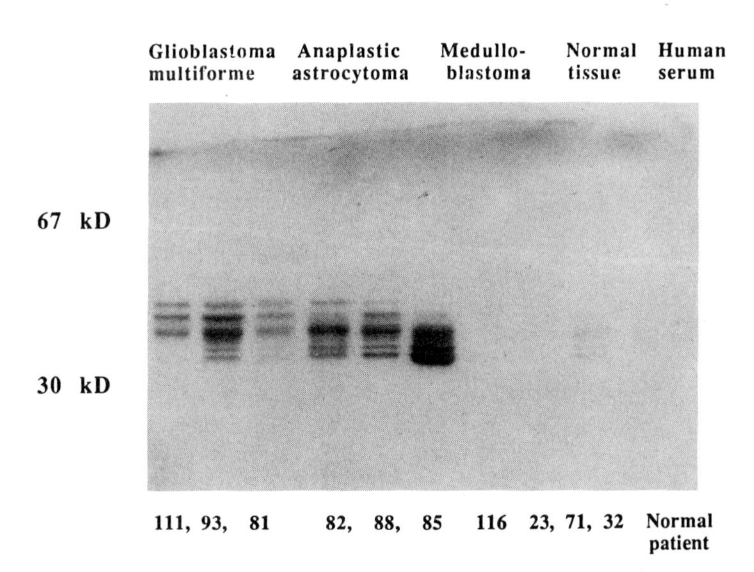

Fig. 5A Western immunoblots using the fully absorbed goat antiserum on multiple glioma tumor extracts, medulloblastoma, and on normal brain and serum protein controls. The IgG fraction of the absorbed antiserum was labeled with ^{125}I, and 8 x 10^5 cpm/ml of the reagent was used in this experiment.

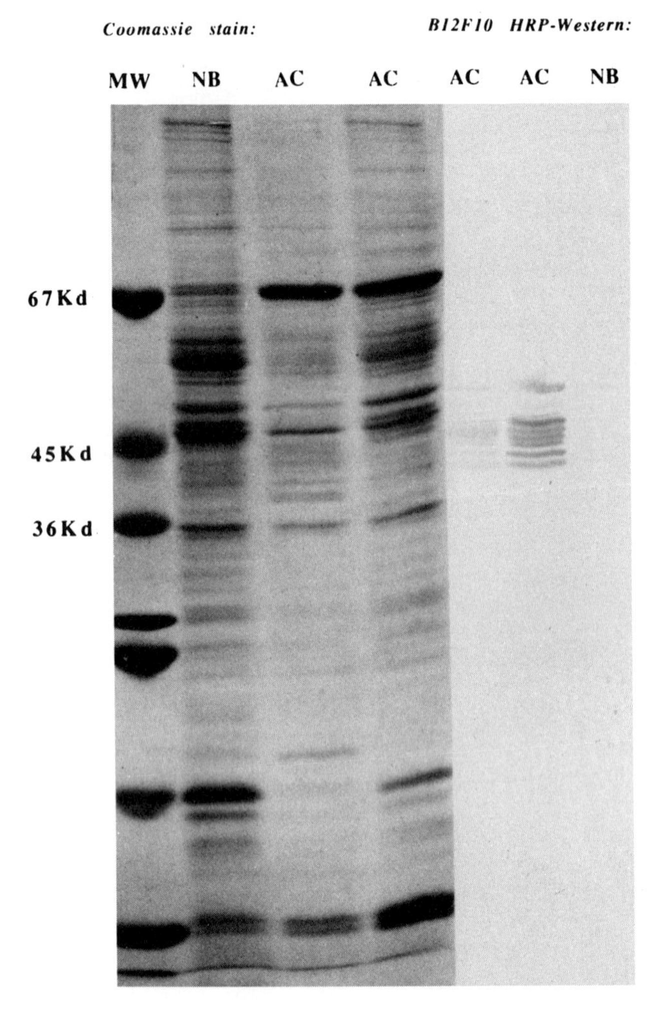

Fig. 5B PAGE and Western HRP-conjugated immunoblot of normal brain (NB) and anaplastic astrocytoma (AA or AC) extracts using the murine MAb B12F10. Note the abundant 67 Kd serum albumin in the tumor extracts. The murine MAb identifies a similar polypeptide pattern, possibly related to human GFAP, as the fully absorbed goat antiserum.

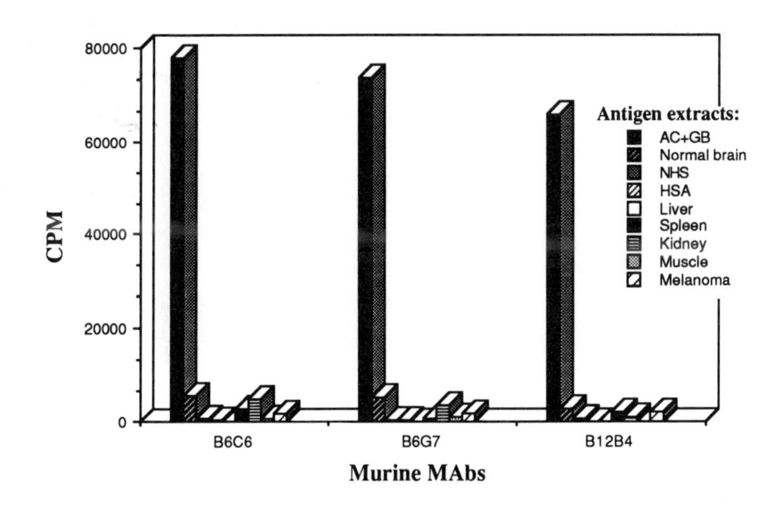

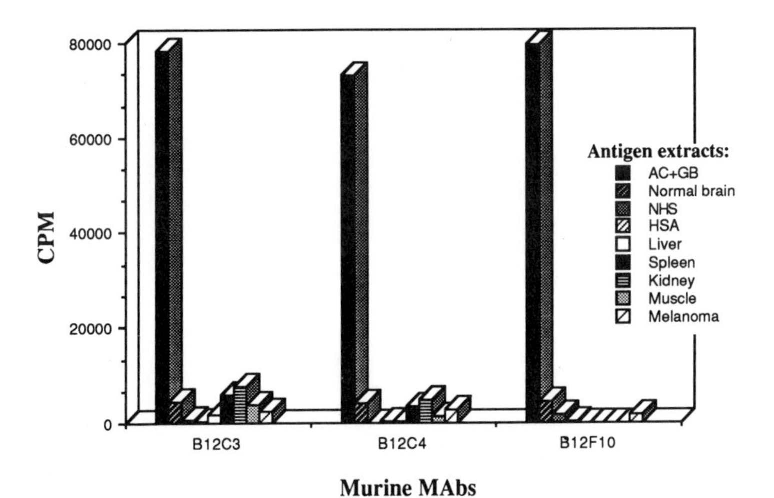

Fig. 6 Specificity of murine monoclonal (MAbs). The MAbs B6C6, B6G7, B12B4, B12C3, B12C4, B12F10 were tested by solid-phase RIA for binding to a mixture of AA + GBM (or AC + GB) extracts, normal brain, normal human serum, HSA, liver, spleen, kidney, muscle, and certain other tumors including melanoma.

Table 1. Specificity of Anti-Glioma Mouse MAbs

					Antibody Bound (cpm)				
MAb	Glioma*	Normal Brain	NHS	HSA	Liver	Spleen	Kidney	Muscle	Melanoma
B_6C_6	77,940	5707	720	297	0	2728	5066	546	1747
B_6G_7	73,636	5384	369	448	0	662	3387	1178	1902
$B_{12}B_4$	65,766	2777	647	0	0	1973	1118	0	2182
$B_{12}C_3$	78,331	4479	661	184	1801	5953	7755	3682	2276
$B_{12}C_4$	73,145	4121	0	191	43	3584	4974	1463	2628
$B_{12}F_{10}$	79,682	4458	1591	341	0	0	0	0	1844

*This is a mixture of extracts of 2 anaplastic astrocytomas and 1 glioblastoma multiforme.

Table 2. Isotype analysis of Murine MAbs

MAb	IgG_1	IgG_2a	IgG_{2b}	IgG_3	Kappa	Lambda
B_6C_6	+	-	-	-	+	-
B_6G_7	+	-	-	-	+	-
$B_{12}B_4$	+	-	-	-	+	-
$B_{12}C_3$	+	-	-	-	+	-
$B_{12}C_4$	+	-	-	-	+	-
$B_{12}F_{10}$	+	-	-	-	+	-

human GFAP or other intermediate filament proteins, as the fully absorbed goat antiserum (Figure 5B). Little reactivity was similarly noted to certain other human tumors, including melanoma (Table 1, Figure 6). The MAbs were all of the IgG_1 Kappa isotype (Table 2).

DISCUSSION

The search for human brain tumor-associated markers is crucial to the development of new diagnostic and therapeutic strategies for these malignancies. Epidemiologic studies estimate about 14,700 new cases of primary intracranial malignancy and 10,900 deaths due to these tumors each year in the United States (15). Despite some advances in radiotherapy and chemotherapy, the 24-month survival for malignant glioma after surgical resection remains at less than 15% to 20% (16). Modern immunological modalities afford the possibility of: (i) high sensitivity in the detection of markers associated with the neoplasm, thereby aiding in the early diagnosis of malignant disease and its recurrence, (ii) using highly specific antibodies directed against those marker antigens to deliver conjugated radioisotopes, cytotoxic agents, and drugs preferentially to the tumor and (iii) subcloning lymphocyte populations to dissect the humoral and T-cell-mediated responses in glioma patients.

These promising immunological approaches have been hampered because specific antigenic markers for central nervous system malignancy have not yet been unequivocally identified (17). Murine monoclonal antibodies generated against astrocytoma, medulloblastoma, and glioma antigens have uncovered broadly shared neuroectodermal and differentiation antigens in astrocytoma, neuroblastoma, melanoma, hematopoietic, and fetal cells (17-22). Clearly, a monoclonal antibody although specific for a particular antigenic site, would nevertheless be broadly cross-reactive if that site were a common region shared by many proteins. In contrast, polyclonal reagents can be extensively absorbed in multiple diverse ways to generate a specific antibody fraction that is free of unwanted reactivities.

Rabbit antisera against an AA cell line and lyophilized GBM tissue, which showed after extensive absorption no cross-reactivity with normal cells, were found to react with 7/7 astrocytoma and 15/15 glioma lines (23-26). Moreover, two primate antisera generated against a glioma line and possessing no reactivities with normal cells after absorptions with adult and fetal brain, detected glioma-associated specificities distinct from normal brain in 7/12 and 3/11 glioma lines, respectively (1). Finally, absorbed rabbit antisera directed against glioma membranes without detectable normal cell cross-reactivity, reacted with high percentages of glioma tissues and lines, and identified candidate protein tumor markers at 10, 30, 55 and 70 kilodaltons (2).

None of the aforementioned studies considered the cross-reactivity with normal serum proteins, in particular human serum albumin, in spite of the fact that the blood-brain barrier is well known to be compromised in these malignancies (27). Indeed, in the present work we have found that tumor extracts contain large amounts of contaminating albumin and other serum proteins (Figures 2 and 5). Therefore, it was decided to undertake careful sequential absorption experiments using the goat anti-glioma antibodies. Furthermore, new immunological technologies are now available which can dissect the complex reactivities of these antisera, as well as decipher the "discordant" reactivity profiles noted for many of the monoclonal reagents (22). Moreover, many of the early studies were unable to take advantage of the sensitive radioimmunoassay or of Western immunoblotting (9,10). The latter technique combines the specificity of a characterized antibody with the molecular resolution of polyacrylamide gel electrophoresis. Finally, the generation of a fully absorbed, specific antibody fraction in a large amount, such as in the goat, would provide ample starting materials for diagnostic applications and biochemical characterization.

The present findings clearly show that it is possible to prepare an antibody fraction that is entirely specific for tumor tissue by careful absorptions of polyclonal anti-tumor antisera on adsorbents of normal brain tissue extracts, normal human serum, and HSA. These results indicate that glioma tissue has potential tumor-associated markers that are not shared in substantial amounts by normal brain tissue or by serum proteins. Indeed, Western immunoblots with this fully absorbed antiserum revealed at least six proteins that are potentially associated with or over-expressed in glioblastoma. The characterization of these candidate polypeptides, including their possible relationship to glial fibrillary acidic protein (GFAP) or other intermediate filament proteins, is currently in progress.

The successful absorptions with polyclonal antisera prompted us to focus our attention on the preparation of MAbs against the tumor markers. The selection and screening strategies enabled the preparation of several murine MAbs that were entirely specific for human glioma tissue, and did not cross-react with normal brain tissue, HSA, normal human serum, kidney, spleen, liver, muscle, or melanoma. Characterization of the tumor-associated markers which bind to these MAbs, and of the humoral response in glioma patients is now in progress.

ACKNOWLEDGEMENTS

This work was supported by the Herbert M. Karol Foundation, the Welch Foundation (Grant Q-994), and by the award to MZA of the Robert A. Welch Chair of Chemistry. DKF is a 1989-1991 American College of Surgeons Scholar. The support and encouragement of Professor Robert G. Grossman are gratefully acknowledged. The technical assistance of Magdalena G. Coronado was invaluable in the execution of these studies.

REFERENCES

1. Wikstrand, C.J. and Bigner, D.D., Surface antigens of human glioma cells shared with normal adult and fetal brain, *Cancer Res.* 39:3235 (1979).
2. Birkmayer, G.D. and Stass, H.P., Humoral immune response in glioma patients: a solubilized glioma-associated membrane antigen as a tool for detecting circulating antibodies, *Int. J. Cancer* 25:445 (1980).
3. Schnegg, J.F., de Tribolet, N., Diserens, A.-C., Martin-Ashard, A., and Carrel, S., Characterization of a rabbit anti-human malignant glioma antiserum, *Int. J. Cancer* 28:265 (1981).
4. Lowry, O.H., Rosenbrough, N.J., Farr, A.L., and Randall, R.J., Protein measurement with the Folin phenol reagent, *J. Biol. Chem.* 193:265 (1951).
5. Atassi, M.Z., Preparation of monoclonal antibodies to preselected protein regions, *Methods in Enzymol.* 121:69 (1986).
6. Hunter, W.M., 1969, V. 608, In: *Handbook of Experimental Immunology*, D.M. Weir, ed., Davis, Philadelphia.
7. Laemmli, U.K., Cleavage of structural proteins during the assembly of the head of bacteriophage T4, *Nature* 227:680 (1970).
8. Merril, C.R., Goldman, D., Van Keuren, M.L., Simplified silver protein detection and image enhancement in polyacrylamide gels, *Electrophoresis* 3:17 (1982).
9. Towbin, H., Staehelin, T. and Gordon, J., Electrophoretic transfer of proteins from polyacrylamide gels to nitrocellulose sheets: procedure and some applications, *Proc. Natl. Acad. Sci. USA* 76:4350 (1979).
10. Burnette, W.N., Western blotting: electrophoretic transfer of proteins from sodium dodecyl sulfate-polyacrylamide gels to unmodified nitrocellulose and radiographic detection with antibody and radioiodinated Protein A, *Anal. Biochem.* 112:195 (1981).
11. Twining, S.S. and Atassi, M.Z., Use of immunoadsorbents for the study of antibody binding to sperm-whale myoglobin and its synthetic antigenic sites, *J. Immunol. Methods* 30:139 (1979).
12. Sakata, S., and Atassi M.Z., Immune recognition of serum albumin. 13. Autoreactivity with rabbit serum albumin of rabbit antibodies against bovine or human serum albumins and autoimmune recognition of rabbit serum albumin, *Mol. Immunol.* 18:961 (1981).
13. Schmitz, H.E., Atassi, H. and Atassi, M.Z., Production of monoclonal antibodies with preselected submolecular specificities: demonstration with sperm-whale myoglobin, *Mol. Immunol.* 19:1699 (1982).

14. Fischer, D.K., Miller, G., Gradoville, L., Heston, L., Westrate, M.W., Wright, J., Brandsma, J. and Summers, W.C., Genome of a mononucleosis Epstein-Barr virus contains DNA fragments previously regarded to be unique to Burkitt's lymphoma isolates, *Cell* 24:543 (1981).
15. Silverberg, E. and Lubera, J.A., Cancer statistics, *Ca-A Cancer J. for Clinicians* 38:5 (1988).
16. Walker, M.D., Green, S.B., Byar, D.P., Alexander, E., Batzdorf, U., Brooks, W.H., Hunt, W.E., MacCarty, C.S., Mahaley, M.S., Mealey, J., Owens, G., Ransohoff, J., Robertson, J.T., Shapiro, W.R., Smith, K.R., Wilson, C.B. and Strike, T.A., Randomized comparisons of radiotherapy and nitrosoureas for the treatment of malignant glioma after surgery, *N. Engl. J. Med.* 303:1323 (1980).
17. Narayan, R.K., Heydorn, W.E., Creed, G.J. and Jacobowitz, D.M., Protein patterns in various malignant human brain tumors by two-dimensional gel electrophoresis, *Cancer Res.* 46:4685 (1986).
18. Schnegg, J.F., Diserens, A.C., Carrel., S., Accolla, R.S. and de Tribolet, N., Human glioma-associated antigens detected by monoclonal antibodies, *Cancer Res.* 41:1209 (1981).
19. Cairncross, J.G., Mattes, M.J., Beresford, H.R., Albino, A.P., Houghton, A.N., Lloyd, K.O. and Old, L.J., Cell surface antigens of human astrocytoma defined by mouse monoclonal antibodies: identification of astrocytoma subsets, *Proc. Natl. Acad. Sci. USA* 79:5641 (1982).
20. Carrel, S., de Tribolet, N. and Mach, J.P., Expression of neuroectodermal antigens common to melanomas, gliomas, and neuroblastomas. I. Identification by monoclonal anti-melanoma and anti-glioma antibodies, *Acta Neuropathol.* 57:158 (1982).
21. Bourdon, M.A, Wikstrand, C.J., Furthmayr, H., Matthews, T.J. and Bigner, D.B., Human glioma-mesenchymal extracellular matrix antigen defined by monoclonal antibody, *Cancer Res.* 43:2796 (1983).
22. Jones, D., Fritschy, J., Garson, J., Nokes, T.J.C., Kemshead, J.T. and Hardistry, R.M., A monoclonal antibody binding to human medulloblastoma cells and to the platelet glycoprotein IIb-IIIa complex, *Br. J. Haematol.* 57:621 (1984).
23. Fischer, D.K., Chen, T.L., and Narayan, R.K., Immunological and biochemical strategies for the identification of brain tumor-associated antigens, *J. Neurosurg.* 68:165 (1988).
24. Coakham, H, Surface antigen(s) common to human astrocytoma cells, *Nature* 250:328 (1974).
25. Wahlstrom, T., Linder, E., Saksela, E. and Westermark, B., Tumor-specific membrane antigens in established cell lines from gliomas, *Cancer* 34:274 (1974).
26. Coakham, H.B. and Lakshmi, M.S., Tumor-associated surface antigen(s) in human astrocytomas, *Oncology* 31:233 (1975).
27. Narayan, R.K., Heydorn, W.E., Creed, G.J., Kornblith, P.L. and Jacobowitz, D.M., 2. Two-dimensional gel electrophoretic protein patterns in high-grade human astrocytomas. In: *Biology of Brain Tumours*, Walker, M.D. and Thomas, D.G., eds,. Martinus Nijhoff Publishers, Boston (1986).

A PEPTIDE ANTIBODY THAT SPECIFICALLY INHIBITS CATHEPSIN L

Clive Dennison and Robert N. Pike

Department of Biochemistry, University of Natal
P. O. Box 375, Pietermaritzburg 3200, South Africa

Cathepsins L is a cysteine proteinase, normally found in lysosomes, but cancer-associated extra-lysosomal variants of cathepsin L, having different molecular weights and pH stability, have been reported and it has been suggested that these may be responsible for the invasive properties of some malignant tumours (Mason et al., 1987; Denhardt et al., 1987; Maciewicz et al., 1989; Rozhin et al., 1989).

It has been suggested (Dennison, 1989) that, in such cases, inhibiting anti-proteinase antibodies, and more especially anti-peptide antibodies (Briand et al., 1985), might be therapeutically useful. In this context anti-peptide antibodies have many advantages. Since a judiciously-chosen sequence of ten or more amino acids has a very high probability of being unique to a particular proteinase, the corresponding anti-peptide antibody is also likely to very specifically inhibit only its target proteinase.

We report here on a peptide, selected from the primary sequence of human cathepsin L, that is able to elicit (rabbit) antibodies which, in turn, are able to inhibit the catalytic activity of the human enzyme. This result suggests that the peptide, or antibodies targeted against the peptide sequence, should be explored for therapeutic usefulness.

MATERIALS AND METHODS

Human cathepsin B was a gift from Dr D Buttle, Strangeways Laboratory, Cambridge, U.K. Human spleen cathepsin L and sheeps' liver cathepsin L were isolated by a modification of the method of Pike and Dennison (1989). Human kidney cathepsin L was from Novabiochem, U.K. Z-Phe-Arg-NHMec and Z-Arg-Arg-NHMec were from Cambridge Research Biochemicals.

A peptide sequence was selected, from human cathepsin L, by consideration of the 3–D structure of the analogous enzyme, papain (Drenth et al., 1971), since it has been deduced, from amino acid sequence information, that the 3-dimensional structures of these two enzymes are comparable (Dufour, 1988). The selected sequence, E-P-D-C-S-S-E-D-M-D-H-G-V, corresponding to residues 153-165 in the amino acid sequence of human cathepsin L (Ritonja et al., 1988), consists of an accessible loop of amino acids, containing the active site histidine (Fig. 1). The presence of histidine in the sequence might increase its antigenicity (Sada, et al., 1990) and the fact that this histidine is part of the active site increases the probability that antibodies targeting this region might be inhibitory. The sequence is largely hydrophilic but has a cluster of hydrophobic residues towards its C-terminus. It is also relatively conserved between species (Ishidoh et al.,1987; Dufour et al., 1987; Ritonja et al., 1988) and antibodies to this sequence should target cathepsin L across species.

Immunobiology of Proteins and Peptides VI
Edited by M.Z. Atassi, Plenum Press, New York, 1991

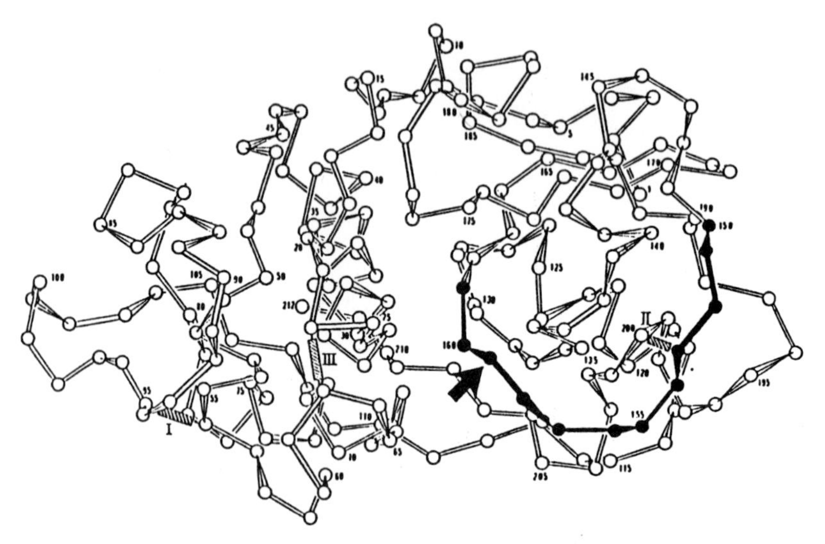

Fig. 1. The 3-dimensional structure of papain, showing the position of the peptide sequence selected from the amino acid sequence of human cathepsin L.
The selected sequence is shaded and the active site histidine is marked with an arrow. Note that human cathepsin L has one amino acid residue more than papain in the selected region. Adapted from Drenth et al. (1971), with permission

The selected peptide was synthesised, in a modified form, by Multiple Peptide Systems, San Diego, California; the cysteine residue was substituted with α-amino butyric acid and an amide group was added to the C-terminus. The peptide was conjugated to KLH (in a ratio of 40:1), through its N-terminus, using 1% (v/v) glutaraldehyde, according to Briand et al. (1985). Two rabbits were each inoculated with peptide according to standard protocols. Anti-KLH antibodies were removed by passage through KLH-Sepharose and serum proteinase inhibitors were removed by purification of the IgG, by the method of Polson et al. (1964). Anti-cathepsin antibodies were assayed, by standard ELISA methods (Tjissen, 1989), using either free peptide or isolated cathepsin L coated to microtiter plate wells. Immunoblots, carried out essentially as described by Towbin et al. (1979), were visualised using either sheep anti-rabbit IgG-HRPO conjugate, or protein A-gold with silver amplification (Moeremans et al., 1984). Assays for the immunoinhibition of cathepsins B and L were carried out using the substrates Z-Arg-Arg-NHMec and Z-Phe-Arg-NHMec, respectively, as described by Barrett and Kirschke (1981). Stopped time assays were carried out over a range of IgG concentrations, and the inhibition by anti-peptide antibodies was calculated in comparison to normal rabbit IgG. Plasma kallikrein activity was controlled by the addition of SBTI, which inhibits kallikrein but not cathepsin L.

RESULTS

The peptide conjugate sucessfully elicited antibodies, which reacted with both the immobilized peptide and immobilized human cathepsin L, in an ELISA (Fig. 2). The antibody titer peaked at about 8-12 weeks. In Western blot analyses (results not shown) the antibodies targeted human cathepsin L strongly and also targeted the sheep enzyme, but relatively weakly. They did not cross-react with human cathepsin B.

The antibodies inhibit human cathepsin L almost completely at high antibody concentrations (Fig. 3) but did not inhibit cathepsin B at all (result not shown). Sheep liver cathepsin L was also inhibited, but to a lesser extent than the human enzyme. Due to the

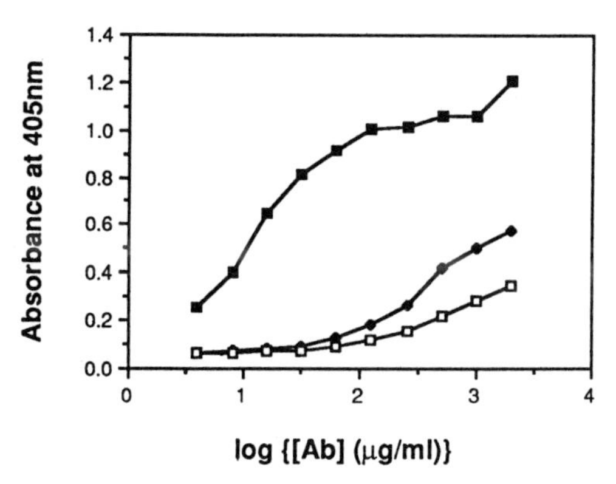

Fig. 2. ELISA of binding of anti-peptide antibodies to whole immobilised cathepsin L.
Reaction of peptide antibodies with human cathepsin L (◆) and human cathepsin L peptide (■). Normal rabbit IgG (❑).

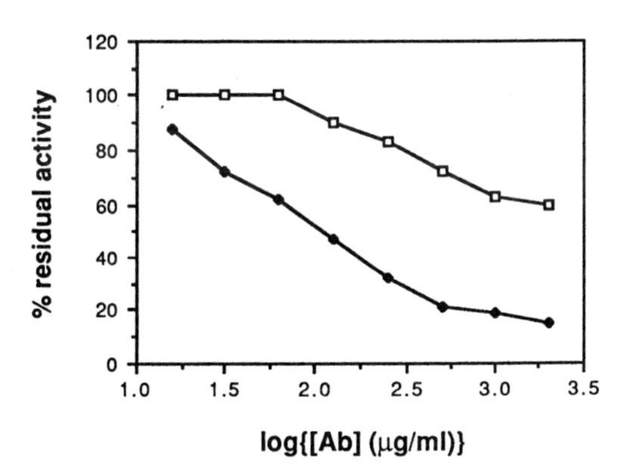

Fig. 3. Immunoinhibition of human and sheep cathepsin L by cathepsin L peptide antibodies, targeted against a human sequence.
Assays were carried out using human (❑) and sheep (▲) cathepsin L and the percentage inhibition was calculated relative to control assays with normal rabbit IgG.

specificity of this inhibition, these anti-cathepsin L peptide antibodies may be useful research tools, since the inhibitors currently in use are unable to qualitatively discriminate between cathepsins B and L (Kirschke et al., 1988). The peptide, used as a vaccine, or monoclonal antibodies, targeted at the peptide sequence, may also be useful as therapeutic agents in pathological conditions caused by excessive cathepsin L activity (Dennison and Pike, 1990).

ACKNOWLEDGEMENTS

This work was supported by grants from the University of Natal Research Fund and the Foundation for Research Development.

REFERENCES

Barrett, A. J. and Kirschke, H., Cathepsin B, cathepsin H and cathepsin L, in: Methods in Enzymology, Vol. 80, L. Lorand, ed., Academic Press, New York (1981).

Briand, J. P., Muller, S. and Van Regenmortel, M. H. V., 1985, Synthetic peptides as antigens: pitfalls of conjugation methods, J. Immunol. Methods, 78:69.

Denhardt, D., Greenberg, A. H., Egan, S. E., Hamilton, R. T. and Wright, J. A., 1987, Cysteine proteinase cathepsin L expression correlates closely with the metastatic potential of H-ras-transformed murine fibroblasts, Oncogene, 2:55.

Dennison, C., 1989, Anti-protease peptide antibodies: a possible mode of tumour immunotherapy, S. Afr. J. Sci. 85:363.

Dennison, C. and Pike, R. N., 1990, Peptides and their use in eliciting immune responses to enzymes. S. A. Provisional Patent 90/6293.

Drenth, J., Janonius, J. N., Koekoek, R. and Wolthers, B. G., 1971, The structure of papain, Advan. Prot. Chem., 25:79.

Dufour, E., Obled, A., Valin, S., Bechet, D., Ribadeau-Dumas, B. and Huet, J. C., 1987, Purification and amino acid sequence of chicken liver cathepsin L, Biochemistry, 26:5689.

Dufour, E., 1988, Sequence homologies, hydrophobic profiles and secondary structures of cathepsins B, H and L, Biochimie, 70:1335.

Ishidoh, K., Towatari, T., Imajoh, S., Kawasaki, S., Kominami, I., Katanuma, N. and Suzuki, K., 1987, Molecular cloning and sequencing of cDNA for rat cathepsin L, FEBS Lett. 223:69.

Kirschke, H., Wikstrom, P. and Shaw, E., 1988, Active center differences between cathepsins L and B: the S_1 binding region, FEBS Lett., 228:128.

Maciewicz, R. A., Wardale, R. J., Etherington, D. J. and Paraskeva, C., 1989, Immunodetection of cathepsins B and L present in and secreted from human pre-malignant and malignant colorectal tumour cell lines, Int. J. Cancer, 43:478.

Mason, R. W., Gal, S. and Gottesman, M. M., 1987, The identification of the major excreted protein (MEP) from a transformed mouse fibroblast cell line as a catalytically active precursor form of cathepsin L, Biochem. J., 248:449.

Moeremans, M., Daneels, G., Van Dijck, A., Langanger, G. and De Mey, J., 1984, Sensitive visualisation of antigen-antibody reactions in dot and blot immune overlay assays with immunogold and immunogold/silver staining, J. Immunol. Methods, 74:353.

Pike, R. N. and Dennison, C., 1989, A high-yield method for the isolation of sheeps liver cathepsin L, Prep. Biochem., 19: 231.

Ritonja, A., Popovic, T., Kotnik, M., Machleidt, W. and Turk, V., 1988, Amino acid sequences of the human kidney cathepsins H and L, FEBS Lett., 228:341.

Rozhin, J., Wade, R. L., Honn, K. V. and Sloane, B. F., 1989, Membrane associated cathepsin L: a role in metastasis of melanomas, Biochem. Biophys. Res. Commun., 164:556.

Sada, E., Katoh, S. and Sohma, Y., 1990, Effects of histidine residues on adsorption equilibrium of peptide antibodies, J. Immunol. Methods, 130:33.

Tijssen, P., 1985, Practice and Theory of Enzyme Immunoassays, Elsevier, Amsterdam.

Towbin, H., Staelin, T. and Gordon, J., 1979, Electrophoretic transfer of proteins from polyacrylamide gels to nitrocellulose sheets: Procedure and some applications, Proc. Natl. Acad. Sci. U.S.A., 76:4350.

IMMUNOMODULATING PROPERTIES OF CORTICOTROPIN-RELEASING FACTOR

Vijendra K. Singh

Molecular Biology Program, Department of Biology and
Developmental Center for Handicapped Persons, Utah State University
Logan, UT 84322-6800, USA

INTRODUCTION

Corticotropin-Releasing Factor (CRF) is a low molecular weight peptide hormone (containing 41 amino acid residues) originally identified in the hypothalamic extracts of the ovine brain (Vale et al., 1981). It is now generally believed that a reciprocal relationship exists between the neuroendocrine system and the immune system (Singh, 1990). Based upon our recent observation that ^{125}I-CRF has specific binding sites on human peripheral blood immunocytes (Singh and Fudenberg, 1988), we hypothesized an immunoregulatory role for this neuropeptide (Singh, 1989 and 1990). In this respsect, our current research (Singh, 1989; Singh et al., 1990; Singh and Leu, 1990) showed that CRF is a stimulator of certain immune functions *in vitro*: (i) stimulation of lymphocyte proliferation in the absence or presence of T cell mitogens, e.g. concanavalin A and phytohemagglutinin which can be blocked specifically by the CRF-antagonist; (ii) increased expression of interleukin-2 receptor (IL-2R) antigen on T cells; and (iii) induction and enhancement of production of interleukin-1 (IL-1) and interleukin-2 (IL-2). Our newer data, as presented in this report, demonstrate that CRF is also a stimulator of natural killer (NK) cell activity but it has a suppressive effect on the production of antibodies or immunoglobulin G (IgG) *in vitro.*

MATERIALS AND METHODS

Heparinized blood was drawn from 6 to 8 young healthy adults aged 25 to 40 years. Mononuclear cells (MNC), isolated by the Ficoll-Hypaque method as described before (Singh, 1989), were resuspended in the growth medium which contained RPMI-1640-penicillin/streptomycin-10% fetal bovine serum.

The NK cell activity was determined in terms of the spontaneous cytotoxicity as measured in the 4-hour [^{51}Cr] -release assay (Edwards et al., 1984). The K562 cell line was used as the source of target cells. The MNC were pretreated with various concentrations of CRF (Sigma code #C2024) for 16 to 18 hours in a 37^{0}C/5% CO_2-incubator, followed by their usage at a effector:target cell ratio of 6.25:1 in the [^{51}Cr]-release assay.

The IgG production *in vitro* was performed by the stimulation of MNC with pokeweed mitogen (PWM) for 7 days. As desired, various concentrations of CRF were supplemented into the MNC cultures which contained 1×10^6 MNC and 25 μg/ml of PWM (Sigma) in a final volume of 1 ml of growth medium. After 7 days of incubation, the cultures were centrifuged at 1,500 rpm for 10 min to collect the supernatant which was subsequently used to quatitate IgG concentration by the microplate enzyme-linked immunosorbant assay (ELISA). The rabbit-anti-human IL-1 (anti-IL-1) was from Endogen and the rabbit-anti-human ß-endorphin (anti-ßE) was purchased from Amersham.

Immunobiology of Proteins and Peptides VI
Edited by M.Z. Atassi, Plenum Press, New York, 1991

RESULTS

As shown in Figure 1, a 16 to 18 hour preincubation of MNC with CRF resulted into approximately 42% increase in the activity of NK cells. This increase was statistically significant ($p=0.002$) at a 0.1 nanomolar concentration of CRF. Moreover, the supplementation of anti-IL-1 or anti-ßE into MNC cultures at the beginning of preincubation caused 60 to 70% inhibtion of NK cell activity stimulated with 0.1 nM CRF (Figure 2). The data in Figure 3 showed that the *in vitro* IgG production was significantly ($p<0.05$) suppressed in the presence of picomolar amounts of CRF. This suppressive effect of CRF was furthermore augmented by anti-IL-1 but not by anti-ßE (Figure 4).

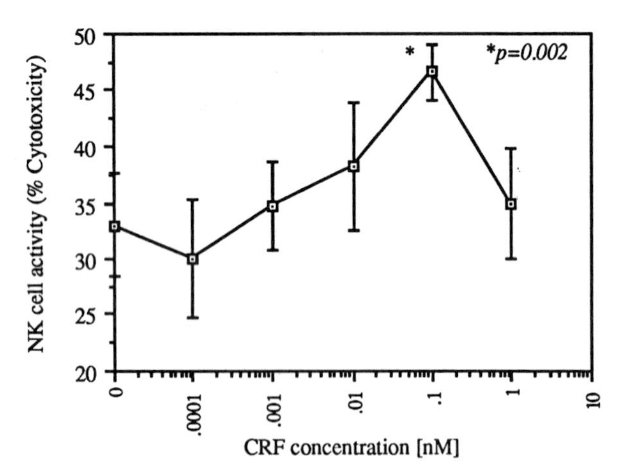

Figure 1. Effect of CRF on the activity of human NK cells
As described in the Methods section, human MNC were preincubated with various concentrations of CRF followed by the assay of NK cell activity against K562 as the target cells. The ratio of effector to target cells was 6.25:1 and the data were obtained from the triplicate analysis of 4 blood donors.

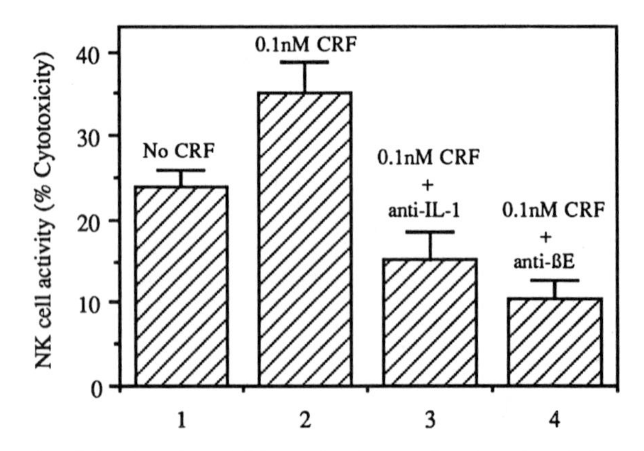

Figure 2. Effect of anti-IL-1 and anti-ßE on the CRF-stimulated activity of NK cells
Twenty microliters of anti-IL-1 (200 NU/ml) or anti-ßE (1 ml reconstituted stock) were added to the MNC cultures which contained 0.1 nM CRF during the preincubation step. Three donors were studied in different experiments.

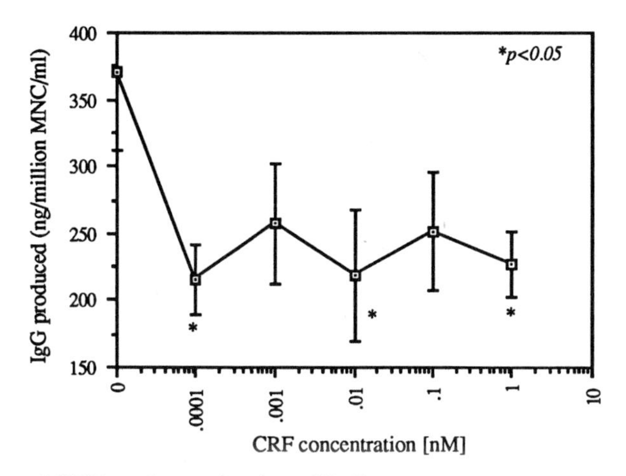

Figure 3. Effect of CRF on the production of IgG
Human MNC were stimulated with PWM in the presence of various concentrations of CRF and the amount of IgG produced was quatitated by the ELISA method as described in the Methods section. The data are the mean ± S.D. (n=4).

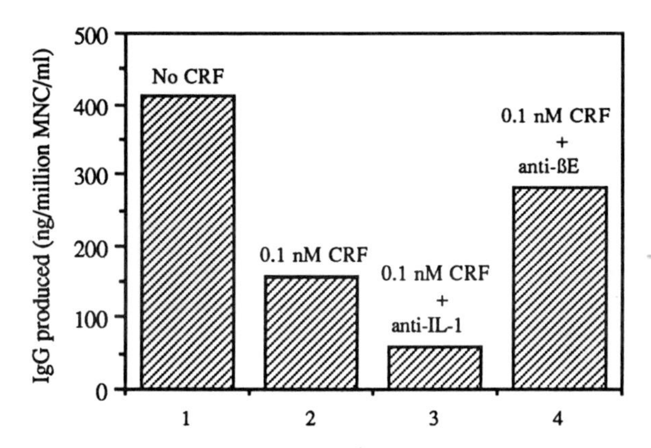

Figure 4. Effect of anti-IL-1 and anti-ßE on the production of IgG
As described in the Methods section, the effect of 0.1 nM CRF on the production of IgG was measured in the presence of 20 µl of either anti-IL-1 or anti-ßE.

DISCUSSION

In the brain, the hypothalamic CRF has been shown to stimulate pituitary gland causing the release of adrenocorticotropin hormone (ACTH), beta-endorphin (ßE), and growth hormone (GH). In addition, CRF is a stress hormone. Our recent studies of immunomodulation by CRF have provided newer data which suggests a more direct interactiveness between neuroendocrine system and immune system (Singh, 1989 and 1990). In the present study, CRF was found to stimulate human NK cell function and this effect was inhibited by anti-IL-1 and anti-BE. The stimulatory effect of CRF on human NK cells is similar to the enhancing effect of CRF on mouse NK cells (Carr and Blalock, 1990). This action, however, required a preincubation step of 16 to 18 hours which yielded a more pronounced effect relative to a preincubation period of 4 to 6 or 22 to 24 hours (data not included). As described previously (Singh, 1989; Singh et al., 1990), CRF did not influence the activity of NK cells if supplemented into the mixture of effector and target cells for the [^{51}Cr]-release assay.

The stimulatory effect of CRF on NK cell function may be mediated indirectly by the activation of monocytes to produce IL-1 which via the release of ßE by B cells may cause the stimulation of NK cell activity. In support of this hypothesis, we have previously shown that CRF induces and enhances the production of IL-1 by human monocytes (Singh et al., 1990; Singh and Leu, 1990) and others have shown that anti-IL-1 blocks the CRF-induced production of ßE by human B cells (Kavelaars et al., 1989). If this type of reaction to occur, then the CRF-stimulated activity of NK cells should also be blocked by anti-IL-1 and/or anti-ßE. Indeed, the supplementation of anti-IL-1 or anti-ßE into the MNC cultures during the preincubation step resulted into a 60 to 70% inhibition of CRF-stimulated NK cell activity as illustrated by the Figure 2.

In terms of antibody (e.g. IgG) production *in vitro*, the CRF was found to significantly suppress this particular immune function. The mechanism of this effect is not clear except that anti-IL-1, but not anti-BE, furthermore decreased the production of IgG. This finding suggested that the CRF is likely to act indirectly via monocyte production of IL-1 and since ßE is probably not involved in the IgG production, therefore anti-ßE has no blocking effect. In addition, our preliminary experiments of the Flow Cytometric analysis suggested another possible mechanism that may involve the activation of either T suppressor cells or a T helper subset having suppressor-inducer function (data not included). Alternatively, the IL-1-induced activation of B cells to generate a specific T suppressor factor may be implicated in the suppression of IgG production by CRF.

In conclusion, we suggest that CRF is a very important soluble messenger or mediator of neuroendocrine-immune pathways (Singh, 1989 and 1990). Furthermore, this neuropeptide hormone has immunomodulating properties based upon our findings of an immunosuppressive effect in the picomolar amounts and an immunostimulatory effect in the nanomolar amounts. The precise mechanism of its action is not fully understood but CRF appears to function on immune cells via IL-1 production by monocytes and may involve functional receptors since human monocytes, as reported previously (Singh and Fudenberg, 1988), are enriched with high-affinity binding sites for CRF.

ACKNOWLEDGMENT

The author is grateful to the USU's Biotechnology Center and Faculty Research program for funding support and to Mrs. S. J. C. Leu for experimental help.

REFERENCES

Carr, D. J. J. and Blalock, J. E. (1990) Corticotropin-releasing hormone enhances murine natural killer cell activity through an opioid-mediated pathway. *Ann. N. Y. Acad Sci.* 594: 371-373.

Edwards, B. S., Merritt, J. A., Jelen, P. A. and Borden, E. C. (1984) Effects of diethyldithiocarbamate, an inhibitor of interferon antiviral activity, upon human natural killer cells. *J. Immunol.* 132: 2868-2875.

Kavelaars, A., Ballieux, R. E. and Heijnen, C. J. (1989) The role of IL-1 in the corticotropin-releasing factor and arginine-vasopressin-induced secretion of immunoreactive ß-endorphin by human peripheral blood mononuclear cells. *J. Immunol.* 142: 2338-2342.

Singh, V. K. (1989) Stimulatory effect of corticotropin-releasing neurohormone on human lymphocyte proliferation and interleukin-2 receptor expression. *J. Neuroimmunol.* 23: 257-262.

Singh, V. K. (1990) Neuroimmune axis as a basis of therapy in Alzheimer's disease. *Prog. Drug Res.* 34: 383-393.

Singh, V. K. and Fudenberg, H. H. (1988) Binding of [125I]-corticotropin-releasing factor to blood immunocytes and its reduction in Alzheimer's disease. *Immunol. Lett.* 18: 5-8.

Singh, V. K. and Leu, S. J. C. (1990) Enhancing effect of corticotropin-releasing neurohormone on the production of interleukin-1 and interleukin-2. *Neurosci. Lett.* (in press).

Singh, V. K., Warren, R. P., White, E. D. and Leu, S. J. C. (1990) Corticotropin-releasing factor-induced stimulation of immune functions. *Ann. N.Y. Acad. Sci.* 594: 416-419.

Vale, W. W., Spies, J., Rivier, C. and Rivier, J. (1981) Characterization of a 41-residue ovine hypothalamic peptide that stimulates secretion of corticotropin and beta-endorphin. *Science* 213: 1394-1397.

INDEX